WITHDRAWN

SULFUR CONCRETE FOR THE CONSTRUCTION INDUSTRY

A Sustainable Development Approach

A. M. O. Mohamed & M. M. El Gamal
UAE University, Al Ain, United Arab Emirates

Copyright ©2010 J. Ross Publishing

ISBN-13: 978-1-60427-005-1

Printed and bound in the U.S.A. Printed on acid-free paper

10 9 8 7 6 5 4 3 2 1

Library of Congress Cataloging-in-Publication Data

Mohamed, Abdel-Mohsen Onsy.
  Sulfur concrete for the construction industry / by Abdel-Mohsen Onsy Mohamed and Maisa El Gamal.
    p. cm.
  Includes index.
  ISBN 978-1-60427-005-1 (hardcover : alk. paper)
  1. Sulfur concrete. I. El Gamal, Maisa, 1966- II. Title.
  TA442.5.M64 2010
  624.1'834--dc22
                                                    2010008163

This publication contains information obtained from authentic and highly regarded sources. Reprinted material is used with permission, and sources are indicated. Reasonable effort has been made to publish reliable data and information, but the author and the publisher cannot assume responsibility for the validity of all materials or for the consequences of their use.

All rights reserved. Neither this publication nor any part thereof may be reproduced, stored in a retrieval system, or transmitted in any form or by any means, electronic, mechanical, photocopying, recording or otherwise, without the prior written permission of the publisher.

The copyright owner's consent does not extend to copying for general distribution for promotion, for creating new works, or for resale. Specific permission must be obtained from J. Ross Publishing for such purposes.

Direct all inquiries to J. Ross Publishing, Inc., 5765 N. Andrews Way, Fort Lauderdale, FL 33309.

Phone: (954) 727-9333
Fax: (561) 892-0700
Web: www.jrosspub.com

*To my divine wife, Amal,
for the 23 years of support she has given me,
and to our beloved daughters, Basant and Dina*
—**A. M. O. Mohamed**

*To my parents
and my little family for their continuous support*
—**M. M. El Gamal**

CONTENTS

Preface xiii
About the Authors xix

Chapter 1. Sustainable Development for the Construction Industry
1.1 Introduction 1
1.2 Sustainable Development 2
    1.2.1 Social Sustainability 4
    1.2.2 Environmental Sustainability 5
    1.2.3 Economic Sustainability 6
    1.2.4 Land Sustainability 7
1.3 Role of Technology 9
    1.3.1 Characteristics of a Sustainable Technology 9
1.4 A Framework for a Sustainable Industry 10
    1.4.1 Current System 10
    1.4.2 Modified System 12
1.5 Sustainability and the Building Construction Industry 13
1.6 Strategies for Implementing Sustainable Design and Construction 15
    1.6.1 Minimizing Consumption 15
    1.6.2 Satisfying Human Needs and Aspirations 17
    1.6.3 Avoiding Negative Environmental Impacts 19
1.7 Sustainability and Project Procurement Life Cycle 21
    1.7.1 Sustainable Business Justification 21
    1.7.2 Sustainable Procurement Process 24
    1.7.3 Sustainable Design 27
    1.7.4 Sustainable Construction Process 29
    1.7.5 Sustainable Management and Operation of the Facility 31
    1.7.6 Sustainable Disposal and Re-Use of the Site 32
1.8 Summary and Concluding Remarks 33

Chapter 2. Sulfur Production and Uses
2.1 Introduction 35
2.2 Global Sulfur Cycle 36
    2.2.1 The Natural Sulfur Cycle 37
    2.2.2 The Anthropogenic Sulfur Cycle 38
2.3 Sulfur Supply 39
    2.3.1 Sulfur Production and Processes 39
2.4 Sulfur Trade 44
2.5 Sulfur Demand 48
2.6 Sulfur Uses 48
    2.6.1 Sulfuric Acid 49
    2.6.2 Agricultural Chemicals 51
    2.6.3 Chemical and Industrial 52
    2.6.4 Construction Industry 54
    2.6.5 Ore Processing 57
    2.6.6 Petroleum Alkylation 57
    2.6.7 Pulp and Paper 58
    2.6.8 Waste Management 59
    2.6.9 Pharmaceutical Industry 60
2.7 Environmental Issues 62
    2.7.1 Production and Processing 62
    2.7.2 Health Effects of Sulfur 63
    2.7.3 Effects of Sulfur on the Environment 64
    2.7.4 Sulfur Waste Management 64
2.8 Summary and Concluding Remarks 67

Chapter 3. Sulfur Properties
3.1 Introduction 71
3.2 Occurrence 72
3.3 Processes 74
3.4 Elemental Sulfur Forms 75
    3.4.1 Liquid Sulfur Allotropes 76
    3.4.2 Solid Sulfur Allotropes 76
3.5 Properties of Elemental Sulfur 77
    3.5.1 Melting and Freezing Points 77
    3.5.2 Viscosity 79
    3.5.3 Density 79
    3.5.4 Color 79
    3.5.5 Strength Characteristics 81
    3.5.6 Thermal Characteristics 82

	3.5.7	Allotropic Transformation	82
	3.5.8	Polymerization	86
3.6	Chemical Properties		88
	3.6.1	Electronic Structure	89
	3.6.2	Oxidation States	90
	3.6.3	Chemical Reactions of Elemental Sulfur	93
	3.6.4	Chemical Reactions of Sulfur with Olefins	95
	3.6.5	Chemical Reactions of Sulfur Compounds	95
	3.6.6	Biological Reactions of Sulfur Compounds	96
3.7	Thermal Properties		99
3.8	Electrical Properties		102
3.9	Isotopes		102
3.10	Potential Ecological Effects of Elemental Sulfur		103
3.11	Products		104
	3.11.1	Product Groups	104
	3.11.2	Practical Applications	104
3.12	Summary and Concluding Remarks		107

Chapter 4. Elemental Sulfur Concrete

4.1	Introduction		109
4.2	History of Sulfur Concrete Development		110
4.3	Terminology		111
4.4	Compressive Strength		111
	4.4.1	Strength Development for Portland Cement Concrete	112
	4.4.2	Strength Reduction for Portland Cement Concrete	114
	4.4.3	Strength Development for Elemental Sulfur Concrete	118
4.5	Material Composition		121
	4.5.1	Elemental Sulfur	121
	4.5.2	Aggregates	122
	4.5.3	ACI Guide for Material Selection	130
4.6	Durability		135
	4.6.1	Effect of Sulfur Loading, Aggregate Type, and Different Admixtures	135
	4.6.2	Effect of Water and Temperature	137
4.7	Summary and Concluding Remarks		140

Chapter 5. Sulfur Cement

5.1	Introduction	143
5.2	Development Background	143
5.3	Terminology	145
5.4	Modified Sulfur	146

		5.4.1	Mechanism	146
		5.4.2	Types	148
		5.4.3	Modification Conditions	151
5.5	Industrial Modified Sulfur Cement			153
		5.5.1	Sulfur Modified with Dicyclopentadiene	153
		5.5.2	Sulfur Modified with DCPD and an Oligomer of Cyclopentadiene	159
		5.5.3	Sulfur Modified with Styrene	164
		5.5.4	Sulfur Modified with a Combination of DCPD and Styrene	167
		5.5.5	Sulfur Modified with Olefinic Hydrocarbon Polymers	168
		5.5.6	Sulfur Modified with Bitumen	169
		5.5.7	Sulfur Modified with 5-ethylidene-2-norbornene	184
5.6	Sulfur Cement			186
5.7	Factors Controlling Formation of Sulfur Cement			187
5.8	Standard Testing of Sulfur Cement			189
5.9	Advantages and Disadvantages of Sulfur Cement			190
		5.9.1	Advantages	190
		5.9.2	Disadvantages	190
5.10	Summary and Concluding Remarks			191

Chapter 6. Sulfur Concrete

6.1	Introduction		193
6.2	Terminology		194
6.3	Development of Sulfur Concrete		194
6.4	Composition		197
	6.4.1	Sulfur	198
	6.4.2	Chemical Additives	199
	6.4.3	Mineral Fillers	199
	6.4.4	Aggregates	200
6.5	Sulfur Concrete Requirements		201
	6.5.1	Binder Requirements	201
	6.5.2	Mix Design Requirements	202
6.6	Manufacturing Equipment and Methods		202
6.7	Recommended Testing		204
6.8	Advantages of Using Sulfur Concrete		205
6.9	Dicyclopentadiene-Modified Sulfur Concrete		207
	6.9.1	DCPD Loadings and Aggregate Type	208
	6.9.2	Storage Time	209
	6.9.3	Thermal Stability	211

6.10	Dicyclopentadiene-Cyclopentadiene Oligomer-Modified Sulfur Concrete		212
	6.10.1	Effect of Mix Composition on Strength	213
	6.10.2	Effect of Freeze-Thaw on Strength	213
6.11	Olefinic Hydrocarbon Polymer-Modified Sulfur Concrete		214
6.12	5-ethylidene-2-norbornene-Modified Sulfur Concrete		216
	6.12.1	Weight Loss in Alkaline Environment	218
	6.12.2	Compressive Strength in Alkaline Environment	218
	6.12.3	Ignition and Biological Oxidation	219
6.13	Bitumen-Modified Sulfur Concrete		220
	6.13.1	Production of Bitumen-Modified Sulfur Concrete	220
	6.13.2	Thermal Stability	221
	6.13.3	Effect of Sulfur Ratio and Loading	223
	6.13.4	Microstructure Characterization	226
	6.13.5	Strength Development	227
	6.13.6	Reaction Products	234
	6.13.7	Durability	237
	6.13.8	Hydraulic Conductivity	241
	6.13.9	Long-Term Hydromechanical Behavior	243
	6.13.10	Leachability	244
6.14	Potential Ecological Effects of Sulfur Concrete		251
6.15	Summary and Concluding Remarks		254

Chapter 7. Technological Aspects of Sulfur Concrete Production

7.1	Introduction		257
7.2	Sulfur Concrete Production		258
	7.2.1	The 1970s	258
	7.2.2	The 1980s	260
	7.2.3	The 1990s	263
	7.2.4	The 2000s	264
7.3	Mix Design		265
7.4	Mixing Process		267
7.5	Equipment		271
	7.5.1	Development	271
	7.5.2	Commercial Scale Application	273
7.6	Manufacturing		274
	7.6.1	Pre-cast Mixing and Production	274
	7.6.2	*In situ* Construction Mixing and Placing Techniques	281
	7.6.3	Placing and Finishing	284
	7.6.4	Cold Weather Placements	284

	7.6.5	Wind and Moisture 285
	7.6.6	Repairing Damages 286
	7.6.7	Joints and Joint Sealing 286
	7.6.8	Forming and Reinforcement 287
7.7	Energy Requirement 287
	7.7.1	Heating Process 287
	7.7.2	Recovery Process 289
	7.7.3	Cooling Process 291
7.8	Durability Issues 292
	7.8.1	Types of Fillers and Aggregates 292
	7.8.2	Water Absorption 300
	7.8.3	Frost Resistance 305
	7.8.4	Service Temperature 306
	7.8.5	Fire Load 307
	7.8.6	Crazing Resistance 308
	7.8.7	Creep 309
	7.8.8	Fatigue Strength 310
	7.8.9	Reinforcement 311
	7.8.10	Abrasion Resistance 313
	7.8.11	Chemical Resistance 314
	7.8.12	Corrosion Potential 314
7.9	Service Life 323
	7.9.1	ISO Approach 324
	7.9.2	Estimation of Sulfur Concrete Service Life 325
7.10	Sulfur Concrete Assessment Protocol 326
7.11	Summary and Concluding Remarks 328

Chapter 8. Sulfur-Modified Asphalt

8.1	Introduction 331
8.2	Asphalt 332
	8.2.1	Asphalt Composition 332
	8.2.2	Asphalt Fractionation 334
	8.2.3	Asphalt Component Interaction 335
	8.2.4	Asphalt Aging 338
	8.2.5	Lime-Modified Asphalt 339
8.3	Sulfur-Modified Asphalt 340
	8.3.1	Beneficial Use of Sulfur-Modified Asphalt 341
	8.3.2	Sulfur Asphalt History 341
	8.3.3	Sulfur Behavior in Liquid State 342
	8.3.4	Sulfur Asphalt Interaction 343

	8.3.5	Sulfur Asphalt Mix Concepts 348
	8.3.6	Rheology of Sulfur Asphalt Binder... 351
8.4	Sulfur Asphalt Processing Technology 356	
	8.4.1	Manufacturing Evaluation... 356
	8.4.2	Preparation Apparatus for Sulfur Asphalt Binders............... 356
	8.4.3	Premixing... 356
	8.4.4	Sulfur Asphalt Module 357
	8.4.5	Mix Production 357
	8.4.6	Construction Procedure... 357
	8.4.7	Safety and the Environment.. 358
	8.4.8	Emissions.. 358
8.5	Improved Performance 359	
	8.5.1	Low Temperature Stiffness.. 359
	8.5.2	Tensile Strength.. 359
	8.5.3	Resilient Modulus of Elasticity.. 359
	8.5.4	Fatigue Life... 360
	8.5.5	Resistance to Stripping.. 360
8.6	Sulfur-to-Asphalt Ratios and Properties 360	
	8.6.1	Sulfur-to-Asphalt Ratio 360
	8.6.2	Sulfur-Modified Asphalt Characteristics................................ 362
	8.6.3	Hydrogen Sulfide Emission Control 363
8.7	Sulfur Asphalt Development 363	
	8.7.1	Sand-Asphalt-Sulfur... 364
	8.7.2	Sulfur-Extended Asphalt.. 364
8.8	Sulfur-Extended Asphalt and Traditional Asphalt Materials.............. 368	
8.9	Plasticized Sulfur.. 370	
	8.9.1	Plasticization Concept... 370
	8.9.2	Chemicals Used for Plasticization of Sulfur........................... 372
	8.9.3	Plasticizing Agent Requirement 373
	8.9.4	Plasticization Perceptions.. 374
	8.9.5	Plasticized Mixing Conditions.. 374
	8.9.6	Case Studies for Plasticization of Sulfur................................. 375
8.10	Potential Ecological Effects... 376	
8.11	Summary And Concluding Remarks 379	

References... 381

Index 411

PREFACE

The recent global attention to the issues and challenges of sustainable development is forcing industries to conduct self-assessments to identify where they stand within the framework for sustainability and, more importantly, to identify drivers, opportunities, strategies, and technologies that support achieving this goal.

In striving to achieve sustainability in the built environment, three themes emerge. First, awareness of the impact that built facilities have on both human and natural systems is essential and should be considered as early as possible in the planning and design of any built facility. Second, the ecological, social, and economic contexts of the facility must be taken into account for all project decision making. Finally, sustainable designers and constructors must be aware of the connectivity of human systems to the natural environment. No human action can take place without affecting the ecological context in which it occurs. All human activity must be undertaken with an awareness of the potential consequences to other humans and nature, especially the construction of built facilities because of its large scale.

Over the last few years, a diverse number of public- and private-sector organizations around the world has given increased attention to the problems of excessive natural resource consumption, depletion, degradation, waste generation and accumulation, and the environmental impact. The construction industry is a major contributor to these problems and now faces: (a) increasingly restrictive environmental conservation and protection laws and regulations to address environmental quality and performance and (b) substantial pressures from civic and private environmental groups. As a result, private- and public-sector owners face new, complex, and rapidly changing challenges imposed by these laws, regulations, standards, and pressures. All life cycle stages of a capital project are affected, from initial planning, design, construction, and operation/maintenance, to ultimate rehabilitation, decommissioning, and/or disposal. Furthermore, traditional approaches to capital projects of environmental regulatory compliance or reactive corrective actions such as mitigation or remediation have proven to be consistently costly, inefficient, and frequently ineffective.

There are strong incentives for the development of a sustainable approach to capital projects. Such an approach goes beyond the traditional focus on cost, time, and quality performance to include the goals of minimal: (a) natural resource consumption, depletion, and degradation, (b) waste generation and accumulation, and (c) environmental impact and degradation; all within the contextual satisfaction of human needs and aspirations. These goals are explicitly and systematically incorporated within the decision-making process at all stages of the life cycle of a capital project, particularly the early funding allocation as well as the planning and conceptual design phases. However, most stakeholders within the capital project delivery process (i.e., owners, planners, designers, vendors/suppliers, constructors, users/operators) face a complex task when attempting to implement a sustainable approach. First, they already face the challenges imposed by increasingly limited resources on the effective and efficient delivery of capital projects. Second, they do not have clear incentives, the proper resources, nor the mechanisms or tools to do so. Finally, there is a lack of awareness and understanding of the actual or potential impact and/or implications of environmental regulations and standards on capital projects as well as ignorance of the opportunities and potential benefits to an organization created by a sustainable approach to its capital projects. Additionally, there is a lack of credible and reliable quantitative indicators, metrics, and/or data on the actual benefits and associated costs.

A sustainable approach to capital projects would allow the construction industry to take a more aggressive role in finding both short- and long-term solutions for a more effective and efficient use of its increasingly limited capital resources.

Consumption of natural resources is at the heart of sustainability. With its large-scale use of material and energy and the displacement of natural ecosystems, the built environment greatly influences the sustainability of human systems as well as the natural ecosystems of which we are a part. Minimizing consumption of matter and energy is essential to achieve sustainability in creating, operating, and decommissioning built facilities.

One strategy for minimizing consumption in the built environment is to improve the technological efficiency of our materials and processes. We need to improve the efficiency with which they meet the needs of their intended use. Reusing buildings, materials, and equipment is a second strategy for making design and construction more sustainable. By reusing what already exists, we save the cost, material, and energy input that would be required to create new facilities from scratch. By using techniques such as adaptive reuse, rehabilitation, or retrofitting, old facilities can be modified or improved to meet new use criteria at a much lower consumptive cost than that required to build a new facility. Creating new technologies is a third strategy for increasing the sustainability of human activity. Consumption of matter and energy can be reduced by developing new technologies that do not rely on traditional types or amounts of materials and energy to meet human needs. Opportunities for new technologies can be found by observing natural ecosystems: what sources of energy and matter are used by these systems? Particularly promising opportunities exist in the area of waste recovery and reuse. Examples of taking artificially-generated waste that would otherwise have been disposed in the natural environment, and using it as input back into the building process are: (a) using waste masonry and concrete from demolished structures as aggregate in new concrete and (b) using sulfur, from the oil industry, fly ash, and aggregates for production of sulfur cement and concrete for the construction and asphalt pavement industries.

Global sulfur demand has been relatively stagnant at approximately 57 million metric tons per year over the last decade. Based on new regulations limiting sulfur content in diesel and gasoline, the current small global surplus in sulfur supply is projected to reach between 6 and 12 million metric tons by 2011 (The Sulphur Institute), or between 10 to 20% of demand. This projected surplus represents obvious challenges to existing producers, potentially leading to drastically reduced sulfur prices and even the possibility of reduced costs to producers for the disposal of the surplus in some regions. On the other hand, the surplus may also represent opportunities for new uses of sulfur driven by these same reductions in sulfur prices that can make new uses more economically feasible and attractive. Even with relatively small surpluses, the oil and gas industry has already experienced strains on sulfur storage facilities. Sulfur is being stored on site in block, granular, or palletized form or in a molten state in rail tank cars at great expense because there is insufficient storage capacity to handle sulfur generation at refineries and gas processing plants. Therefore, new markets must be found for sulfur to avoid a disposal crisis. A promising potential market is the construction, asphalt pavement, and sewerage management industries.

In view of these emerging challenges, this book is designed to evaluate the potential applications of sulfur in the construction, asphalt pavement, and sewerage management industries by providing:

1. A brief overview of the wide range of technological issues at the construction industry level while emphasizing the need for an integrated approach and an understanding of the various components of a sustainable system (Chapter 1). In this chapter, we stress the importance of adopting a new paradigm that considers industry as a total system rather than focusing on individual components of processes and operations. To achieve sustainability for society as a

whole, and for construction in particular, intelligent decision making is required that includes full consideration and knowledge of the many trade-offs and impacts associated with each available alternative.
2. Basic information related to sulfur cycle, supply, trade, uses, and environmental issues (Chapter 2). There is a noticeable trend in the global sulfur industry over the last decade that signals a dramatic change. Sulfur consumption is dominated by a number of large-scale uses with several complex interactions that complicate the future outlook for demand.
3. A brief overview of sulfur properties (chemical, electrical, mineralogical, physical, and thermal) (Chapter 3). Sulfur products for industrial uses are also discussed in this chapter.
4. Fundamental aspects of manufacturing elemental sulfur concrete (Chapter 4). We have looked into what constitutes elemental sulfur concrete. What are the factors that contribute to its strength development? What are the impacts of the environmental variables on elemental sulfur concrete strength?
5. Basic ingredients of sulfur cement and its physicochemical properties (Chapter 5). Also, we define terminologies used in manufacturing sulfur cement and a clear distinction was made between sulfur modifiers and sulfur cement. The performance of sulfur cement made with modifying chemical reagents such as dicyclopentadiene, dicyclopentadiene plus oligomer of cyclopentadiene, styrene, dicyclopentadiene plus styrene, bitumen, RP220, RP020, CTLA, 5-ethylidene-2-norbornene, etc., are discussed. Controlling parameters such as the type and percentage of modifiers and thermal history, including reaction and curing temperatures, curing time, cooling rate, and storage are discussed and evaluated.
6. Basic ingredients of sulfur concrete (Chapter 6). We discuss the performance of sulfur concrete prepared with the addition of various chemical additives used to modify sulfur. These additives are: (a) dicyclopentadiene, (b) dicyclopentadiene and cyclopentadiene oligomer, (c) olefinic hydrocarbon polymers, (d) bitumen, and (e) 5-ethylidene-2-norbornene. For each additive, we discuss the performance of the sulfur concrete product in view of: (a) biological stability (resistance to sulfur-oxidizing bacteria), (b) chemical stability (chemical reactions, leachability due to acid, base and salt attack, and control of hydrogen sulfide gas generation), (c) hydrodynamic stability (hydraulic conductivity and water absorption), (d) mechanical stability (compressive, flexural, and creep), (e) mineralogical sta-

bility (formation of new minerals), and (f) thermal stability (phase transformation, fire resistance, and cyclic freezing). Additionally, we discuss the use of various chemical additives as fire retardants and techniques to control hydrogen sulfide gas emission during sulfur modification and mixing to form sulfur concrete articles.
7. Sustainable manufacturing technology for sulfur concrete production (Chapter 7). We discuss the durability of sulfur concrete in view of mix design; aggregate composition and amount of binder; choice of aggregates that resists harsh environment conditions; choice of filler that influences workability and thermal stress; composite (sulfur binder, fillers, and aggregates) resistance to acidic, basic, and salt environments; water permeability and absorption; casting procedures and binder compensation; frost resistance; service temperature; fire load; creep; fatigue load; steel and other reinforcements; abrasion resistance and electrochemical stability (corrosion resistance).
8. Basic concepts of sulfur-modified asphalt pavements (Chapter 8). We detail: (a) asphalt composition, properties, and modes of interactions, (b) sulfur-modified asphalt interaction, performance, and processing technology, and (c) sulfur plasticizers.

The authors are indebted to their families and colleagues for valuable and timely input to the development of much of the materials contained in this book.

Indeed, few concepts had been developed to explain sulfur cement and concrete properties and behavior by the early 1970s. The concepts grow gradually as more discussions occur and more research materials become available. This is true in evaluating the physical, chemical, mechanical, and thermal properties of modified-sulfur cement and concrete for their potential use in the construction, asphalt pavement, and sewerage management industries. Undoubtedly, much remains to be done and significant milestones lie ahead in bridging theory and engineering practice.

A.M.O. Mohamed
M.M. El Gamal

ABOUT THE AUTHORS

Prof. Abdel-Mohsen Onsy Mohamed (known as A. M. O. Mohamed) earned his M. Eng. in 1983 and Ph.D. in 1987 from the Department of Civil Engineering and Applied Mechanics, McGill University, Montreal, Canada. From 1987 to 1998, Prof. Mohamed was employed by McGill University and was the associate director of the former Geotechnical Research Centre (GRC) and adjunct professor in the Department of Civil Engineering and Applied Mechanics. In 1998, he joined the United Arab Emirates (UAE) University where he is currently the Director of Research for the university and professor of Civil and Environmental Engineering.

Prof. Mohamed is currently the president of the Gulf Society for Geoengineering (GSGE: http://www.engg.uaeu.ac.ae/gsge) and the Editor-in-Chief of Developments in Arid Regions Research (DARE) Series Published by Taylor and Francis (http://balkema.ima.nl/Scripts/cgiBalkema.exe). He is a board member of several scientific journals and served as a member of the international advisory board for various international conferences.

Professor Mohamed's present research activities contribute to soil properties and behavior, ground improvement, soil-pollutant interactions, transport processes, multiphase flow, remediation of polluted soils, monitoring of subsurface pollutants, solid waste management, environmental impact assessment, risk management, sulphur cement and concrete, and sustainable geoengineering technologies for carbon management.

For publications in refereed journals and conference proceedings, Professor Mohamed published over 200 scientific papers, and for the training of highly qualified personnel, he supervised 29 graduate students (M. Sc., M. Eng. and Ph.D.) in UAE and McGill Universities.

Professor Mohamed's research activities have resulted in co-authoring the following patents:

- Mohamed, A. M. O. and El Gamal, M. New use of surfactant, U.K. Patent Application No. 0807612.7, filed by J. A. KEMP & Co., U.K., dated 25 April 2008
- Mohamed, A. M. O. and El Gamal, M. Method for treating cement kiln dust, U.S. Patent Application No. 12/119525, filed by , Low Hauptman Ham & Berner, LLP, U.S.A., dated 13 May 2008
- Mohamed, A. M. O. and El Gamal, M. New use of surfactant, International Patent Application No. PCT/IB2009/005338, filed by J. A. KEMP & Co., U.K., dated 21 April 2009
- Mohamed, A. M. O. and El Gamal, M. New use of surfactant, GCC Patent Application No. GCC/P/2009/13350, filed by J. A. KEMP & Co., U.K., dated 25 April 2009
- Mohamed, A. M. O. and El Gamal, M. Method for treating particulate material, International Patent Application No. PCT/IB09/005579, filed by J. A. KEMP & Co., U.K., dated 8 May 2009
- Mohamed, A. M. O. and El Gamal, M. Method for treating particulate material, GCC Patent Application No. GCC/P/2009/13456, filed by J. A. KEMP & Co., U.K., dated 12 May 2009

Professor Mohamed's research activities have resulted in co-authoring the following books:
- Yong, R. N., Mohamed, A. M. O., and Warkentin, B. P. (1992) *Principles of Contaminant Transport in Soils*, Elsevier, The Netherlands, 327 pages; The book was translated into Japanese in 1995
- Mohamed, A. M. O., and Antia, H. E. (1998) *Geoenvironmental Engineering*, Elsevier
- Mohamed, A. M. O., Chenaf, D., and El-Shahed, S. (2003) *DARE's Dictionary of Environmental Sciences and Engineering: English-French-Arabic*, A. A. Balkema Publishers
- Mohamed, A. M. O. (2006) *Principles and Applications of Time Domain Electrometry in Geoenvironmental Engineering*, Taylor & Francis Group

Professor Mohamed has edited/co-edited the following books:
- Yong, R. N., Hadjinicolaou, J., and Mohamed, A. M. O. (1993) *Environmental Engineering, Volumes 1 and 2*, ASCE-CSCE
- Mohamed, A. M. O. and Al-Hosani, K. (2000) *Geoengineering in Arid Lands*, A. A. Balkema
- A. M. O. Mohamed (ed.) (2006) *Arid Land Hydrogeology: In Search of a Solution to a Threatened Resource*, Taylor & Francis/Balkema

- A.M.O. Mohamed (ed.) (2006) *Reclaiming the Desert: Towards a Sustainable Environment in Arid Lands*, Taylor & Francis/Balkema
- The proceedings of the UAE University Annual Research Conference for:
    a. The 10th ARC, April 13-16, 2009, Research Affairs, Al Ain, UAE
    b. The 9th ARC, April 23-25, 2008, Research Affairs, Al Ain, UAE
    c. The 8th ARC, April 23-25, 2007, Research Affairs, Al Ain, UAE
    d. The 7th ARC, April 23-25, 2006, Research Affairs, Al; Ain, UAE
    e. The 6th ARC, April 25-27, 2005, Research Affairs, Al Ain, UAE
    f. The 5th ARC, April 25-27, 2004, Research Affairs, Al Ain, UAE
    g. The 4th ARC, April 25-27, 2003, Research Affairs, Al Ain, UAE

Professor Mohamed awarded:
- The Best Performance Award for Excellence in Scholarship, College of Engineering, UAE University (2007)
- The Outstanding Performance and Distinction Award in Research in the College of Engineering, UAE University (2004) and (2001)
- The Outstanding Performance and Distinction Award in University and Community Services in the College of Engineering, UAE University (2004)
- The Best Interdisciplinary Research Project Award in the College of Engineering, Research Affairs Sector, UAE University (2004)

Dr. Maisa El Gamal is an expert in polymer science and technology. Dr. El Gamal received her M.Sc. degree in material science from Alexandria University, Egypt in 1993, and her Ph.D. in polymer science from Tanta University, Egypt, in 1999. From 1993 to1999, Dr. El Gamal was employed as an assistant researcher by the Agriculture Research Centre (ARC), Ministry of Agriculture and Land Reclamation, Cairo, Egypt. From 1999 to 2000, she was employed by ARC as a researcher. In 2000, she joined the Research Affair Sector, United Arab Emirates University, where she is currently employed as a senior research associate.

Dr. El Gamal's research interests are related to soil stabilization, waste solidification and stabilization, polymer science (synthesis, structure, morphology, chemical, thermal, and rheological properties), polymer technology (design, processing, and mechanical characterization), material science, and controlled release formulations.

For publications in journals and conference proceedings, Dr. El Gamal published over 40 papers in the field of soil stabilization, nano-composite, waste solidification and stabilization, sulfur concrete manufacturing, and carbon capture and storage by solid wastes.

Dr. El Gamal's research activities have resulted in co-authoring the following patents:

- Mohamed, A. M. O. and El Gamal, M. New use of surfactant, U.K. Patent Application No. 0807612.7, filed by J. A. KEMP & Co., U.K., dated 25 April 2008
- Mohamed, A. M. O. and El Gamal, M. Method for treating cement kiln dust, U.S. Patent Application No. 12/119525, filed by Low Hauptman Ham & Berner, LLP, U.S.A., dated 13 May 2008
- Mohamed, A. M. O. and El Gamal, M. New use of surfactant, International Patent Application No. PCT/IB2009/005338, filed by J. A. KEMP & Co., U.K., dated 21 April 2009
- Mohamed, A. M. O. and El Gamal, M. New use of surfactant, GCC Patent Application No. GCC/P/2009/13350, filed by J. A. KEMP & Co., U.K., dated 25 April 2009
- Mohamed, A. M. O. and El Gamal, M. Method for treating particulate material, International Patent Application No. PCT/IB09/005579, filed by J. A. KEMP & Co., U.K., dated 8 May 2009
- Mohamed, A. M. O. and El Gamal, M. Method for treating particulate material, GCC Patent Application No. GCC/P/2009/13456, filed by J. A. KEMP & Co., U.K., dated 12 May 2009

SUSTAINABLE DEVELOPMENT FOR THE CONSTRUCTION INDUSTRY

1.1 INTRODUCTION

Over the last few years, a diverse number of public- and private-sector organizations around the world have given increased attention to the problems of excessive natural resource consumption, depletion, and degradation; waste generation and accumulation; and environmental impact and degradation. Since the construction industry is a major contributor to these problems, it now faces increasingly restrictive environmental conservation and protection laws and regulations, the emergence of international standards to address environmental quality and performance such as ISO 14000, and substantial pressures from civic and private environmental groups. As a result, private and public sector owners face new, complex, and rapidly changing challenges imposed by these laws, regulations, standards, and pressures at all life cycle stages of a capital project from initial planning, design, construction, and operation/maintenance to ultimate rehabilitation, decommissioning, and/or disposal. Furthermore, traditional approaches to capital projects of mere environmental regulatory compliance or reactive corrective actions such as mitigation or remediation have proven to be consistently costly, inefficient, and, many times, ineffective.

There are strong incentives for the development of a sustainable approach to capital projects. Such an approach goes beyond the traditional focus on cost, time, and quality performance to include the goals of minimal (1) natural resource consumption, depletion, and degradation, (2) waste generation and accumulation, and (3) environmental impact and degradation, all within the contextual satisfaction of human needs and aspirations. These goals are explicitly and systematically incorporated within the decision-making process at all stages of the life cycle of a capital project, particularly the early funding allocation and the planning and conceptual design phases. However, most stakeholders within the capital project delivery process (i.e., owners, planners, designers, vendors/suppliers, constructors, users/operators) face a complex task when attempting to implement a sustainable approach. First, they face the challenges imposed by increasingly limited resources on the effective and efficient delivery of capital projects. Second, they do not have clear incentives, the proper resources, or the mechanisms or tools to do so. Finally, there is a lack of awareness and understanding of the actual or potential impact and/or implications of environmental regulations and standards on capital projects; a lack of awareness and understanding of the opportunities and potential benefits to an organization created by a sustainable approach to its capital projects; and a lack of credible and reliable quantitative indicators, metrics, and/or data on the actual benefits and associated costs.

A sustainable approach to capital projects would allow the construction industry to take a more aggressive role in finding both short- and long-term solutions for a more effective and efficient use of its increasingly limited and tight capital resources. The anticipated beneficiaries include: owners who would directly accrue the economic benefits of the implementation of specific strategies for investment, execution, and management of capital resources within a sustainable framework; designers and constructors who could significantly enhance environmental quality and performance of capital projects as a result of the application of specific guidelines for sustainable design and construction; and vendors and suppliers who would have a strong incentive to develop and supply sustainable construction technologies, systems, products, and materials.

This chapter discusses the framework of sustainability and implication for the building design and construction industry; how the construction industry can move toward sustainable development in view of energy, material, waste, and pollution; the role of technology in sustainable development; sustainable technology characteristics; and strategies for implementing changes for a more sustainable future.

1.2 SUSTAINABLE DEVELOPMENT

Sustainable development offers a new way of thinking that reconciles the ever-present human drive to improve the quality of life with the limitations imposed on us by our global context. It requires unique solutions for improving our welfare that do not come at the cost of degrading the environment or impinging on the well-being of other people. Sustainable development has been defined by various organizations as:

1. The 1987 World Commission on Environment and Development (WCED) defined sustainable development as:
   Development that meets the needs of the present without compromising the ability of future generations to meet their own needs.
2. The 1989 Tokyo Conference on Conservation of the Global Environment stated that:
   Sustainable development calls for a review of not only the conventional framework of the world economy, such as trade, direct investment, international financing and official development aid, but each country's domestic economic, financial and monetary policies.
3. The 1992 Rio Conference on Environment and Development stated in Principle 3 that:
   The right to development must be fulfilled so as to equitably meet developmental and environmental needs of present and future generations.
   and in Principle 4 further stated that:
   In order to achieve sustainable development, environmental protection shall constitute an integral part of the development process and cannot be considered in isolation from it.
4. Weston (1995) stated that:
   Sustainable development is a process of change in organizing and regulating human endeavors so that humans can meet their needs and exact their aspirations for current generations without foreclosing the possibilities for future generations to meet their own needs and exact their own aspirations.

Although there is no general agreement regarding the precise meaning of sustainability, beyond respect for the quality of life for future generations, most interpretations and definitions of the term *sustainable* refer to the viability of natural resources and ecosystems over time and to the maintenance of human living standards and economic development (National Science and Technology Council, 1994).

Sustainable development is a relationship, or balancing act, between many factors (social, environmental, and economic realities and constraints) that are constantly changing. Because sustainable development is a dynamic concept, it requires decision makers to be flexible and willing to modify their approaches according to the changes in the environment, human needs and desires, or technological advances. This means that actions that contribute to sustainable development today, either in perception or in reality, may be deemed detrimental tomorrow if the context has changed:

Ensuring sustainability over time means maintaining a dynamic balance among a growing human population and its demands, the changing capabilities of the physical environment to absorb the wastes of human activity, the changing possibilities revealed by new knowledge and technological changes and the values, aspirations, and institutions that channel human behavior. Thus, visions of a sustainable world must naturally change in response to shifts in any part of this dynamic relationship (Pirages, 1994).

The basic elements of sustainable development are (Mohamed and Antia, 1998):

1. Education
2. Determining environmental limits
3. Efficient use of natural resources
4. Integrated environmental management systems
5. New technologies and technology transfer
6. Perception and attitude changes
7. Population stabilization
8. Refining market economy
9. Social and cultural changes
10. Waste reduction and pollution prevention

Fundamentally, sustainable development aims for the satisfaction of human needs, the maintenance of ecological integrity, the achievement of equity and social justice, and the provision of social self-determination (Mohamed and Antia, 1998). The real challenge lies in finding ways of putting sustainable development into practice. In the following sections, the social, environmental, and economic issues that are essential to sustainable development are discussed.

1.2.1 Social Sustainability

Sustainability is inherently anthropocentric since it is the welfare of humans with which we are concerned. More than a concern for mere survival, sustainability is a desire to thrive and have the best life possible. There are many socio-cultural issues that influence sustainability. The most prominent issue is intergenerational equity in which we must insure that we leave our children with the tools and resources they need to survive and enjoy life. Therefore, we must strive to raise the standard of living of those people who today lack the most basic requirements such as clean water and adequate food. Other issues in this dominion are: environmental justice, population growth, human health, cultural needs, and personal preferences. These elements have a great deal to do with our quality of life and should not be ignored in favor of the more easily measurable economic elements.

Ways in which social sustainability can be promoted include:
1. Encouraging systematic community participation
2. Emphasizing full cost accounting and *cradle-to-grave* pricing, including social costs
3. Promoting qualitative improvement of social organization patterns and community *well-being* over quantitative growth of physical assets
4. Using resources in ways that increase equity and social justice while reducing social disruptions

1.2.2 Environmental Sustainability

Environmental concerns are also important for sustainable development. The natural environment is the physical context within which we live. Sustainable development requires that we recognize the limits of our environment. There are limits to the quantities of natural resources that exist on the planet. Some of these resources, including trees and wildlife, are renewable so long as we leave enough intact to regenerate. Other resources, such as minerals, are renewed at such slow rates that any use whatsoever depletes the total stock. We need to minimize our consumption of all resources, renewable and depletable.

Another key environmental issue is to minimize our impact on global ecosystems. The earth is like an organism and we must maintain it in a healthy state. Natural ecosystems can survive some impacts, but these must be small enough so that the earth can recover. In some cases, there are particular resources or elements of an ecosystem that are essential to its health. Protecting ecosystem health may involve the protection of an endangered species, the preservation of a wetland, or the promotion of biodiversity in general.

To maintain the natural resources, Goodland and Daly (1995) developed the following practical rules:

1. Input rules:
   a. *Renewable*: Harvest rates of renewable resources inputs would be within the regenerative capacity of the natural system that generates them.
   b. *Nonrenewable*: Depletion rates of nonrenewable resources inputs should be equal to the rate at which renewable substitutes are

developed by human invention and investment. Part of the proceeds from liquidating nonrenewable resources should be allocated to research in pursuit of sustainable substitutes (El Sarafy, 1991).
2. Output rule:
   a. Waste emission from a project should be within the assimilative capacity of the local environment to adsorb without unacceptable degradation of its future waste absorption capacity or other input services.

1.2.3 Economic Sustainability

Economics, as it pertains to sustainable development, does not simply refer to Gross National Product (GNP), exchange rates, inflation, profit, etc. Economics is important to sustainable development because of its broader meaning that explains the production, distribution, and consumption of goods and services. The exchange of goods and services has a significant impact on the environment since the environment serves as the ultimate source of raw material inputs and the repository for discarded goods.

Economic gain has been the driver for much of the unsustainable development that has occurred in the past. A shift to sustainability will only occur if it is shown not to be excessively costly and disadvantageous. Part of sustainability is changing the way things are valued to take into consideration the economic losses due to lost or degraded natural resources and expand our scope of concern from short- to long-term impacts. Once this is done, sustainable development will be revealed to be a more economically beneficial option than current development patterns.

In view of Principle 4, environmental protection is considered to be an integral part of the development process. This is different from the traditional pattern of making economic decisions and then correcting the environmental impacts which may result. This is illustrated in Figure 1.1 in which the natural system includes the ambient physical environment, ecosystem, and natural resources. The economic system refers to the factors of production of goods and services. Utilization of the natural system by the economic system results in a decrease in the natural resources and produces additional environmental problems such as solid wastes, air and water pollution, and greenhouse gases. The importance of these impacts on the natural system varies geographically, depending on the existing states of both the natural environment and the economy.

Economic sustainability involves the consumption of interest rather than capital and is defined by Goodland and Daly (1995) as:

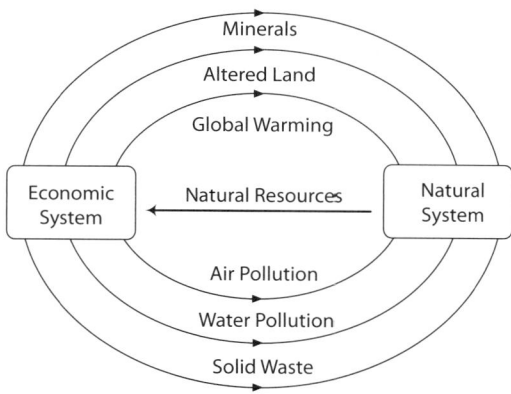

Figure 1.1. Conceptual interactions between economic and natural system (Mohamed and Antia, 1998).

The amount one consumes during a period and still be as well off at the end of the period.

Though this definition has focused solely on man-made capital, the principle can be broadened to include natural capital. Rather than viewing the economy as an isolated system, we must begin to see it as linked to the environmental system. Therefore, the economic sustainability can be defined as *the maintenance of capital in general, both man made and natural* (Mohamed and Antia, 1998).

Nowadays, the concept of *environmental economics*, which seeks to integrate the environmental system into a broader economic system in which current economic principles will still apply, is growing. Much research is being done on internalizing environmental values, which are external to the classic economic system, as well as finding ways of placing monetary values on intangible non-market (and nonmarketable) components of the environment.

1.2.4 Land Sustainability

Land sustainability can be achieved by integrating the ecological, economical, and social objectives (Serageldin, 1993). Figure 1.2 describes this concept. Ecologists stress the preservation of the integrity of the ecological system that is critical to the overall stability of our global ecosystem and that deal in measurement units of physical, chemical, and biological entities.

Economists seek to maximize human welfare within the existing capital stock and technologies and use economic units (i.e., money or perceived value) as a measurement standard. Sociologists emphasize that the key factors in sustainable

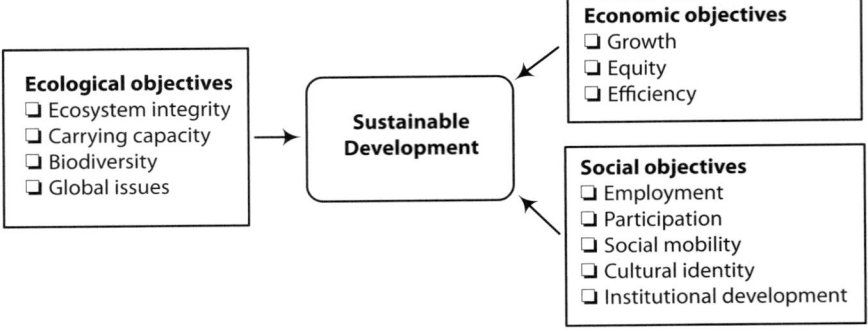

Figure 1.2. Integrated set of objectives for sustainable development (Mohamed and Antia, 1998).

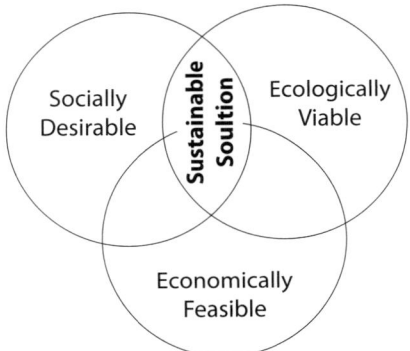

Figure 1.3. Sustainable solution for the development of a sustainable land (Mohamed and Antia, 1998).

development are people with a range of needs and desires and use units that are often intangible such as well-being and social empowerment.

Sustainable solutions for land development fall at the intersection of the spheres, as shown in Figure 1.3, that represent the three key ingredients for sustainable development (Campbell and Heck, 1997). Sustainable development occurs only when management goals and actions are simultaneously *ecologically viable*, *economically feasible*, and *socially acceptable*. These imply environmental soundness and political acceptability. Imbalance among the three components, due to failure in one or more of the spheres, will likely result in failure to achieve sustainable development.

1.3 ROLE OF TECHNOLOGY

Technology plays an important role in sustainable development because it is one of the most significant ways in which we interact with our environment; we use technologies to extract natural resources, to modify them for human purposes, and to adapt our man-made living space. It is through the use of technology that we have seen drastic improvements in the quality of life for many people. Unfortunately, many of these short-term improvements in the immediate quality of life have also exacted a great toll on the environment. In order to proceed toward sustainability, we will have to be more deliberate and thoughtful in our employment of technology. We need to develop and use technologies with sustainability in mind. We need *sustainable technologies*.

To avoid confusion and ambiguity, it is necessary to establish a working definition of *technology*. In this book the term technology is taken to mean *the application of knowledge to the achievement of particular goals or to the solution of particular problems* (Moore, 1972). Thus, technologies include not only the physical tools we use to interact with our environment, but also symbols, processes, and other nontangible effectors such as language and economic transactions that serve as interfaces between humans and enable actions toward the solution of problems to occur.

1.3.1 Characteristics of a Sustainable Technology

A sustainable technology is one that promotes a societal move toward sustainability and one that fits well with the goals of sustainable development. Sustainable technologies are practical solutions to achieve economic development and human satisfaction in harmony with the environment. These technologies serve to contribute, support, or advance sustainable development by reducing risk; enhance cost effectiveness; improve process efficiency; and create processes, products, or services that are environmentally beneficial or benign, while benefiting humans (National Science and Technology Council, 1994). To qualify as sustainable technologies, these solutions must have the following characteristics, in addition to meeting pre-existing requirements and constraints (e.g., economic viability):

- Minimize use of nonrenewable energy and natural resources
- Satisfy human needs and aspirations with sensitivity to cultural context
- Minimize the negative impact on the earth's ecosystems

Minimizing Consumption

The use of nonrenewable energy and natural resources should be minimized because consumption of resources inherently involves increasing the disorder of materials and energy, rendering them of lower utility for future use (Roberts, 1994; Rees, 1990). By subjecting materials and energy to consumption processes, we decrease their potential utility to current and future generations. Therefore, consuming as little matter and energy as possible, or *doing more with less* is a fundamental objective of sustainability.

Maintaining Human Satisfaction

A sustainable technology must fulfill the needs of the population that it is intended to serve. In fulfilling those needs, the technology must account for human preferences and cultural differences. In some cases, these preferences may conflict with environmental and economic criteria and a compromise will have to be found. This does not mean that human preferences should be ignored; fulfillment of our desires means the difference between merely surviving and truly living.

Minimizing Negative Environmental Impacts

Finally, causing minimal negative environmental impacts (as well as maximizing positive impacts) is an important objective of sustainability since the environment consists of ecosystems whose ongoing health is essential for human survival on earth (Goodland, 1994). Sustainability of the human race requires that ecosystems are protected and preserved in a reasonable state of health through maintaining biodiversity, adequate habitat, and ecosystem resilience.

1.4 A FRAMEWORK FOR A SUSTAINABLE INDUSTRY

1.4.1 Current System

In order to understand the changes that need to be made to develop sustainable technologies, it is useful to look at the paradigm that is currently being used. Despite a wide range of positions and opinions on the subject of sustainability, there is general agreement that the current paradigm that disregards constraints to material or energy consumption is unsustainable (Roberts, 1994). In Figure 1.4, a model of the unsustainable development approach that has prevailed over the last few centuries is shown. In this model, several systems are linked. It begins with both renewable and nonrenewable natural resources such as air, water, soil, mineral, or biological resources.

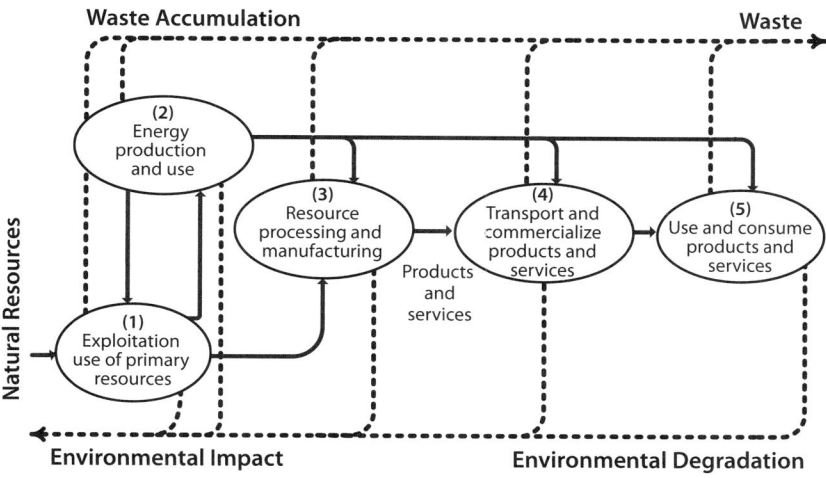

Figure 1.4. Unsustained sustainable development process (adapted from Roberts, 1994).

The basic components are:

1. **Subsystem 1**: *Exploitation and use of primary natural resources*: This system provides inputs for industrial processes. Its outputs are the principal inputs for two other systems.
2. **Subsystem 2**: *Production and use of energy*: The output of this system is a critical input to all the systems.
3. **Subsystem 3**: *Resource processing and manufacturing*: The output of this system is a set of industry-specific products or services.
4. **Subsystem 4**: *Transport and commercialize*.
5. **Subsystem 5**: *Use and consumption*: Products or services are generated by the industrial system across all segments of society.

This process has two additional outputs from each of its systems that are at the core of many problems facing the world today: (a) increasing amounts of hazardous and nonhazardous waste and (b) increasing levels of environmental impact.

It is worth noting that during this process inputs enter at Subsystem 1 and move in one direction through the system to Subsystem 5 and then are disposed, going through the system only once with no cycling of materials. To aggravate the situation even more, this process is fueled by continuous increases in the demand for and the use and consumption of products and services, creating pressures for further exploitation of natural resources, and for continued expansion of energy production, resource processing, and manufacturing capabilities. This unrelent-

12 Sulfur Concrete for the Construction Industry

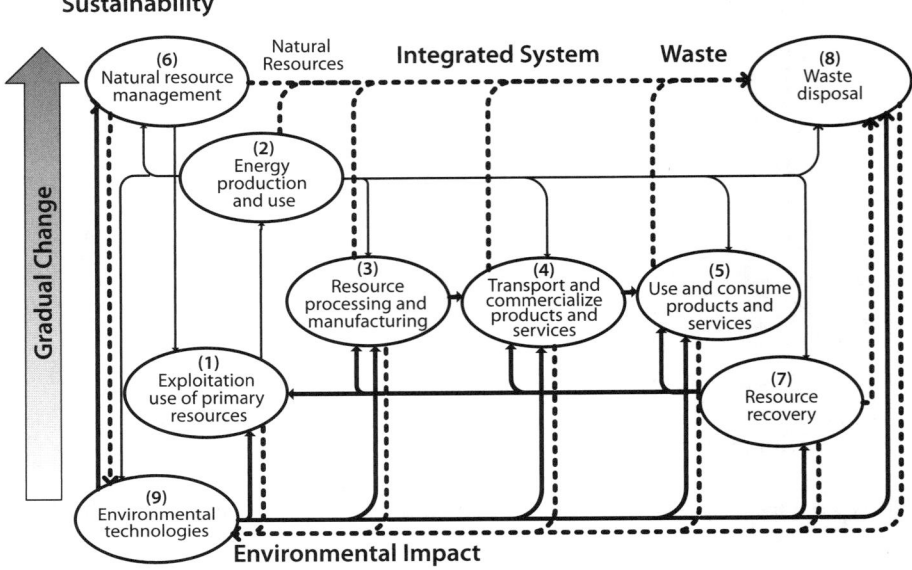

Figure 1.5. Sustained sustainable development process (adapted from Roberts, 1994).

ing growth has created three serious problems: (1) natural resource depletion, (2) accumulation of waste, and (3) environmental degradation. These challenges must be addressed to achieve sustainability.

1.4.2 Modified System

A new way of thinking must be adopted to redirect our development toward sustainability. This cyclic sustainable process is a direct response to the challenges and problems posed by the unsustainable process described in Figure 1.4 and offers a mechanism to gradually overcome the problems of unsustainability.

The framework for a sustainable system presented in Figure 1.5 highlights one way of looking at this new approach (Roberts, 1994). This system shows how to implement two of the three criteria for sustainable technology: (1) economical in use of nonrenewable energy and natural resources and (2) minimal negative impact on the earth's ecosystems. The criteria regarding the satisfaction of human needs and aspirations is not represented explicitly in this figure but nonetheless remain important.

First, the framework represents a closed cyclical system. The total integrated system includes the same five systems described earlier and, in addition, it incorporates four new subsystems, each a response to a specific sustainability challenge:

6. **Subsystem 6**: *Natural resource management:* Addresses the need to manage the exploitation of renewable natural resources in a way that ensures that the supply will always exceed the demand. At the same time, this management system monitors and controls the use of nonrenewable natural resources to prevent their total depletion.
7. **Subsystem 7**: *Resource recovery:* Addresses the need to recover and recycle selected resources and products from waste. These recovered resources would then become inputs to the five basic subsystems in the described framework. They also would contribute to reducing the amount of waste that requires disposal.
8. **Subsystem 8**: *Waste disposal*: Recognizes that a certain amount of waste is inevitable and, thus, will require disposal in ways that are not detrimental to the environment.
9. **Subsystem 9**: *Environmental technologies:* Addresses the need to incorporate proactively, in every subsystem within the framework, strategies and mechanisms that mitigate environmental impacts at the root before the impact happens, through the application of preservation, pollution prevention, avoidance, monitoring, assessment, and control strategies and mechanisms. This subsystem also takes into account that some damage already has been done to the environment and that corrective actions such as remediation or restoration are necessary.

Sustainable technologies should adopt this cyclic closed loop system that mimics natural systems. In this system, the generation of waste is avoided; instead, all byproducts are used as inputs back into production or as inputs into some other process. By minimizing waste, environmental impact is lessened. Because the scale of impact is kept low in this system, change to the environment will be gradual and the surrounding environment will be able to adapt and remain healthy.

1.5 SUSTAINABILITY AND THE BUILDING CONSTRUCTION INDUSTRY

Whereas traditional design and construction focus on cost, performance, and quality objectives, sustainable design and construction adds to these criteria a minimization of resource depletion and environmental degradation and, therefore, creating a healthy built environment (Kibert, 1994). Figure 1.6 illustrates the

14 Sulfur Concrete for the Construction Industry

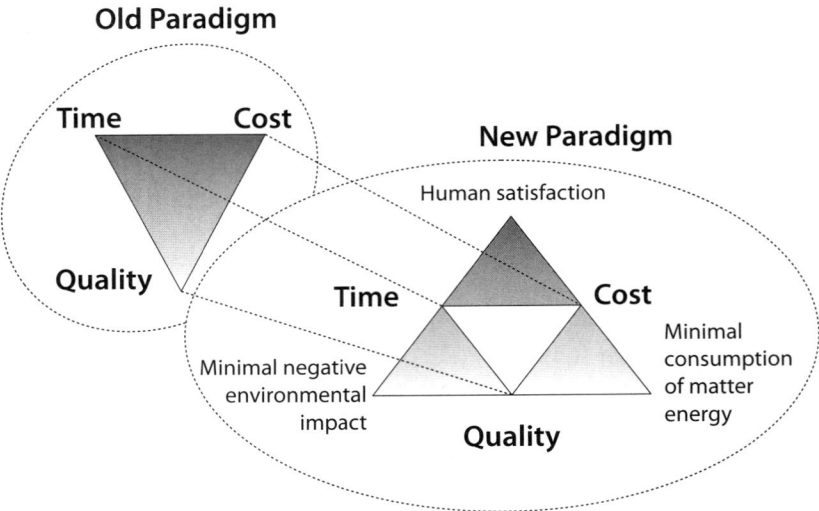

Figure 1.6. Paradigm shift from traditional to sustainable design construction (adapted from Kibert, 1994).

primary paradigm shift to sustainability within the building design and construction industry. This model of the new sustainability paradigm shows issues that must be considered for design making at all stages of the life cycle of the facility.

Sustainable designers and constructors will approach each project with the entire life cycle of the facility in mind, not just the initial capital investment. Instead of thinking of the built environment as an object separate from the natural environment, it should be viewed as part of the flow and exchange of matter and energy that occurs naturally within the biosphere. In addition to the nonliving components that make up the built environment, sustainable designers and constructors must also consider the living components of the built environment (flora, fauna, and people) that operate together as a whole system in the context of other ecosystems in the biosphere (Yeang, 1995).

Life cycle considerations are particularly important with respect to the design and construction of built facilities because the life cycle of a facility involves more than just constructing the facility itself. Operation, maintenance, and decommissioning or disposal of the facility also consume matter and energy and are largely constrained by the design and construction decisions made in the early phases of the facility's life. Not only are changes easier to make during the design of the facility, but the costs of the changes are lower also, since the facility exists only *on paper* as opposed to being a physical artifact that exists in reality after construction begins and ends. Additionally, choices of more costly design features made during facility and design construction may be offset by costs, resources, and energy savings realized over the life cycle of the facility. Thus, the primary responsibility for creating sustainable built facilities falls to the designers and constructors of such facilities.

People who make project decisions with sustainability as an objective will need to evaluate the long-term as well as short-term impacts to the local and global environments. And those who take a sustainability approach to design and construction will be rewarded with reduced liability, new markets, and an earth-friendlier construction process that will help future and current generations to achieve a better quality of life (Kinlaw, 1992; Liddle, 1994).

1.6 STRATEGIES FOR IMPLEMENTING SUSTAINABLE DESIGN AND CONSTRUCTION

In the creation of built facilities, there are many opportunities to improve how design and construction are currently accomplished to make them more sustainable. Three general objectives should shape the implementation of sustainable design and construction while keeping in mind the three categories of sustainability issues discussed earlier (social, environmental, and economics). These objectives are:

1. Minimizing consumption of matter and energy over the whole life cycle of consumption
2. Satisfying human needs and aspirations with sensitivity to cultural context
3. Avoiding negative environmental impact

In the following subsections, we present specific strategies for approaching each of the three objectives, along with examples of technologies and opportunities related to each of the strategies.

1.6.1 Minimizing Consumption

Consumption of natural resources is at the heart of sustainability. With its large scale use of materials and energy and the displacement of natural ecosystems, the built environment greatly influences the sustainability of human systems as well as the natural ecosystems of which we are a part. Minimizing consumption of matter and energy is essential to achieve sustainability in creating, operating, and decommissioning built facilities. The following sections highlight several strategies for minimizing consumption of natural resources over the life of the built facilities.

Improving Technological Efficiency: Doing More with Less

One strategy for minimizing consumption in creating the built environment is improving the technological efficiency of our materials and processes. For materials, we need to improve the efficiency with which they meet the needs for which they are used. An example of this is improving the technology of windows to reduce unwanted thermal losses and air leakage in climate-controlled applications. With respect to processes, technological efficiency means reducing the amounts of input matter and energy required to generate the desired outcome of the process. In construction, improving the site layout to reduce the travel distance of excavating equipment is an example of improving process efficiency, resulting in fewer equipment hours, less fuel used, and lower maintenance costs.

Reuse, Rehabilitation, and Retrofitting

Reusing buildings, materials, and equipment is a second strategy for making design and construction more sustainable. By reusing what already exists, we save the cost, material, and energy input that would be required to create new facilities *from scratch*. The primary reason for disposal of facilities and materials is that those artifacts do not meet the present needs of humans. By using techniques such as adaptive reuse, rehabilitation, or retrofitting, old facilities can be modified or improved to meet new use criteria, at a much lower consumptive cost than building a new facility. An example of adaptively reusing existing facilities are loft apartments developed in the structures formerly used for factories. Materials and equipment can also be reused or rehabilitated to varying degrees. The greatest impediments to this strategy are artifacts that are designed for obsolescence, with short life cycles, or where economic constraints have forced sub-quality construction or manufacturing.

Creating New Technologies

Many opportunities exist to increase the sustainability of human activity by creating new technologies. Consumption of matter and energy can be reduced by developing new technologies that do not rely on traditional types or amounts of materials and energy to meet human needs. Photovoltaic (PV) panels, which generate electricity from solar radiation, are one example of such a technology. Instead of using finite reserves of coal or oil to make the electricity used by humans, PV panels use the essentially infinite resource of solar energy. Opportunities for new technologies can be found by observing natural ecosystems: what sources of energy and matter are used by these systems? Particularly promising opportunities exist in the area of waste recovery and reuse. Examples of taking artificially-generated waste that would otherwise have been disposed in the natural environment, and using it as input back into the building process are: (a) using waste masonry and concrete from demolished structures as aggregate in new concrete and (b) using sulfur, from the oil industry, fly ash, and aggregates for production of sulfur cement and concrete.

Modifying Historical Technologies

Technologies have been used over the course of human history to meet the needs of people. Many of these technologies have been forgotten or abandoned as new technologies were developed to replace them. Whereas most of these technologies may appear to be obsolete, some may prove to be useful in and of themselves, or to suggest ideas for new technologies. Traditional construction techniques such as rammed earth have found new applications in structures constructed from waste automobile tires filled with compacted earth. By combining a knowledge of historical building techniques with consideration of the dangerous problem of waste tire disposal, builders have developed a low-cost system that helps to deal with waste disposal while creating a useful and durable structure.

Reshaping Human Desires

A more fundamental strategy for minimizing consumption is to attempt to change human desires and tastes. Whereas fundamental human needs such as food, shelter, and water are not greatly adaptable, other human wants are often significantly responsive to external influences. The obvious architectural trends in built facilities from decade to decade are an example of how designers can influence consumer demand and thus the consumption of matter and energy. Other mechanisms for changing human consumptive patterns are education and awareness. If people are aware of the impacts of their choices on the ecosystems of which they are a part, they may make more enlightened choices.

1.6.2 Satisfying Human Needs and Aspirations

The quality of the facility as a man-made environment for people is determined by how well it meets human needs and aspirations for such things as security, nontoxicity, shelter, aesthetics, and other functional requirements. Other human needs that are indirectly met by built facilities include economic profitability for those who participate in the design and construction of the facility. Since sustainability is meaningless without reference to humans and their continued survival, the second objective of applied sustainability is satisfaction of human needs and aspirations.

Improving Economic Viability

In today's world, economic viability is an important consideration for any building project. Indeed, a facility design that is sustainable but too expensive to construct has little value in and of itself. Thus, increasing cost effectiveness of facilities is a critical strategy for creating sustainable built facilities. Economic viability often follows from achieving the objectives of minimizing cost and negative environmental impacts, since less consumption means less cost, and reduced environmental impact means lower liability and remediation costs. However, tradeoffs usually exist with respect to economic viability. While sustainable choices save money in the long-term, they are often more expensive initially, making these choices seem unattractive from a short-term perspective. To accurately identify the economic viability of sustainability choices, we need technologies that assist in cost-benefit analysis, financial forecasting, and long-term predictions. In addition, revised economic valuation schemes that assign meaningful values to reserves of natural resources and ecological habitats are essential in assessing the economic viability of construction projects.

Matching User Needs with Facility Design

In creating a facility that is sustainable based on the human satisfaction criteria, the first step must be to identify the needs of the people who will use the facility. These needs shape the basic functional requirements of the facility and must be met in order for the facility to be considered sustainable. The facility design process has been described by one architect as *"establishing a fit between the pattern of needs and use: the patterns of built form, servicing systems, technological factors, and environmental factors"* (Yeang, 1995). Opportunities exist in the area of systematic human needs assessment, and adapting those needs as input to the design process. Additionally, technologies such as decision support systems can help designers and project decision makers to match user needs with appropriate building functionalities within the design.

Creating a Healthy Built Environment

In addition to the basic functional requirements of users that must be met by the facility, designers, and constructors must also strive to include factors that create a healthy environment both inside and outside of the facility. Nontoxic materials are an essential component of a healthy built environment, as well as design features that convey aesthetic or spiritual values conducive to the tasks and activities that occur within the facility. Besides the requirements for creating a healthy indoor environment, sustainable design also requires consideration of the interfaces between the built environment and the natural environment. Nontoxic materials and processes are essential technologies for achieving sustainability throughout the facility life cycle.

Empowering People to Meet their Own Needs

A final strategy for satisfying human needs in the built environment is empowerment. By including users in decision making for the planning and design of facilities, the final facility will be more likely to meet the needs of those users. Allowing user participation at all phases of the facility life cycle creates an awareness among the users of the interfaces of the facility with its environmental context and a respect for the flows of energy and material through the built system over time. Strategies such as owner/builder programs in which people are taught techniques for constructing their own homes that invite a respect for the final outcome that might not exist for manufactured or contractor-built houses. This respect and understanding can only lead to more sustainable design and construction.

1.6.3 Avoiding Negative Environmental Impacts

Built facilities impact the natural environment in many ways over their entire life cycles. Yeang (1995) lists four categories of impacts that built facilities have on the earth's ecological systems and resources:

1. Spatial displacement of natural ecosystems, and modification of surrounding ecosystems as a result
2. Impacts resulting from human use of the built environment, and the tendency for that use to encourage further human development of the surrounding ecosystems
3. Depletion of matter and energy resources from natural ecosystems during the construction and use of the facility
4. Generation of large amounts of waste output over the whole life cycle of the facility, which is deposited in and must be absorbed by natural ecosystems

Given their large scale and long life cycles, built facilities have particularly large and long-lasting effects on the environment as a whole. The following strategies are examples of approaches that can be taken to improve the sustainability of built facilities by avoiding negative environmental impacts over their life cycle.

Recovering Waste: Reduce, Reuse and Recycle

Various approaches exist to help recover waste from building construction and operation processes. Pollution prevention, for example, is a strategy that advocates anticipating and eliminating pollution before it is produced and has been used successfully in factory fabrication applications. Material recycling is also commonly used in prefabrication processes where careful planning can eliminate waste or enable it to be directly recycled back into the manufacturing process or to other complementary processes. Construction and demolition waste recycling is also becoming increasingly popular as disposal options become more expensive. Promising applications include recycling construction and demolition waste into new composite materials for construction such as the concrete aggregate.

Reusing Existing Development

Another way of minimizing impacts on the natural environment is by making better use of sites and facilities that have already been used. Rehabilitation of existing structures for similar or adaptive uses, as well as using retrofitted existing sites rather than green field sites for new development, are examples of strategies that reduce negative impacts on the natural environment. By reusing existing sites and/or facilities, we save costs and avoid negative impacts by avoiding the need to *start from scratch*. Additionally, peripheral costs such as extending utility and transportation systems to green field facilities, as well as travel savings for users, are reduced or eliminated. Thus, not only is reuse of existing development more sustainable because of its reduced environmental impact, it can also be economically beneficial. Finally, redeveloping unsavory components of built systems can lead to improvements in the human system, as well, by providing better environments for living and by encouraging further development.

Integrating the Built Environment into Ecological Systems

Sustainability must occur within the context of natural ecological systems because these systems provide the resources for all human activity. The built environment can be integrated into the natural environmental context of its site and bioregion by designing material and energy flows into and out of the built system to fit within the yield and assimilative capacities of that context. Grey water systems are an example of a technology that has been successfully used to facilitate the processing and absorption of human wastewater back into the natural environment. Rather than collecting the wastewater and using artificial chemical treatment processes to eliminate contaminants, grey water systems take advantage of the naturally purifying processes of ecosystems in their operation. As an added bonus, the grey water relationship is symbiotic since the plants, which purify the wastewater, use the contaminants as a nutrient. Thus, integration of built systems into the surrounding ecological context can be mutually beneficial to humans and nature provided that humans do not exceed the assimilative capacity of natural systems.

1.7 SUSTAINABILITY AND PROJECT PROCUREMENT LIFE CYCLE

Construction project procurement life cycle encompasses the following critical phases:

a. Business justification
b. Project brief and procurement process
c. Design brief
d. Construction process
e. Operation and management
f. Disposal and reuse

1.7.1 Sustainable Business Justification

Defining a business case establishes the need for the project. An effective business case sets out the range of solutions that would meet the business objectives and justify the proposed project. This includes whether the delivery of a construction project, or of a service, is the most appropriate way to meet the business need. Questions around sustainability must be an integral part of this process. Examining these issues will help the project team identify the range of options and deliver the best value, whole-life solution—one that not only meets the business need but also makes informed decisions about the feasibility and nature of the project.

The key issues that should be considered in the development of the business case are: (a) whether construction is the best solution to meet a recognized business need, (b) the impact that alternative service, delivery, and construction options have on the effectiveness and efficiency of the business operation, (c) the impact that options (in delivering and operating the facility) will have on the stakeholders, and (d) including staff and the local community, as well as the social, economic, and environmental impacts of the various options.

The key areas of sustainability that should be considered during the development of the business case are:

A. Economic Aspects:

- *Whole-Life Costing (WLC) and Value for Money:* Assessment should be carried out on a whole-life cost basis. (a) Is construction the best choice? (b) Can the business need be delivered in a different way? (c) Have the options and cost assessment been undertaken for refurbishment versus new build?
- *Economic Regeneration:* How will it affect the local economy?

- *Function:* The project should offer flexibility and have the ability to adapt to future changes.
- *Investment and Project Delivery:* WLC must be assessed against the benefits that will be delivered; higher WLC may deliver greater business benefits.

B. Environmental Aspects:

- *Location:* (a) Does the construction utilize previously developed land and buildings? (b) If so are there any issues of contamination? (c) are there any planning issues? (d) Has an environmental impact assessment (EIA)/site appraisal been carried out?
- *Transport Infrastructure:* (a) How is the area currently serviced? (b) Are public services readily available? (c) Is the development reliant on energy-intensive forms of transport? (d) Is there an infrastructure in place to support the community? (e) How would the project impact on current infrastructure? (f) Has a transport impact assessment (TIA) been carried out?
- *Biodiversity:* (a) Has a detailed survey and EIA been undertaken? (b) Does the development create opportunities to enhance green spaces and nature conservation? (c) Are there planning conditions relating to biodiversity?
- *Energy and Water:* (a) Will the development actively contribute to the government targets to reduce emissions of CO_2? (b) Will the development use energy efficiently? (c) How will the development impact on local gas and distribution networks? (d) How will the development impact on the local sewage system? (e) Can the local water supply cope with additional demand from the development? (f) Will the development pose risks to water pollution?

C. Social Aspects:

- *Stakeholders:* The local community should be consulted during the decision-making process. Have their needs been identified and taken into consideration?
- *Culture/Heritage:* The project should enhance or preserve the existing culture and heritage and should address any negative visual impact. Issues to be considered are: (a) Will the project be sympathetic to the local styles of architecture? (b) Is there a possibility of uncovering archaeological remains? (c) Are there planning constraints in place?

Sustainable Development for the Construction Industry 23

Table 1.1. Sustainability and the project procurement lifecycle

Project activity	Items to be considered to achieve sustainability		
	Economic	Environmental	Social
Sustainable business justification	• Life cycle assessment • Economic regeneration • Investment & project delivery	• Location • Transport infrastructure • Biodiversity • Energy • Water	• Stakeholders • Culture/heritage • Health & safety
Sustainable procurement process	• Life cycle assessment	• Biodiversity • Energy • Water • Waste minimization & management • Materials • Pollution (air, noise, land, & water) • Environmental performance standards	• Respect for people • Health & safety • Stake holder/ local community • Culture/heritage • Project team • Contractor selection • Supply team
Sustainable design	• Function: adaptability & re-use	• Location • Enhancing biodiversity • Energy efficiency • Materials • Waste minimization & management • Transport & travel • Water • Pollution (air, noise, land & water) • Climate change	• Internal environment & accessibility • Culture/heritage • Health & safety • Local communities
Sustainable construction process	• Performance monitoring • Cost management • Logistics	• Biodiversity • Energy efficiency • Waste minimization & management • Pollution (air, noise, land & water)	• Respect for people • Health & safety • Local communities
Sustainable management & operation of the facility	• Cost management	• Water • Energy • Biodiversity • Waste minimization & management • Environmental performance standard • Environmental management system	• Post occupancy evaluation • Health & safety • Education
Sustainable disposal & re-use of the site	• Adaptation for new use	• Waste minimization & management • Materials • Pollution (air, noise, land & water)	• Remove any hazardous materials such as asbestos

- *Health and Safety:* Appropriate resources should be allowed for within the business case to comply with government policy of health and safety.

1.7.2 Sustainable Procurement Process

The procurement process consists of: (a) the establishment of the procurement route, (b) finalizing tender documents, including the project brief and the output based specification, and (c) the tender selection process for an integrated supply team.

The project brief sets out the vision, strategy, and requirements of the project. It must clearly highlight the importance of sustainability considerations to the client to ensure that all parties involved in the project are conscious of the client's needs and requirements.

Before beginning the tender process, the client must develop the project brief into an output-based specification that defines the objectives that they wish to achieve but not how these should be met. Within this, the client must set out those objectives that are considered essential and those that are desirable. This specification must include the sustainable performance objectives for the project, covering both the construction and operation of the facility so that prospective companies who are submitting tenders can fully respond to these requirements. Equally, it is vital to assess which risks are more appropriately managed by the client and which should be managed by the supply team and to develop the detail of the specification taking this analysis into account.

Companies submitting tenders should also be asked to provide full details of how they will respond to the required sustainability objectives. The importance of this element in the tender appraisal process should be made clear. This will encourage them to suggest innovative approaches and alternatives that offer better value for money and/or whole-life cost performance. As part of the tender evaluation process, the client should explicitly appraise the responses to the sustainability criteria defined within the tender documentation. This appraisal should also consider the supply team's knowledge and experience of sustainable projects.

The key areas of sustainability that should be considered during the development of the project brief and the output-based specification are:

A. Economic Aspects:
- *Whole Life Costing/Value for the Money*: Tender documents should emphasize the importance of life-cycle-assessment costs (LCAC) in delivering a value for money, sustainable project. Project briefs should clearly set out the benefits the facility is intended to deliver

and should seek to link their delivery to supply team rewards. In a long-term concession, does the contract ensure that the residual assets value at the end of the concession reflects life cycle assessment costing constraints?

B. Environmental Aspects:

- *Biodiversity*: A requirement for a biodiversity management plan (BMP) for current development and long-term management should be included in the brief. The BMP should encompass the following: (a) consultation and scoping study, (b) detailed surveys and impact assessment, (c) design of development to incorporate biodiversity objectives, (d) enhancement, mitigation, and compensation, and (e) management and after care. We have to make sure that biodiversity interest has been identified on the site.
- *Energy*: The brief should define targets for energy consumption during construction and operation, and how they will be monitored. The brief should identify minimum requirements for energy performance for both new facilities and major renovations. Will the brief include the requirement to procure only buildings in the top quartile of energy performance for the government estate?
- *Water*: The brief should include targets and how they will be monitored for water consumption both during construction and when the facility is in operation.
- *Waste Minimization and Management*: A requirement for suppliers/contractors to provide a waste management plan should be included in the brief. Targets should be specified for reuse/recycling during construction and in operation and how they will be monitored.
- *Materials*: The brief should include a requirement to use materials that contribute to the sustainability goals the project is aiming to achieve. Examples are: (a) use of preferred standards, (b) reuse of materials that can be recycled or reclaimed on-site, (c) avoidance of environmentally damaging materials, (d) use of natural materials, and (e) avoidance of materials that are potentially harmful to humans.

The brief should include an outcome-based requirement for overall material efficiency such as a minimum requirement for recycled content in the project. For example, one may state that a minimum requirement such as 10% of the materials value of the project should derive from recycled or reused content.

- *Pollution (Air, Noise, Land, and Water)*: The brief should define targets to minimize or reduce pollution where possible. Risks should be identified and a plan to mitigate potential sources of pollution should be stated in the brief. Targets for minimizing the pollution should be set out by the client and key performance indicators (KPIs) should be used to benchmark performance.
- *Environmental Performance Standards*: The brief should include a requirement for the use of a performance measurement target set by various professional agencies. The brief should take account of governmental or departmental strategy and targets.

C. Social Aspects:
- *Respect for People*: The respect for people toolkit addresses the following six action themes: (a) equality and diversity in the workplace, (b) working environment, (c) health, (d) safety, (e) career development and lifelong learning, and (f) worker satisfaction.
- *Health and Safety*: The brief should take full account of government policies.
- *Stakeholders/Local Community*: The views, interests, and requirements of stakeholders should be addressed within the brief. The brief should include a provision for future consultations on design, construction, and operating issues.
- *Culture/Heritage*: Does the brief identify the client's commitment to preserving and maintaining the culture and heritage of the local community?
- *Project Team and Contractor Selection*: The brief should identify clearly the sustainability criteria those tendering will be measured against. Does the tender documentation encourage supplies to innovate and offer higher sustainability solutions?
- *Supply Team*: The supply team should have the following characteristics:
    1. *Knowledge and Competence*: (a) Can the integrated supply team provide examples of successful sustainable projects they have completed? (b) Does the integrated team have the relevant experience?
    2. *Commitment and Motivation*: (a) Is the integrated supply team enthusiastic and interested in sustainability issues? (b) Does the integrated supply team actively promote sustainability?
    3. *Resources*: Does the integrated supply team have suitably experienced resources to implement a sustainable project?

4. *Training*: (a) Is the integrated supply team willing to educate all stakeholders? (b) Is the integrated supply team able to provide handover training?
5. *Team Building and Communication*: Will the integrated supply team share knowledge and best practice?

1.7.3 Sustainable Design

Developing the Sustainable Design

The design brief should be developed in partnership with the integrated project team. It should expand on the project brief by providing greater detail but still be flexible enough to allow for alternative solutions. This can then be used as the base for developing the scheme and for detailed design proposals. During the design phase, the client and the supply team continue to have an opportunity to influence the sustainability performance of a development. A key focus is to identify those construction materials that best meet sustainability targets. These sustainability considerations, such as the requirement for energy efficiency or accessibility, will have a significant influence on how the final design is reached.

The key areas of sustainability that should be considered during the development of both the outline and detailed design briefs are:

A. Economic Aspects:

- *Function*: Does the design encompass adaptability and reuse?

B. Environmental Aspects:

- *Location*: The design should take account of a facility's orientation, solar radiation levels, wind speed, and wind direction.
- *Enhancing Biodiversity*: Biodiversity objectives should be drawn up to reflect both opportunities and constraints for conservation within the design. Will the building materials utilized in the design be benign to local species?
- *Energy Efficiency*: The design should incorporate energy-saving features. (a) Does the design make use of renewable energy sources? (b) Does the design make use of alternative means of heating/cooling? (c) Does the design utilize products or apply processes that allow the facility to utilize energy efficiently?
- *Materials*: (a) Have considerations been given to using materials with low-embodied energy? (b) Can the material be sourced locally reducing the energy used in delivery to site? (c) Does the specified materials comply with government policy? (d) Does the design maximize the

cost-effective use of recycled products? (e) Does the design incorporate materials with a long life and low-maintenance requirements? (f) Does the design include specifications for low-energy use materials?

Examples:

Crop-based materials are increasingly being used by the construction industry as an alternative to more traditional materials. A number of crop-derived materials are available that provide significant benefits through reduced environmental impact and cost savings during disposal. In many cases, the additional cost of renewable material is offset over the whole life of the product.

a. *Insulation Materials*: Have the benefits of low-embodied energy in manufacture, naturally good performance when damp, and renewable feedstock been identified?
b. *Paints*: Have the benefits of low-embodied energy in manufacture, reduced toxicity and disposal issues, and renewable feedstock been identified?
c. *Floor Covering*: Have the benefits of reduced health and allergy issues, disposal benefits, and renewable feedstock been identified?
d. *Biomass Heat Boilers*: Have the benefits of renewable energy and efficient and carbon neutral been identified?
e. *Geotextiles for Landscaping and Roadside Use*: Have the benefits of much lower-embodied energy, degrades naturally at end of life, and renewable feedstock been identified?

- *Waste Minimization and Management*: (a) Has thought been given to a design that minimizes waste both during construction, operation, refit, and demolition? (b) Does the design make use of prefabricated components? (c) Can waste off-cuts be returned to suppliers for recycling? (d) Can standard components be used? (e) Does the design take account of the segregation and storage of waste during operation of the facility? (f) Does the design facilitate both routine maintenance and component life cycle replacement?
- *Transport and Travel*: Does the design promote the use of public transport?
- *Water*: (a) Does the design incorporate water-saving features both for consumption and discharge of waste? (b) Has a sustainable urban drainage system been considered? (c) Does the design incorporate grey water recycling and rain water harvesting?

- *Pollution (Air, Noise, Land, and Water)*: Does the design mitigate against any possible risks to pollution?
- *Climate Change*: The facility should be robust enough to cope with future climate change. (a) What practical measures can be adopted within the design?

C. Social Aspects:

- *Internal Environment and Accessibility*: Does the design provide for a healthy and comfortable environment? i.e., the design should be accessible to the young, elderly, or disabled.
- *Culture/Heritage*: Does the design enhance the historic or local environment?
- *Health and Safety*: The design should consider the health and safety requirements of the end user. (a) Will the materials, products, and furniture be assessed for safety? (b) Does the design minimize the risk of crime?
- *Local Communities*: (a) Has the local and wider community been consulted on the design? (b) Have design quality indicators been used to help stakeholders evaluate the design quality?

1.7.4 Sustainable Construction Process

Before Construction Begins

Before the construction process begins the client must be satisfied that the proposals meet or exceed the original project and design brief.

During Construction

Construction sites often have a negative impact on the local environment and community through noise, air, water and land pollution. The client and the integrated supply team should make provisions to minimize pollution and disruption and to ensure the health and safety of local residents as well as construction site staff. However, during the construction process there are opportunities to make cost savings and reduce the environmental impact through waste recycling and recovery. For example, the client can cut the costs of disposal and reduce the pressure on the landfill by utilizing reclaimed materials and recycled aggregates.

After Construction: The Handover

The client should see the commissioning and handover of a project as an important and final phase of the construction process. As part of the handover, it is essential to provide the client with training, facility operations and maintenance information, health and safety files, and procedures for reporting defects.

Key areas of sustainability that should be considered both before and during the construction process are:

A. Economic Aspects:

- *Performance Monitoring*: Is economic, social, and environmental performance being monitored, recorded, and reported on site?
- *Cost Management*: Have whole-life costs been reassessed?
- *Logistics*: Has the use of a central delivery-handling center been considered?

B. Environmental Aspects:

- *Biodiversity*: (a) Has a biodiversity management plan been implemented? (b) Are construction techniques sympathetic to the local habitat and species?
- *Energy Efficiency*: Are there plans in place to minimize energy use during construction?
- *Waste Minimization and Management*: Hazardous waste on-site should be disposed of in the correct manner. (a) Has a site waste management plan been implemented? (b) Is there a provision for waste segregation and auditing? (c) Have steps been taken to minimize construction waste? (d) Can any construction waste be recycled or sold? (e) Who is responsible for disposing of the waste and are they licensed to do so?
- *Pollution (Air, Noise, Land, and Water)*: (a) Are there plans to minimize and monitor pollution? (b) Are there plans to conserve and minimize water usage on-site?

C. Social Aspects:

- *Respect for People*: (a) Are the contractors, suppliers and designers committed to their workforce? (b) Are they committed to achieving the respect for people standard? (c) Are the site staff and subcontractors educated and trained in environmental awareness?
- *Health and Safety*: The site should be secure from theft and vandalism. (a) Are the contractors registered to and committed to the considerate contractors scheme or something similar? (b) Are the contractors registered to and committed to the construction skills certification scheme or something similar?
- *Local Communities*: Are the relevant stakeholders being kept informed of the progress?

1.7.5 Sustainable Management and Operation of the Facility

It is essential that, as far as possible, the facility is monitored and maintained according to the predefined sustainability criteria set out in the project and design brief. Following the handover of the completed facility, the client must ensure that its end-users are educated and trained in how to use the facility efficiently. This is not only an ideal opportunity to promote the sustainability aims (and achievements) of the facility itself, but it will encourage end-users to play their part in meeting those goals as well as instilling a sense of belonging. Once the end-users have had time to adjust to their surroundings, a post-occupancy evaluation should be carried out to identify not only how satisfied people are with the building, but to examine how the facility is meeting its environmental objectives.

This feedback is extremely useful. It can help iron out any problems with the facility and the information gathered could be used to inform on future projects. Those key performance indicators that were identified for the site in-operation should be assessed and verified appropriately throughout the operational life of the facility. All major facilities should operate under an environmental management system that provides a framework for setting, implementing, and monitoring environmental targets. This will help deliver cost savings and demonstrates, through verification, an organization's positive environmental achievement.

Key areas of sustainability that should be considered throughout the management and operation of a facility are:

A. Economic Aspects:

- *Cost Management*: Is there a process in place to validate whole-life costing?

B. Environmental Aspects:

- *Water*: Plans should be in place to monitor and reduce water usage during operation.
- *Energy*: Plans should be in place to monitor and reduce energy usage during operation. Are there plans in place to undertake regular inspections of boilers and air conditioning systems (in buildings)?
- *Biodiversity*: Plans should be in place to manage and care for areas of conservation both on-site and within the facility.
- *Waste Minimization and Management*: There should be guidance on waste minimization and recycling for users and facility management. Are recycling facilities provided?
- *Environmental Performance Standard*: Has a performance measurement been carried out?

- *Environmental Management System*: Has an environmental management system been put in place? Many organizations are now compliant with ISO 14000 that provides a framework for the development of an environmental management system and the supporting audit program.

C. Social Aspects:

- *Post Occupancy Evaluation*: Feedback should be obtained from the occupants on user satisfaction. Has feedback been obtained on the performance of the facility?
- *Health and Safety*: (a) Is there a health and safety file? (b) Are risk assessments carried out regularly?
- *Education*: End-user training should be completed as part of the handover. (a) Has environmental training been given? (b) Are end-users aware of the sustainability aims and achievements of the facility? (c) Have procurement advisors been trained in the need for securing sustainable procurement contracts, e.g., buying recycled, buying energy efficient, etc.

1.7.6 Sustainable Disposal and Re-Use of the Site

There are two distinct areas a client must be aware of at this phase of the construction project life cycle. These are: (a) disposal of a surplus facility and (b) reuse of an existing facility. Each area has its own set of considerations that will apply, depending on the individual circumstance of the facility in question.

Disposal

Where a facility is identified as surplus to requirements, it is important to dispose of it on best terms bearing in mind the sitting tenants, biodiversity, and historical context.

Reuse

Occasionally facilities reach the end of their life or no longer fulfill the function for which they were built. There are a number of options a client can choose to adopt. However, it is important that the solution represents an efficient, affordable, and sustainable use of an existing built asset. The options for re-use are: (a) adapt for a small change in use, (b) refurbish and alter for a major change in use, and (c) demolish and recycle if the facility cannot be reused.

It is always preferable to adapt or refurbish an existing facility rather than choosing to construct anew. However, there will be circumstances in which a facility has come to the end of its life. Demolition of an existing facility can create a large and complex waste stream that covers a wide array of materials. This waste stream, if managed carefully, can provide materials for re-use in new structures, can lessen the associated environmental impacts, and deliver cost savings through avoided waste disposal fees (for example, design for deconstruction).

Key areas of sustainability that should be considered before commencing disposal or re-use are:

A. Economic aspects:

- *Adaptation for New Use*: Can the facility be adapted to meet future needs?

B. Environmental Aspects:

- *Waste Minimization and Management*: (a) Has a waste management plan been put in place? (b) Can any demolition waste be recycled or sold?
- *Materials*: Have existing building materials been identified for re-use?
- *Pollution (Air, Noise, Land, and Water)*: Plans should be in place (a) to control and minimize emissions to air, noise and vibration, and (b) to avoid contaminating land and water.

C. Social Aspects:

- The demolition process should be fully planned in order to minimize risk to health and safety. Is there a need to remove any hazardous materials such as asbestos?

1.8 SUMMARY AND CONCLUDING REMARKS

The recent global attention to the issues and challenges of sustainable development is forcing industries to conduct self-assessments to identify where they stand within the framework for sustainability and, more importantly, to identify drivers, opportunities, strategies, and technologies that support achieving this goal. However, in order to understand the changes that need to be made to achieve sustainability, it is useful to look at the paradigm that is currently being employed. Despite a wide range of positions and opinions on the subject, there is general agreement that the current paradigm of development that disregards constraints to material resources and/or energy consumption is unsustainable. This paradigm, which has prevailed over the last few centuries, is based on an unsustainable approach to development that begins with the extraction and use of primary natural resources, both renewable and nonrenewable such as air, water, soil, mineral, or biological resources. These resources are then utilized for energy production and use and as inputs to resource processing and manufacturing processes. The results of these processes are industry-specific products or services that are eventually transported and commercialized, and ultimately used and consumed across all segments of society. In the process, all inputs and outputs move in one direction until disposed, going through the system only once with no recovery of materials. Aggravating this situation even more is a continuous increase in the demand, use, and consumption of products and services that creates pressures for further extraction of natural resources and for continued expansion of energy production, resource processing, and manufacturing capabilities. This unrelenting growth has created three serious problems: (1) excessive natural resource consumption, depletion, and degradation (both renewable and nonrenewable); (2) waste generation and accumulation (including organic and inorganic; hazardous and nonhazardous); and (3) environmental impact and degradation (on the air, water, land, biota). These are the challenges that must be overcome to achieve sustainability.

In striving to achieve sustainability in the built environment, three themes emerge. First, awareness of the impacts that built facilities have on both human and natural systems is essential and should be considered as early as possible in the planning and design of any built facility. Second, the ecological, social, and economic contexts of the facility must be taken into account for all project decision making. Finally, sustainable designers and constructors must be aware of the connectivity of human systems to the natural environment. No human action can take place without affecting the ecological context in which it occurs. All human activity must be undertaken with an awareness of the potential consequences to other humans and nature, especially the construction of built facilities because of its large scale.

The principal conclusion is that the area of sustainable technologies is one of increasing interest that has many levels and complex dimensions. In this chapter we have tried to provide a brief overview of the wide range of technological issues at an industry level while emphasizing the need for an integrated approach and understanding of the various components of a sustainable system. We have stressed the importance of adopting a new paradigm that considers industry as a total system rather than focusing on individual components of processes and operations. In order to achieve sustainability for society as a whole, and for construction in particular, intelligent decision making is required that includes full consideration and knowledge of the many trade-offs and impacts associated with each alternative available as an option. Sustainability is a desirable state toward which to strive, but the journey is not easy.

2

SULFUR PRODUCTION AND USES

2.1 INTRODUCTION

Sulfur is different from other minerals with respect to mining issues and supply concerns. The sustainable production of necessary sulfur supplies is not in question as long as the world economy is petroleum-driven. Sulfur mining has stopped, not because sulfur resources have become depleted, but because sufficient sulfur can be recovered as a byproduct at petroleum refineries and natural gas processing plants. The amount of sulfur that is presently being produced is more than the demand for sulfur worldwide. Sulfuric acid, as a byproduct from metal smelters, contributes additional supplies.

Whereas the production of recovered sulfur has been increasing around the world, as a result of growing environmental awareness and concerns, the use of sulfur in a multitude of end uses has increased at a slower pace. Huge quantities of sulfur and sulfuric acid are consumed in many industries, but not as much as is produced. Expansive world trade transfers large quantities of sulfur from the major producing areas to the large consumers. Stocks, however, are accumulating at recovery operations in remote locations in which the distances to the market make its transportation prohibitively expensive.

The cumulative effects of sulfur production also differ from the results of other mineral production. Modern sulfur production actually results in improved environmental conditions rather than having to repair mining processes and outcomes. Sulfur recovery prevents the emissions of SO_2 and other harmful compounds into the atmosphere, avoiding the detrimental effects of those emissions.

The challenges facing the sulfur industry are unique. In other mineral industries, the major concerns are how to continue to produce necessary materials and protect the environment as much as possible at the same time. The sulfur industry is confronting the question of what to do with all the sulfur it produces.

Current production exceeds consumption, but environmental regulations continue to increase with little growth in sulfur uses. The 21st century sulfur industry will need to expand sulfur consumption in nontraditional markets and find acceptable ways to dispose of unneeded sulfur without compromising environmental protection.

This chapter discusses some basic information related to sulfur cycle, supply, trade, uses, and environmental issues.

2.2 GLOBAL SULFUR CYCLE

Although most chemical elements have a global cycle, the global sulfur cycle is unusually active and pervasive with inputs from natural and man-made sources. Much of the cycle is difficult to quantify. The amount of sulfur that is produced through mining or as environmental byproducts at oil refineries, natural gas processing plants, and nonferrous metal smelters is reasonably well defined; but the quantity of SO_2 released from electric power plants and industrial facilities in developing countries is harder to measure. Estimates of sulfur emissions from natural sources are even more difficult to determine because of the variety of sources, the variability of emissions over time, the wide range of compounds involved, and the difficulties in measuring in remote locations (Ober, 2003).

The important reactions of the sulfur pathways into the environment (Figure 2.1) include:

- *Assimilative sulfate reduction*: Sulfate (SO_4^{2-}) is reduced to organic sulfhydryl groups by plants, fungi, and various prokaryotes. Sulfur oxidation state is +6.
- *Desulphurization*: Organic molecules containing sulfur can be desulphurized, producing hydrogen sulfide gas (H_2S). Sulfur oxidation state is -2.
- *Oxidation*: Oxidation of hydrogen sulfide produces elemental sulfur ($S°$). Sulfur oxidation state is 0. This reaction is completed by the photosynthetic green and purple sulfur bacteria. Further oxidation of elemental sulfur by sulfur oxidizers produces sulfate.

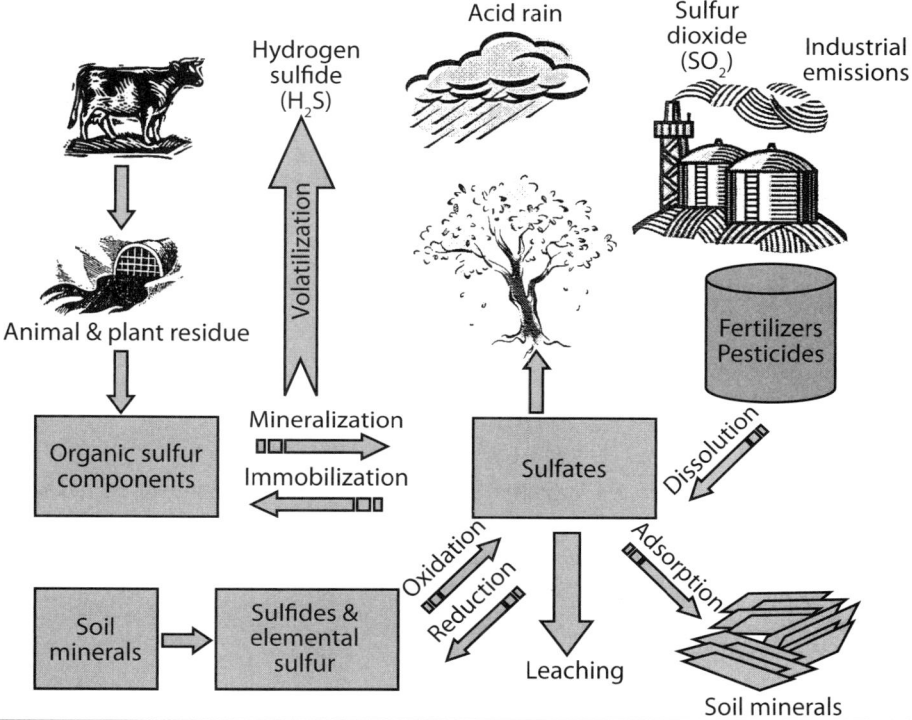

Figure 2.1. Sulfur pathways into the environment.

- *Dissimilative sulfur reduction*: Elemental sulfur can be reduced to hydrogen sulfide.
- *Dissimilative sulfate reduction*: Sulfate reducers generate hydrogen sulfide from sulfate.

2.2.1 The Natural Sulfur Cycle

The natural sulfur cycle is extremely complex and difficult to measure. Sulfur occurs in the Earth's crust in a wide range of minerals and is one of the few elements that can be found in nature in its elemental state. It is found also in coal, crude oil, and natural gas in the form of many different compounds and in varying quantities. Sulfur is essential in all living things, both plants and animals (Moss, 1978).

Sulfide minerals in the lithosphere are weathered to sulfates, some of which eventually reach the oceans through river runoff and erosion. The sulfates that reach the oceans become components of marine sediment. Other weathered sulfates react with bacteria to form compounds that are incorporated into the soil

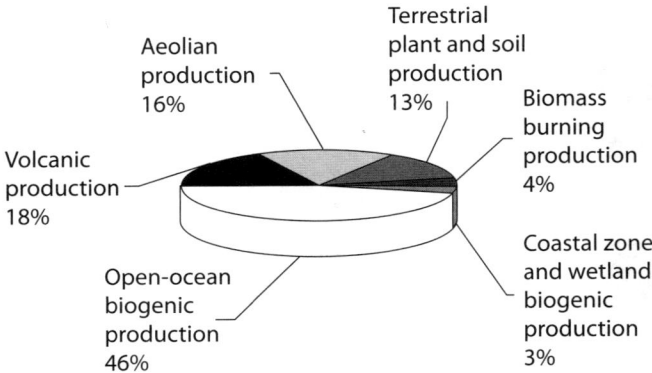

Figure 2.2. Estimated natural sulfur sources in the atmosphere.

and plant systems. Animals may then ingest the plants and the sulfur compounds are ultimately returned to the environment as sulfates (Moss, 1978).

Perhaps the best known, and certainly the most dramatic, natural sources of sulfur are volcanoes. Most volcanic emissions of sulfur compounds occur during noneruptive periods of volcanic activity, but sulfur also is emitted during eruptions (Whelpdale, 1992).

Seawater contains about 2.65 milligrams of sulfate per gram of water. As bubbles of seawater break, particles of sea salt are formed and emitted into the atmosphere. This sea spray is one of the largest sources of sulfur in the atmosphere, especially over open oceans (Ober, 2003). The sulfate in the seawater may come from the weathered minerals or through the decay of oceanic organisms.

Whelpdale (1992) estimated the natural sulfur sources in the atmosphere (Figure 2.2) as:

a. 46% of total natural sulfur in the atmosphere due to open-ocean biogenic production derived from the physiological activities of organisms
b. 18% due to volcanic production
c. 16% due to aeolian production from dust
d. 13% due to terrestrial plant and soil production
e. 4% due to biomass burning production
f. 3% due to coastal zone and wetland biogenic production

2.2.2 The Anthropogenic Sulfur Cycle

The amount of sulfur entering the atmosphere through human activity, the anthropogenic sulfur cycle, is easier to define than the natural sulfur cycle, but significant uncertainties remain to its size in less developed areas of the world (Ober, 2003). The majority of anthropogenic sulfur emissions are in the form of SO_2 resulting from the burning of fossil fuels (coal, petroleum, and natural gas), the smelting of nonferrous metal ores, and other industrial processes and burning (Whelpdale, 1992). Nearly all of this sulfur is emitted into the atmosphere in the Northern Hemisphere.

Globally, man-made sulfur inputs to the atmosphere began to increase significantly early in the 20th century, and the trend continued until the mid-1970s when environmental regulations in North America and Western Europe began to limit allowable sulfur emissions. Environmental regulations have reduced the quantity of SO_2 emitted into the atmosphere, but they have not eliminated the problem. Significantly decreased emissions have been realized at many industrial facilities, but statistics on actual emissions are not readily available (Ober, 2003).

2.3 SULFUR SUPPLY

At the turn of the 20th century, the only significant production of elemental sulfur was in Sicily, Italy. Sulfur ores were mined there using conventional mining methods, but the processes for recovering the sulfur from the ore were unusual (Day, 1885). Sulfur recovery was always poor, ranging from 30 to 50%. The remainder of the sulfur that was not retained in the ore upon completion of the process was released into the atmosphere as SO_2. As early as 1885, the environmental damage caused by emissions of sulfur was recognized. The Italian government restricted the processing of the ore to certain times throughout the year, opposite the primary growing season. During the time that Italian sulfur dominated the global sulfur industry, many companies were searching for other ways to meet the growing industrial demand for sulfur. Pyrites were being burned to recover their sulfur content while sulfuric acid and other methods were being investigated to melt or dissolve the sulfur from the ores. Sulfur was known to be contained in petroleum and natural gas, but none was recovered; the sulfur either remained in the final products or was emitted into the atmosphere.

2.3.1 Sulfur Production and Processes

Sulfur is mined from both surface and underground deposits and is recovered as a byproduct from a number of industrial processes. In sulfur mining, three techniques are applied: (1) conventional underground methods, (2) conventional open pit methods, and (3) the Frasch mining method. About 90% of all sulfur mined is obtained through Frasch mining.

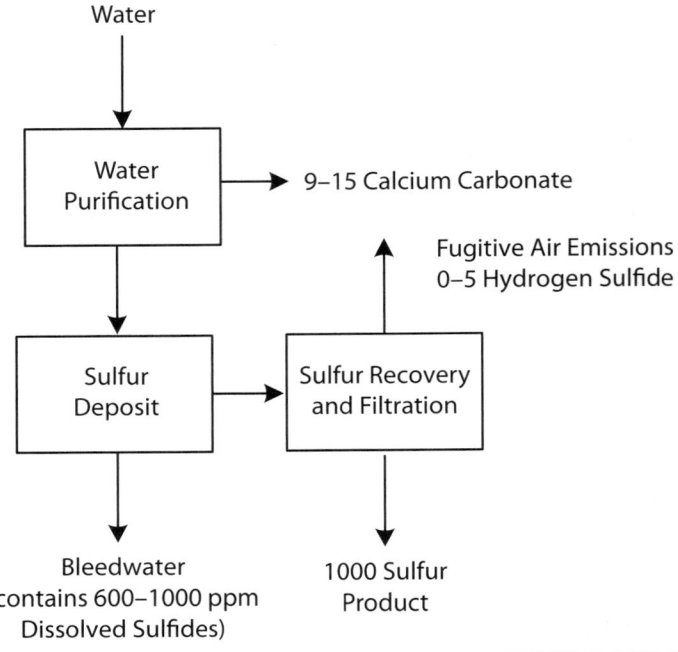

Figure 2.3. Frasch process (adapted from the U.S. Environmental Protection Agency, 1980).

Frasch Mining

The Frasch mining process uses hot water to melt sulfur trapped in salt domes. The sulfur is then pumped to the surface and is either sold as a liquid or cooled and solidified into a number of forms for market. Figure 2.3 presents a process flow diagram for the Frasch process.

Early in the century, at the point in the process when the sulfur reached the surface, it was pumped into wooden forms or molds where it cooled and solidified. Modern facilities use insulated pipes to move the sulfur to heated storage tanks where it is held until transfer to a terminal from which it is shipped to customers. Currently, when large quantities of excess sulfur are stockpiled, the process is similar to that of the early days of the industry. Molten sulfur is poured into molds where it cools and solidifies, creating huge blocks of solid sulfur.

The Frasch sulfur process only works under specific geologic conditions. Deposits amenable to the process have proven to be either salt domes or bedded evaporite deposits in which permeable native sulfur deposits are enclosed in impermeable formations (Morse, 1985). The elemental sulfur results from the bacterial alteration of anhydrite or gypsum, in the presence of hydrocarbons, into limestone and H_2S. The H_2S then oxidizes to sulfur through contact with oxygenated water that percolates through the formation or through the action of anaerobic, sulfur-reducing bacteria (Bodenos and Nelson, 1979).

Combined Sulfur from Smelting of Nonferrous Sulfides

Combined sulfur can be recovered during the smelting of nonferrous sulfides. Sulfur dioxide in the smelter gases is converted to sulfuric acid and liquid sulfur dioxide. In the United States, byproduct sulfuric acid from nonferrous metal smelters and roasters supplied about 11% of the total domestic production of sulfur in all forms in 1990. Sulfur may also be recovered from sulfur dioxide emissions. Regenerative or throw-away flue-gas desulphurization methods may be used either to recover sulfur in a useful form or to dispose of it as solid waste. Both recovery methods may employ wet or dry systems and use a variety of compounds such as limestone, sodium carbonate, and magnesium oxide to neutralize or collect the sulfur dioxide. End products include gypsum, sulfuric acid, liquid sulfur dioxide, and elemental sulfur, all of which can be used if a local market exists. If no local markets exist, large quantities of gypsum or sulfuric acid may have to be neutralized or otherwise disposed.

Natural Gas

H_2S must be removed from natural gas because of its extreme toxicity. It also causes corrosion problems; but these problems compared to elemental sulfur by means of the Claus are minor when compared to the hazards of breathing even low concentrations of H_2S. At natural gas processing facilities, the first step in recovery of sulfur is the separation of H_2S from the rest of the natural gas stream. The natural gas containing H_2S, known as sour gas, is passed through a solvent in which the H_2S dissolves and the desirable portions of the natural gas are insoluble. The solvent is then heated and the H_2S is expelled from the solution (Fischer, 1984). The most common solvents used are amines, which are organic derivatives of ammonia (Schmerling, 1981). After the various components of the gas have been separated, H_2S is converted using the Claus process, named after its inventor, British chemical engineer C. F. Claus (Zwicker, 1990).

Petroleum

During the early days of the petroleum industry, simple distillation was the only process used to refine crude oil. Crude oil is a complex mix of hydrocarbon compounds, ranging from simple compounds with small molecules of low densities and very dense compounds with extremely large molecules. Any of these compounds can contain sulfur. An average crude oil contains approximately 84%

carbon, 14% hydrogen, 1 to 3% sulfur, and less than 1% each of nitrogen, oxygen, metals, and salts (Occupational Safety and Health Administration, 1999).

As demand increased for lighter fuels such as gasoline and jet fuel, more complicated processes were developed to break larger particles apart to produce increasing quantities of the more desirable, lower density compounds. These processes often involved reactions that could also break the sulfur apart from the organic compound as H_2S; the same sulfur compound often found in natural gas with all the same undesirable properties. Until about 1970, any H_2S produced in the refining process was used as refinery fuel and burned along with the other gaseous hydrocarbons released during the process. Burning the H_2S produced SO_2 that was released into the atmosphere and was considered a nuisance rather than a hazard.

As refining increased and the atmosphere around refineries became more noxious, SO_2 emissions came under fire as a significant contributor to air pollution and acid rain. Oil refineries were then required to install equipment to reduce the amount of SO_2 allowable quantities or process crude oils with low sulfur content, resulting in low SO_2 emissions. Concern was not limited to SO_2 emissions at the refinery; automobile exhaust was recognized as a large source of SO_2 in the atmosphere. As environmental concerns increased, environmental regulations have further limited emissions. These factors resulted in tremendous growth in sulfur recovery capacity and production across the world.

Oil Sands

Oil sands represent a large source of recovered sulfur, especially in Alberta, Canada. Alberta has huge deposits of oil sands with estimated reserves of 300 billion barrels of recoverable crude oil that also contain 4 to 5% sulfur (Stevens, 1998). The Athabasca oil sands are a mixture of sand, water, clay, and bitumen, a naturally occurring viscous mixture of heavy hydrocarbons. Because of its complexity, bitumen is difficult, if not impossible to refine at most oil refineries. It must be upgraded to a light-oil equivalent before further refining or it must be processed at facilities specifically designed for processing bitumen. Oil sands with more than 10% bitumen are considered rich; those with less than 7% bitumen are not economically attractive (*Oil and Gas Journal*, 1999). During the upgrading and refining of the oil sands, significant quantities of sulfur must be recovered using processes described for conventional oil refining.

Recovered Elemental Sulfur (Claus Process)

Recovered elemental sulfur is a nondiscretionary byproduct of petroleum refining, natural gas processing, and coking plants. Recovered sulfur is produced primarily to comply with environmental regulations applicable directly to

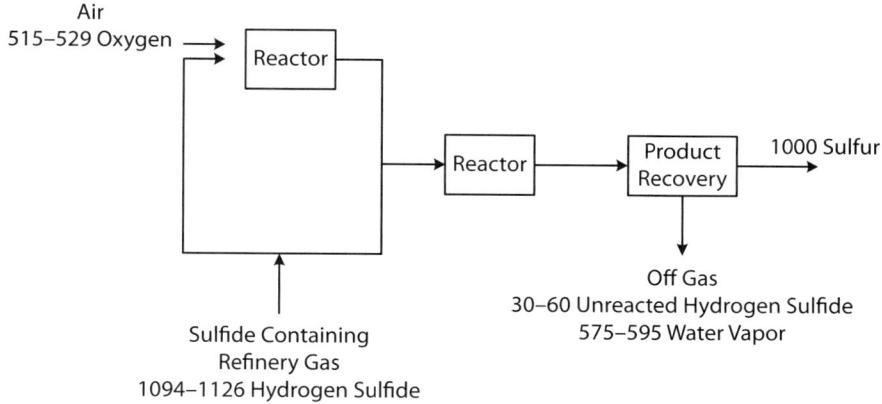

Figure 2.4. Claus process (adapted from the U.S. Environmental Protection Agency, 1980).

processing facilities or indirectly by restricting the sulfur content of fuels sold or used. The principal sources of recovered sulfur are hydrogen sulfide in sour natural gas and organic sulfur compounds in crude oil. Recovery is mainly in the elemental form although some is converted directly to sulfuric acid. Sulfur in crude oil is recovered during the refining process. Organic sulfur compounds in crude oil are removed from the refinery feed and converted to hydrogen sulfide by a hydrogenation process. The sulfur in natural gas is already in the form of hydrogen sulfide. Hydrogen sulfide from both sources is converted to elemental sulfur by the Claus process (Figure 2.4).

In this process, concentrated hydrogen sulfide is fired in a combustion chamber connected to a waste heat boiler. Air is regulated to the combustion chamber so that part of the hydrogen sulfide is burned to produce sulfur dioxide, water vapor, and sulfur vapor. The high temperature gases are cooled in a waste heat boiler and sulfur is removed in a condenser. The efficiency of the process is increased by adding as many as three additional stages in which the gases leaving the sulfur condenser are reheated and passed through catalytic converters and additional condensers. Finally, the total gas stream is incinerated to convert all remaining sulfur-bearing gases to sulfur dioxide before release into the atmosphere. The sulfur is then collected in a liquid form.

Formed Sulfur

Formed sulfur may be made into one of several forms, including flakes, slates, prills, nuggets, granules, pastilles, and briquettes. To produce flakes, the sulfur is cooled and solidified on the outside of large, rotating drums, from which it peels off into small flakes. To produce slates, molten sulfur is cast onto a continuous

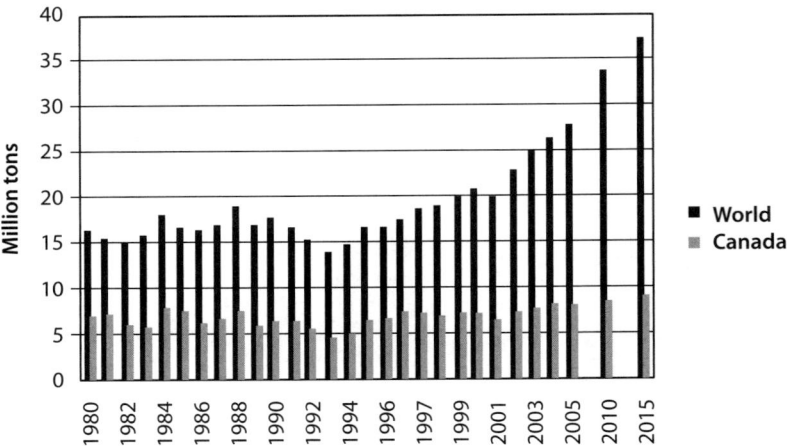

Figure 2.5. Chart showing world sulfur trade (adapted from "Market Prospects for Sulphur Recovered from the Oil Sands," CHOA Meeting, Calgary, Alberta, Canada, April 12, 2006).

conveyer belt and is cooled with air or water so that it solidifies into a thin sheet. As the slate reaches the end of the belt, the sheet breaks into smaller pieces. Sulfur prilling can be accomplished with air or water. In air prilling, molten sulfur is sprayed from the top of a tower against an upward flow of air. As it falls, the sulfur breaks into small droplets and cools into prills. In water prilling, the sulfur is sprayed into tanks containing water from which the prills are collected and dried. Minor modifications to prilling techniques are used to produce nuggets. Granulation involves applying successive coats of sulfur to solid particles of sulfur in a granulator until the particle size reaches the required diameter. In the Procor GX granulation process, liquid sulfur is sprayed into a rotating drum in which small seed particles of sulfur are recycled from the end of the process. Pastilles are individual droplets of molten sulfur that have been dropped on a steel belt and cooled by conduction.

2.4 SULFUR TRADE

Sulfur has become an important internationally traded commodity. It has become more widely traded since recovered sulfur production increased. The dominance of many countries in the petroleum and natural gas industries, and increased environmental awareness and regulations around the world, have created a tremendous increase in recovered sulfur production throughout the world.

An example of a country with significant production and limited consumption is Canada. As many other developed countries, Canada limits the level of

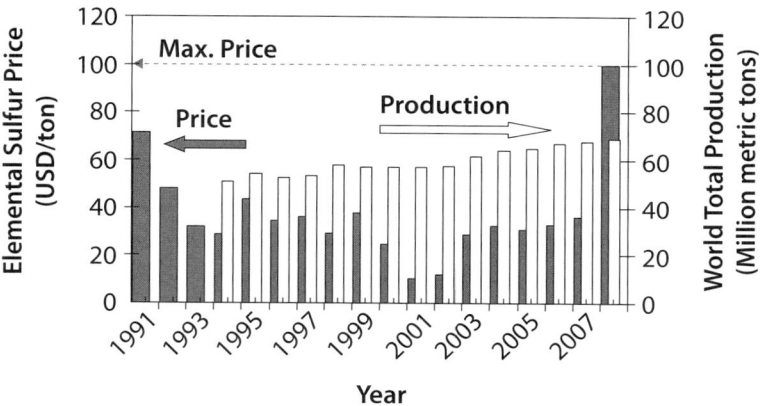

Figure 2.6. Chart showing world sulfur production and the change in prices of sulfur from 1980 through 2008.

sulfur allowable in gasoline and other automotive fuels as well as SO_2 emissions at petroleum refineries. Canada's major contribution to the world sulfur supply, however, comes from its large sour natural gas deposits in western Canada, especially in Alberta (see Figure 2.5). Many operating gas wells contain large quantities of H_2S, some as high as 45%, that must be removed prior to selling the natural gas. Canada is the world's largest producer of recovered sulfur; it is also the largest exporter of elemental sulfur.

Although Canada exports an average 7 Mt/yr, supply exceeds demand to the extent that more than 15 Mt of elemental sulfur has been stockpiled in Alberta. Canadian companies are actively exploring methods for long-term storage and disposal of sulfur such as re-injection and underground storage. In some operations, it can no longer be considered a valuable byproduct but rather a waste material with costs attributed to its storage or disposal. The processing of heavy crude oil associated with tar sands extraction is also a significant and growing source of sulfur, nearly 3 Mt in 2005.

In contrast to Canada, Morocco is a large consumer of sulfur with little domestic supply, necessitating the import of approximately 2.5 Mt/yr. Morocco possesses huge deposits of phosphate rock from which it produces phosphoric acid and phosphate fertilizers. Because the demand in Morocco significantly exceeds domestic supply, the Moroccan phosphate producers import sulfur from many other countries. Even with the huge global sulfur market, production exceeds consumption worldwide. The global sulfur market has become quite competitive, resulting in a trend of continually lower prices. Regional supply constraints have caused brief price increases but, overall, prices have been on a downward trend, especially in constant dollars for the past 20 years (see Figure 2.6).

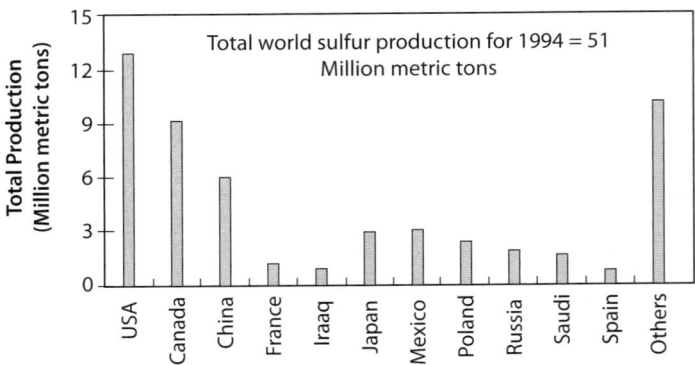

Figure 2.7. Total world sulfur production in 1994.

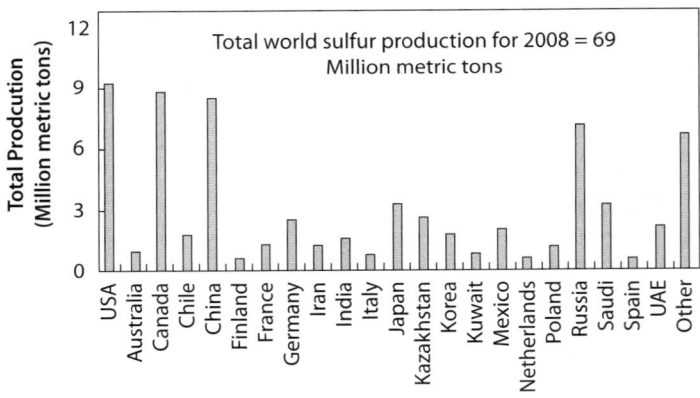

Figure 2.8. Total world sulfur production in 2008.

Detailed descriptions are of major contributing countries is shown in Figures 2.7 and 2.8 for 1994 and 2008, respectively. In 1994, only eleven countries were the main contributors whereas in 2008, the number increased to 22 countries because of increase of oil production. The total world production has increased from 51 Mt in 1994 to 69 Mt in 2008 with a rate of increase of approximately 2.5% per year.

Because many of the countries that produce sulfur are not significant consumers and many of the substantial consumers do not have large sulfur production capabilities, an increasing global sulfur market has emerged. Figures 2.9 and 2.10 illustrate how the global sulfur trade has changed since 1981. Perhaps the most striking change in sulfur trade since 1981 is the number of countries that have entered the market. In 1981, six countries that exported a total of 15 Mt dominated sulfur trade. They were, in descending order, Canada, Poland, the

Sulfur Production and Uses 47

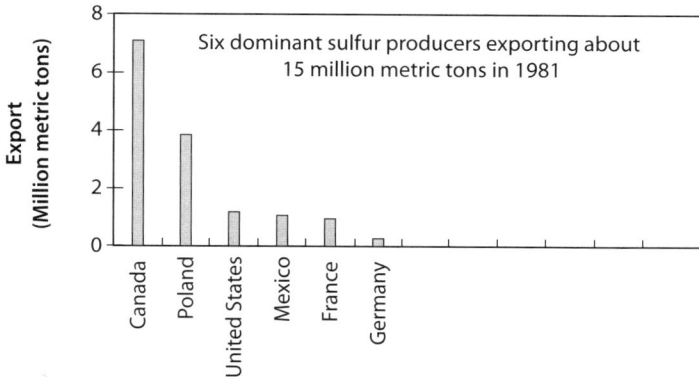

Figure 2.9. Chart showing worldwide sulfur trade in 1981 (data obtained from Ober, 2003).

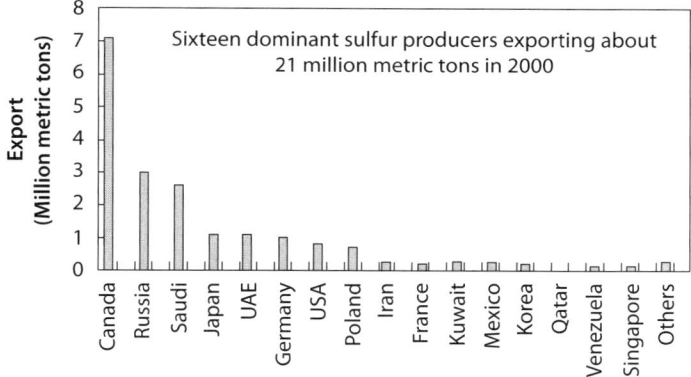

Figure 2.10. Chart showing worldwide sulfur trade in 2000 (data obtained from Ober, 2003).

United States, Mexico, France, and the Federal Republic of Germany. By 1990, Saudi Arabia joined the leaders and the U.S. was decreasing in importance.

By 2000, 16 countries were exporting more than 100,000 t of sulfur, and even more were exporting small quantities for a global total of approximately 21 Mt. Throughout the period, Canada was the leader and Saudi Arabia became more important. Mexico, Poland, and the United States became less important as the Frasch industry in each of those countries dwindled and operations closed. Of these three countries, only Poland continues to mine sulfur. Russia, as part of the Soviet Union, was a major importer of sulfur for use in its fertilizer industry and has become a major exporter of sulfur. This dramatic change is due, in part, to increased recovered sulfur from large sour natural gas operations in Russia, the downturn of the Russian economy, and the subsequent cutbacks at fertilizer operations that resulted from the break-up of the Soviet Union in 1991.

48 Sulfur Concrete for the Construction Industry

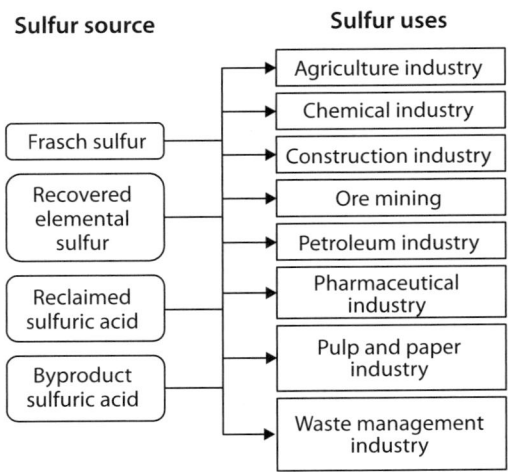

Figure 2.11. Flow diagram that shows the possible sources of sulfur and some of the possible end-use categories.

2.5 SULFUR DEMAND

Sulfur is used in such a wide variety of industries that listing them individually is nearly impossible. Figure 2.11 shows sulfur sources and the categories of industries in which it is consumed, either as elemental sulfur or H_2SO_4. Most sulfur is converted to H_2SO_4 before it is used. Sulfur is unusual in that for many of its end uses, especially those that require sulfur as H_2SO_4, the sulfur is not part of the final product. Although needed for the synthesis of many compounds and the successful accomplishment of numerous industrial processes, sulfur often circulates continuously through the process or is incorporated in its waste stream.

2.6 SULFUR USES

Elemental sulfur is used in the chemical industry to produce H_2SO_4, but it is also used directly in plant nutrient sulfur, petroleum refining, pulp and paper processing, and synthetic rubber production. Some of the major uses for H_2SO_4 are copper ore leaching, phosphate fertilizer and other agricultural chemical production, petroleum refining, and other chemical manufacturing. An example of sulfur end-uses in the United States are shown in Figure 2.12 (Matos and Ober, 2005 and USGS website). It should be noted that until 2003, various sulfur utilizations were reported; however, from 2004 to 2008, reported data were related only to the total sulfur utilization. A brief description of the various utilizations of sulfur is discussed in the following sections.

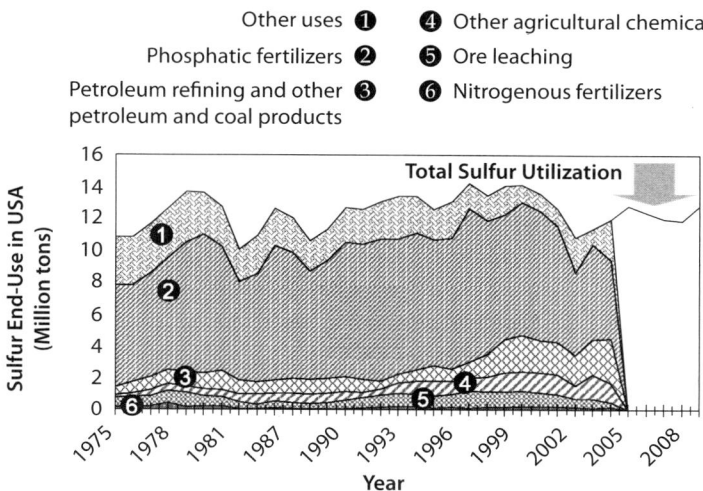

Figure 2.12. Sulfur end-use in the United States from 1975 to 2008.

2.6.1 Sulfuric Acid

Sulfuric acid (H_2SO_4) is a strong acid that, when in dilute form, will dissolve all base metals. A chemical property that makes H_2SO_4 important in industrial uses is its ability to decompose salts of most other acids. Sulfuric acid is the most important inorganic chemical used in commerce due to its low cost, widespread availability, and versatility. The most important application of this property is in the reaction with phosphate rock to produce phosphoric acid and calcium sulfate (Sander et al., 1984).

Fertilizer production is by far the largest consumer of sulfuric acid where it is an important sulfur source or facilitator. As a plant nutrient, sulfur may be applied directly or as part of a fertilizer where it increases the availability of other nutrients, improves crop yield and quality, and fights disease (fertilizer). At the same time, sulfur controls pests, mites, and mildews (insecticide, pesticide, and herbicide) and lowers the pH of saline and alkali soils (soil conditioner).

Sulfuric acid is generally used to:

1. Oxidize metals, convert insoluble oxides, sulfides, carbonates, and silicates to soluble compounds (copper and uranium ore leaching, metallurgy, and mineral processing)
2. Oxidize and remove tars and organic sulfides in petroleum and serve as a catalyst in the alkylation of isoalkanes with alkenes and in the refining of raw paraffin petroleum refining

3. Pickle steel to remove mill scale, rust, dirt, etc, prior to processing (iron and steel)
4. Act as a drying and dehydrating agent because of its avidity for water (explosives, celluloid, acetates, and lacquers)
5. React with aromatic compounds to form sulfonic acids (organic chemicals, rubber, and plastics)
6. Produce reagents such as:
   a. Aluminum sulfate or alum (antiperspirant, clarifier for fats and oils, deodorizer and decolorizer in petroleum processing, fireproofing, leather tanning, water clarifier in papermaking, and water treatment)
   b. Ammonium sulfate (fire retardant, fertilizer, viscose rayon, and chemical feedstock)
   c. Antimony and sulfate
   d. Explosives
   e. Barium sulfate (pigment in paints)
   f. Boric acid (antiseptics, borosilicate glass, catalyst, fiberglass, nuclear applications, pharmaceuticals and cosmetics, photography, porcelain enamels, frits, glazes, liquid SO_3 stabilizer, and textile treating)
   g. Chromic acid (chrome plating, green pigment, ceramic colorant, metallurgy, refractory additive, and chemical feedstock)
   h. Lead sulfate (ceramics, pigment, and vinyl plastics stabilizer)
   i. Lithium sulfate (pharmaceuticals)
   j. Magnesium sulfate (pharmaceuticals, chemicals, dyes, paper sizing, explosives, and fertilizers)
   k. Manganese sulfate (animal feed additive, fertilizer, food supplement, fungicide, paint additive, textile dye, and intermediate manganese chemical)
   l. Nickel sulfate (catalyst, nickel plating, and textiles)
   m. Sodium sulfate (sulfate chemical pulping of wood, Kraft paper, glass, ceramic glazes, dyeing wool textiles, tanning, spinning baths for viscose rayon, nickel smelting, animal feed supplements, photography, water treatment, sulfonated oils, and chemical feedstock)
   n. Zinc sulfate (mordant, aniline black, and ceramic coloring)

2.6.2 Agricultural Chemicals

Phosphoric Acid and Phosphate Fertilizers

H_2SO_4 is an essential component of these manufacturing processes; more H_2SO_4 is consumed in the conversion of phosphate rock into phosphoric acid (H_3PO_4) and other phosphate fertilizers than is consumed in all other industries. Because of this, the fortune of the sulfur industry is closely tied to the fortune of the phosphate industry. Phosphorus is one of the three essential nutrients (along with nitrogen and potassium) for plant growth. Naturally occurring phosphate minerals are insoluble and for plants to efficiently use this nutrient, the phosphate minerals must be processed into soluble compounds. Most phosphate fertilizer producers' use H_2SO_4 that is produced at plants located at or near the phosphate processing facilities used to convert the phosphate ores. Although other processes for producing H_3PO_4 and phosphate fertilizers have been developed, H_2SO_4 remains the preferred acidulating agent because of its easy availability, low cost, and well-established technology.

Nitrogen Fertilizers

Nitrogen fertilizers require a fraction of the sulfur that is consumed in the production of phosphate fertilizers. Ammonium sulfate, the only significant sulfur-containing nitrogen fertilizer, is a small portion of the entire nitrogen fertilizer market. In 2000, it represented about 2% of the market (Kramer, 2002). Most ammonium sulfate is produced as a byproduct of other industrial processes. Production of caprolactam, for example, which is an intermediate in the production of the nylon-6 fiber, results in an ammonium sulfate byproduct.

Plant Nutrient Sulfur

Nitrogen, phosphate, and potassium are recognized as the most important plant nutrients. Increasingly, however, agronomists have emphasized the importance of sulfur in the nutrition of plants. Ammonium sulfate's fertilizer value is related to its nitrogen content (21%) as well as its sulfur content (24%) as sulfate. The sulfate sulfur in ammonium sulfate is a form that is readily available for crop uptake. Other benefits to maintaining the proper nitrogen sulfur ratio to plant growth are improved protein and amino acid content and chlorophyll seed formation.

Over the past 20 years, soils have become increasingly deficient in sulfur for several reasons. Larger crop yields have removed greater quantities of sulfur. Environmental regulations limiting SO_2 have significantly reduced atmospheric deposition of sulfur. Reduced sulfur content of transportation fuels has limited SO_2 emissions from vehicles, reducing deposition from those sources. Finally, increased conservation tillage in which crop residues are not plowed back into the soil and left to decompose on the surface has reduced the mobility of the sulfate ion in the subsoil, making it less available to crops (H.J. Baker and Bro., 2001).

In addition to ammonium sulfate and phosphate-sulfate fertilizers, elemental sulfur products have been developed to provide elemental sulfur for crop nutrition. Because large particles of elemental sulfur oxidize to the sulfate form slowly, limiting the crop nutrient value of the sulfur, plant nutrient sulfur products have been formulated to deliver small sulfur particles that decompose more readily into the sulfate form. The particles can be agglomerated into granules with a binding material that makes the sulfur easier to handle, limiting the sulfur dust and improving the safety of handling finely divided sulfur. Other products incorporate elemental sulfur with 10% bentonite clay. After application, the clay swells as it absorbs soil moisture, breaking off the sulfur in a variety of particle sizes.

Other factors affect the conversion of sulfur to sulfate besides particle size. Rising soil temperature increases bacterial activity and conversion. Moist soils also promote transformation. Specific types of bacteria (*thiobacillus oxidans*) in the soil are necessary for the conversion to occur, and colonies expand with regular application of elemental sulfur. The sulfur must be in close contact with the soil organisms for sulfate formation to occur (Tiger Industries, 1999).

Other Agricultural Uses

Other agricultural uses require elemental sulfur. Sulfur is used as a supplement in animal feeds and in salt blocks. Animal feed sulfur is of high purity and small particle size. Finely divided sulfur is also used in agriculture as fungicides, insecticides, and miticides. Wettable sulfur powders can be mixed with water and sprayed on plants to act as a fungicide or a miticide, reducing fungus formation in crops and killing mites. Dusting and flowable sulfur are used as fungicide on creeping plant crops such as grapes, tomatoes, and peanuts. Sulfur sprays are used to rid the interior of poultry houses from depluming mites, tiny organisms that can cause poultry to lose their feathers. Many of these products can also be used as soil amendments to correct high alkalinity in the soil (Georgia Gulf Sulfur, 2000).

2.6.3 Chemical and Industrial

In the United States, agricultural uses, ore processing, and petroleum refining represented 93% of identified sulfur consumption in 2000 (Ober, 2003). Smaller quantities of sulfur are used in a myriad other industrial processes. Some of these are listed individually and others are discussed briefly.

Caprolactam

Another industrial use of H_2SO_4 is the source of one sulfur compound used in agriculture. Ammonium sulfate is a byproduct from the production of caprolactam ($C_6H_{11}NO$), an organic chemical that is a precursor for the nylon-6 polymer that is used in carpets, other textile products, and plastics. The production process is complex, requiring numerous complicated reactions among inorganic and organic chemicals.

Caprolactam production is one of the few industrial uses for sulfur in which sulfur does not end up as a component of waste. Variations on the caprolactam production process result in different byproduct production volumes of ammonium sulfate through the actual reaction and subsequent neutralization of the reaction products. The amount of ammonium sulfate produced in the caprolactam processes varies from 1.7 to 4.4 t of ammonium sulfate per ton of caprolactam. World capacity for caprolactam production is nearly 4.0 Mt. It is not known, however, which processes are used at various operations, making estimates of byproduct ammonium sulfate production difficult (Fisher and Crescentini, 1992).

Hydrofluoric Acid

Hydrofluoric acid (HF) is a strong, inorganic acid that is used in several industrial processes. Sulfuric acid is used to produce HF through its reaction in a rotating steel kiln with finely ground, acid-grade fluorspar (the mineral calcium fluoride). Approximately 2.4 ton of fluorspar and 2.7 ton of 96 to 98% H_2SO_4 produce about one ton of HF.

Titanium Dioxide

Titanium is well known as a high-tech metal used in aerospace equipment and 21st century sporting goods. The most common use of titanium, however, is in the form of titanium dioxide (TiO_2) pigment, which is used in paints, paper, plastics, and other materials. TiO_2 is widely used because of its whiteness, high refractive index, and light scattering ability. Ilmenite ($FeTiO_3$) and rutile (TiO_2) are ores used for producing pigment-grade TiO_2 through the sulfate and chloride processes. The sulfate process requires sulfur in the form of H_2SO_4. The sulfate process has been in use longer than the chloride process and uses the lower cost ilmenite that contains 35 to 54% TiO_2, or titanium slag with 75 to 79% TiO_2 as raw materials. Most chloride process plants require a high TiO_2-content ilmenite with approximately 64% TiO_2, slag containing 85 percent or greater TiO_2, rutile, and synthetic rutile. The chloride process, however, has become preferred by most TiO_2 producers because technical and environmental considerations of the sulfate process are more complicated and expensive.

Vulcanization of Rubber

An important use of sulfur, although not in large amounts, is the vulcanization of rubber. Rubber is a polymer with the empirical formula $(C_5H_8)_x$ with one C=C double bond to each unit. Sulfur cross-links the polymer chains, taking advantage of these double bonds. However, many good rubber-like materials have been produced as artificial polymers. Sulfur is an important fuel in pyrotechnic mixtures, because it is cheap and stable. It occurs in match heads, the most common pyrotechnical device, and was an ingredient of black powder. Black powder is a special mixture of 75% potassium nitrate, 15% charcoal, and 10% sulphur, more or less. The dry sulfur and charcoal are ground together so that the thixotropic sulfur thoroughly coats the active surface of the charcoal. The nitrate is then added and the mix is wet ground until homogenized. The dried mix is formed into grains of the desired size and the powder is ready to use.

Other Uses

H_2SO_4 is used to make explosives and other organic and inorganic chemicals. It is used in steelmaking and in the production of the synthetic fiber rayon. The electrolyte in lead-acid batteries is H_2SO_4. Many other processes use sulfur and H_2SO_4 but listing all of them would be nearly impossible.

2.6.4 Construction Industry

Although not widely used currently, sulfur construction materials can offer improvements over more traditional materials, especially in specific applications. Sulfur construction materials include sulfur concrete and sulfur-extended asphalt pavements as well as pre-cast concrete components, extrusions, and cast-in-place forms. About 3.2 million kilometers (2.0 million miles) of paved highways, roads, and streets comprise the majority of the ground transportation network in the United States, 93% of which is paved with asphalt-type materials (Weber and McBee, 2000).

Sulfur Extended Asphalt

Asphalt is a heavy, organic byproduct of petroleum refining. It was typically high-volume and low-cost, making it attractive to the road construction industry. As demand for such high-quality petroleum products as gasoline grew, pressure mounted for the petroleum industry to improve refining technology to produce more high-quality products and less asphalt or other low-value byproducts and wastes. These events, coupled with the increasing availability and decreasing costs of recovered sulfur, have made sulfur-extended asphalt cements a more attractive alternative to the traditional asphalt paving materials.

Table 2.1. **STARcrete™** sulfur polymer concrete properties compared with those of Portland cement concrete (**STAR**crete™ Technologies, 2000)

Property	Compared with 34.5 MPa Portland Cement Concrete	Testing Laboratory
Abrasion resistance	Much greater	11
Bond strength to concrete	Much greater	11
Bond strength to reinforcing steel	Greater	2
Coefficient of linear expansion	Equivalent	2
Compressive creep	Less	2
Compressive strength	Greater	1, 6, 7
Corrosion resistance	Much greater	4, 9
Durability under thermal cycling	Equivalent or higher	3, 4
Fatigue resistance	Much greater	5
Fire resistance	Slightly less	4, 8
Flexural strength	Greater	1, 6
Modulus of elasticity	Greater	2
Splitting tensile strength	Greater	2, 6
Thermal conductivity	Less	3
Water permeability	Much less	10

Testing Laboratories:
1. EBA Engineering Consultants, Canada
2. R. M. Hardy & Associates, Canada
3. Ontario Research Foundation, Canada
4. Sulphur Innovations, Canada
5. Iowa State University, Ames, USA
6. J. A. Smith & Associates, Canada
7. Bernard & Hoggan Engineering, Canada
8. Wamock Hersey, Vancouver, Canada
9. Mellon Institute, Pittsburgh, USA
10. Chemical & Geological Laboratories, Calgary, Canada
11. Dow Chemical, Texas Division, Freeport, USA

Tests initiated in 1975 showed that sulfur-extended asphalt was at least comparable with traditional asphalt concrete; some properties being superior. Sulfur-extended asphalt was stiffer, making it more resistant to rutting damage in warm climates. Transverse cracking was also less of a problem than in more traditional asphalt paving. The addition of sulfur to the asphalt mix just prior to pouring the roadway reduces the viscosity of the hot paving material, making the pavement easier to install. Equipment for preparing traditional asphalt requires little or no adaptation to handle sulfur-extended asphalt, an important factor in the acceptance of sulfur-extended asphalts on a large scale (Weber and McBee, 2000).

Even with the demonstrated advantages of sulfur-extended asphalt over more traditional paving materials, no large-scale highway projects have used sulfur-extended asphalt. A period of elevated sulfur prices in the late 1980s eliminated the financial incentives for substitution (Weber and McBee, 2000). Recent

Table 2.2. Potential applications of sulfur solidified concrete

Activity	Applications
Pre-casting	Sewer pipe; railway ties; highway barriers; a range of agriculture products; offshore drilling platforms; construction blocks, slabs, and other building components
Extruding	Bricks and paving stones; curbs and gutters; roof tiles
Cast-in-place	Bridge decks; marine installations; ship hulls and structures exposed to marine environments; drilling platforms; food processing plants; agricultural applications, including barns and effluent systems; sewage treatment plants; acid plants-drainage canals, sumps, tanks, flooring, walls and beams; fertilizer plants; foundations

increases in low-cost sulfur availability have reinvigorated interest, and large-scale testing of sulfur-extended sulfur as a viable, economic alternative to more traditional asphalt formulations are underway.

The main competitor to asphalt pavements is concrete made with Portland cement and aggregates. Sulfur can substitute for Portland cement in roads and other construction uses. The 7% of United States roads not made with asphalt cement use Portland cement concrete, representing a significant market for sulfur.

Corrosion resistance materials

Elemental sulfur incorporating a small quantity of polymer modifiers that stabilize the sulfur structure, especially during thermal and hydraulic cycling, has replaced Portland cement effectively in industrial applications exposed to corrosive environments.

Several companies and researchers have developed slightly different formulations for polymer sulfur concretes, but all share the improved properties of extremely high resistance to corrosion, high physical strength, high resistance to fatigue, low water permeability, and fast curing time, which are listed in more detail in Table 2.1. Molten sulfur is the binder that holds the aggregates together to form the pavement or other concrete forms when it cools and solidifies. Additives contribute special properties to the sulfur binder, the most important of which is to produce a protective burn surface when exposed directly to a flame. The burn surface, when combined with the naturally low thermal conductivity of elemental sulfur, results in a slow penetration of heat. In the event of a fire near any form of sulfur concrete, tests have shown zero flame spread, fuel contribution, and smoke density. Because sulfur concrete sets on cooling, extreme weather conditions do not hamper its installation. Recycling is simply accomplished by crushing and reheating with no loss in strength.

Potential applications

Sulfur concrete can be pre-cast easily into various shapes, such as construction blocks and slabs, railway ties, sewer pipes, support beams, tanks for holding corrosive liquids, and any number of other items. Sulfur concrete can be extruded to form bricks and paving stones, roofing tiles, and street curbs and gutters. Table 2.2 summarizes these applications. In most uses, such as agricultural effluent systems, bridge decks, drilling platforms, and fertilizer plants, sulfur concrete can replace cast-in-place Portland cement. Sulfur concrete has found its greatest acceptance in facilities that must withstand corrosion from process acids, although it could be used in many other applications, including road construction (STARcrete Technologies, 2000).

2.6.5 Ore Processing

Smelting of nonferrous metals has been a major source of sulfur in the form of byproduct acid since early in the 20th century, but more recent developments in the metals industry make them major consumers of H_2SO_4 also. Processes have been developed that dissolve the various metals from the host rock with solvents, frequently H_2SO_4, and then recover the desired metals from the solutions. Copper has been produced through variations of this method since about 1967. Nickel can be produced using a pressure acid leach process. Technically, it could be possible to produce other metals through variations of H_2SO_4 leaching processes, but no commercial processes for other metals have been developed.

2.6.6 Petroleum Alkylation

Crude oil is a complex mix of hydrocarbons that must be separated and recombined to optimize the products manufactured from each barrel of crude; the ultimate goal is to obtain the most gasoline possible from each barrel of oil (Leffler, 2000).

In petroleum refining, several different processes, including distillation, vacuum flashing, and catalytic cracking, are used to separate the various hydrocarbon compounds contained in crude oil. In general, the lighter the product (lower density) and the fewer carbon atoms per molecule, the lower the boiling point and the more volatile the compound. As the number of carbon atoms that molecules contain increases, the compounds become heavier and the boiling point goes up. This makes it possible to separate the various hydrocarbon compounds into products within specific ranges of boiling points and densities.

The densest, largest molecules are the least desirable products, and processes have been developed to *crack* the molecules into smaller, more manageable, and more desirable products. Many refineries have thermal or catalytic crackers, the apparatus that refiners use to break some of the larger petroleum molecules into smaller, more desirable molecules. Refiners use the products of crackers and simpler distillers in alkylation facilities to combine some of the smaller compounds that result from cracking and distillation into compounds that meet gasoline fuel requirements (Schmerling, 1981). The product from this process is known as alkylate, a high quality gasoline blending component with such desirable properties as high octane number, low vapor pressure, and high heat of combustion. Alkylate also blends easily with other gasoline components and additives and optimizes the quantity of gasoline that can be produced from a barrel of oil. Because it contains saturated hydrocarbons and few impurities, it burns cleanly, promoting long engine life and limiting atmospheric emissions of harmful compounds (Liolios, 1989).

Alkylation reactions typically combine isobutane (a saturated, branched hydrocarbon with 4 carbon atoms and 10 hydrogen atoms) with any of several light olefins (unsaturated hydrocarbons containing one carbon-carbon double bond and an equal number of carbon and hydrogen atoms). The olefins used in alkylation are usually propylene, butylene, or any combination of the two. For the reactions between the isobutane and either propylene or butylene to occur, a catalyst is required. Most alkylation plants use H_2SO_4 or HF as the catalyst (Schmerling, 1981). HF-based operations, however, are of interest because H_2SO_4 is required for the production of HF.

2.6.7 Pulp and Paper

During pulp and paper processing, wood is treated to separate the cellulose fibers from the lignin and extractive substances, such as resinous and fatty acids, that make up the wood particles. The cellulose fiber that has been treated is known as pulp and is further processed by bleaching and drying to produce a material that is used to make paper. The quality and type of paper produced is determined by the amount of processing the pulp undergoes.

Wood is treated in mechanical or chemical pulping processes. In mechanical processes, wood particles are torn apart with all of the materials contained in the wood remaining in the pulp. Paper made from mechanical pulp is used for lower quality papers, such as newsprint and paperboard. Chemical pulping dissolves and separates the lignin and the extractive substances from the wood fibers by means of chemical reactions. Chemical pulp is used to produce higher quality paper. Chemical pulp from softwood produces strong paper owing to the wood's long fibers. Hardwood chemical pulp is ideal for making fine papers because of its short fibers. Waste paper has become an increasingly important source of pulp for paper making, but chemical pulping of virgin fibers is the more important in relation to the sulfur industry.

Two chemical processes are used to produce pulp, the sulfate and the sulfite processes, both of which use sulfur chemicals. Several sulfur chemicals are used in pulp and paper processing, including sulfuric acid, sulfur, sulfur dioxide, magnesium sulfate, aluminum sulfate, and sodium-based sulfur-containing chemicals. In the sulfate process, which is also known as the Kraft process and is the more widely used of the two processes, the pulp is cooked with alkaline cooking liquor; in the sulfite process, the cooking liquor is acidic (Ober, 2003).

2.6.8 Waste Management

Sulfur polymer cement (SPC), also called modified sulfur cement, is a relatively new material in the waste immobilization field, although it was developed in the late 1970s by the U.S. Bureau of Mines. The physical and chemical properties of SPC are interesting (e.g., development of high mechanical strength in a short time and a high resistance to many corrosive environments). Because of its very low permeability and porosity, SPC is especially impervious to water, which, in turn, has led to its consideration for immobilization of hazardous or radioactive waste.

Because it is a thermosetting process, the waste is encapsulated by the sulfur matrix, therefore, little interaction occurs between the waste species and the sulfur (as there can be when waste prevents the set of Portland cement-based waste forms).

At present, only a limited number of studies have been performed with SPC for waste immobilization, most of which were by Brookhaven National Laboratory (BNL) on four waste streams: incinerator ash, sodium sulfate salt, boric acid, and ion exchange resins.

Laboratory waste forms containing these wastes passed all Nuclear Regulatory Commission requirements for radioactive waste forms; however, full-scale testing continues at Idaho National Engineering Laboratory where operations and engineering aspects of the process are being studied. Reports indicate that while some new problems are being uncovered significant progress is also being made. SPC has been studied in Europe by researchers from The Netherlands. In general, their finding corroborates those of BNL, showing a low leachability of radioactive tracers from the sulfur matrix.

Whereas SPC does have promising properties, some restrictions do exist that can limit its use, especially since a prospective waste must contain less than 1% water. In addition, the question of long-term durability of this material remains unanswered because regulatory tests do not yet include data on bacteria attacking the sulfur. The fact that elemental sulfur exits only in a narrow thermodynamic stability field, and is also not found at the surface of the earth, indicates that the question of long-term stability for SPC must be answered in considering this material for immobilization of low-level or mixed wastes.

Table 2.3. Most prevalent element in the human body

Composition of the Human Body			
Element	Percent	Element	Percent
Hydrogen	63.0	Potassium	0.06
Oxygen	25.5	Sulfur	0.05
Carbon	9.5	Chlorine	0.03
Nitrogen	1.4	Sodium	0.03
Calcium	0.31	Magnesium	0.01
Phosphorus	0.22	All others	0.01

(Dharmananda: http://www.itmonline.org/arts/sulfa.htm)

2.6.9 Pharmaceutical Industry

Sulfur is essential to life and is the eighth most prevalent element in the human body (Table 2.3). No one is allergic to sulfur itself. Sulfur is not present as an isolated element in the body, but rather occurs in combination with other elements and, most often, in complex molecules. The primary placement of sulfur in the human body is in the sulfur-containing amino acids (SAAs), which are methionine, cysteine, cystine, homo-cysteine, homo-cystine, and taurine. Dietary SAA analysis and protein supplementation may be indicated for vegan athletes, children, or patients with HIV because of an increased risk for SAA deficiency in these groups. Methyl-sulfonyl-methane (MSM), a volatile component in the sulfur cycle, is another source of sulfur found in the human diet. Increases in serum sulfate may explain some of the therapeutic effects of MSM, DMSO, and glucosamine sulfate (Dharmananda: http://www.itmonline.org/arts/sulfa.htm).

Organic sulfur, as SAAs, can be used to increase synthesis of S-adenosyl-methionine (SAMe), glutathione (GSH), taurine, and N-acetyl-cysteine (NAC). MSM may be effective for the treatment of allergy, pain syndromes, athletic injuries, and bladder disorders. Other sulfur compounds such as SAMe, dimethylsulfoxide (DMSO), taurine, glucosamine or chondroitin sulfate, and reduced glutathione may also have clinical applications in the treatment of a number of conditions such as depression, fibromyalgia, arthritis, interstitial cystitis, athletic injuries, congestive heart failure, diabetes, cancer, and AIDS.

Sulfa Drugs

One of the more common drug reactions is an allergy to sulfa drugs. Sulfa drugs are more appropriately labeled sulfonamides and are derivatives of para-amino benzoic acid. Table 2.4 lists common medications that contain a sulfonamide component (Dharmananda: http://www.itmonline.org/arts/sulfa.htm).

The sulfa drugs are usually not allergenic by themselves, but when a sulfonamide molecule is metabolized in the body, it is capable of attaching to proteins,

Table 2.4. Common medications that contain a sulfonamide component

Sulfonamide drug classes/individual drugs that may cause allergic reactions				
Sulfonamide Antibiotics	Thiazide Diuretics	Loop Diuretics	Sulfonylureas	Carbonic Anhydrase Inhibitors
• Sulfadiazine	• Hydrochlorothiazide	• Furosemide	• Chlorpropamide	• Acetazolamide
• Sulfamethoxazole	• Chlorthiazide		• Tolbutamide	
• Sulfasalazine	• Metolazone		• Tolazamide	
• Sulfisoxazole	• Chlorthalidone		• Glipizide	
• Sulfacetamide	• Indapamide		• Glyburide	
• Sulfanilamide	• Methyclothiazide			
• Sulfathiazole				
• Sulfabenzamide				

(Dharmananda: http://www.itmonline.org/arts/sulfa.htm)

thus forming a larger molecule that could serve as an allergen. Therefore, the allergy is not to the original drug, but to a drug-protein complex. It is estimated that a skin rash occurs in about 3.5% of hospitalized patients receiving sulfonamides, but people with HIV infection seem to have a considerably higher sensitivity to them.

The benefits of sulfur compounds used in health products are often mentioned. Popular items include alpha-lipoic acid (thiotic acid), methyl-sulfonylmethane (MSM), allicin (the sulfur compound that is the main active ingredient of garlic), glucosamine sulfate (and its natural polymer, chondroitin), SAMe (S-adenosylmethionine), and several important antioxidants such as glutathione, N-acetylcysteine (NAC), and dimethyl-sulfoxide (DMSO).

Examples for sulfur use in cosmetics are:

1. Sulfur is particularly necessary for the body's production of collagen that helps to form connective tissue.
2. Sulfur is an antibacterial agent. It can be a potent skin irritant and sensitizer. It also has a high pH that can encourage the growth of bacteria on skin.
3. Sulfur accelerates peeling of the skin (micro-exfoliant) and has an antiseptic action on the skin surface when applied topically. It is particularly suitable for skin that is prone to breakout and congestion.
4. Sulfur is an element used in anti-acne products.
5. Sulfur is used to manufacture cysteine, an amino acid that contains sulfur and is found in the protein of the hair; the bond it forms strengthens the hair.

2.7 ENVIRONMENTAL ISSUES

2.7.1 Production and Processing

In any industrial endeavor, attempts are made to minimize losses during production and processing in order to maximize the economic benefit of the operations. This is certainly the case in sulfur mining in which producers strive to optimize sulfur extraction.

The nature of sulfur mining differs from more conventional mining methods in ways that reduce losses during the Frasch mining process. Because ores are not processed by mechanical methods to obtain the elemental sulfur product, virtually no product is lost during handling. Sulfur leaks may occur at various locations of Frasch operations, but in small quantities and are insignificant in relation to the total sulfur production and environmental considerations at those locations. With the demise of the Frasch industry, these losses are inconsequential to the sulfur industry.

In petroleum refining and natural gas processing, recovery of sulfur can be an economic liability rather than an asset. Sulfur recovery equipment and operations add significantly to the cost of oil and gas processing. Costs vary considerably by location, size of operation, and the type of equipment that must be installed. For example, a relatively small petroleum refinery in Oklahoma contracted an engineering firm to double its sulfur recovery capacity at a cost of $22 million in 2000 (*Sulphur*, 2000). In the same year, a deal to build a new desulphurization unit in Kuwait was estimated to cost $125 million (Cunningham, 2000). Environmental concerns related to sulfur and SO_2 emissions, however, prohibit virtually all losses to the atmosphere, making these kinds of installations necessary. Although recovered sulfur sales can be profitable to producers in some locations, others incur expenses in order to have sulfur removed from their premises. Discontinuing sulfur recovery operations while continuing to process oil and gas, however, is never an option. If a refinery or natural gas operation cannot recover sulfur, for whatever reason, then that facility must close, at least temporarily, until the situation has been resolved and sulfur recovery is again possible.

Environmental protection legislation is also a more influential factor to H_2SO_4 byproduct production than economics. Regulations mandate that smelter operators recover the majority of SO_2 liberated during the smelting process. These mandates include monitoring programs to ensure that the emissions regulations are not violated. If excess SO_2 emissions are detected, then fines may be imposed. H_2SO_4 plants that burn elemental sulfur to produce acid for their own use or for sale also must meet environmental regulations limiting SO_2 emissions, significantly reducing sulfur losses at these operations.

Losses during handling and shipping solid bulk sulfur have been significant in the past, although difficult to quantify. When sulfur was poured to block in the early days of the industry, the solid material was broken up and moved with bulldozers, creating a tremendous problem with fine sulfur dust. The fine particles were difficult to contain and could be transported great distances in the air. In addition to contaminating the area adjacent to the production locations, contamination was a problem along rail lines and at port facilities. Of even more concern than the dust contamination was the hazardous nature of the sulfur dust. Finely divided sulfur presents explosive and fire hazards, and the SO_2 generated by such a fire is toxic (West, 1966).

Because of these issues, regulations were established to limit shipments of crushed and broken sulfur. Several processes have been developed to minimize the loss and lessen the hazards in handling solid sulfur. In the U.S., nearly all sulfur is shipped molten, avoiding any of the dusting problems associated with bulk sulfur (Ober, 2003). To make this material acceptable for bulk transport, the molten sulfur is processed in forming apparatus that solidifies the sulfur into distinct particles such as granules, pastilles, prills, and slates that resist breakage, significantly reducing the problems during handling. Additional dust suppression techniques include covered conveyors systems, dust collectors, and enclosed railcar unloading facilities equipped with water sprays. These innovations have reduced losses to a minimum during handling.

2.7.2 Health Effects of Sulfur

All living things need sulfur. It is especially important for humans because it is part of the amino acid methionine, which is an absolute dietary requirement for human beings. The amino acid cysteine also contains sulfur. The average person takes in around 900 mg of sulfur per day, mainly in the form of protein.

Elemental sulfur is not toxic, but many simple sulfur derivates are such as sulfur dioxide (SO_2) and hydrogen sulfide. Sulfur can be found commonly in nature as sulfides. During several processes, sulfur bonds are added to the environment that are damaging to animals, as well as humans. These damaging sulfur bonds are also shaped in nature during various reactions, mostly when substances that are not naturally present have already been added. They are unwanted because of their unpleasant smells and they are often highly toxic.

Globally, sulfuric substances can have the following effects on human health:

1. Neurological effects and behavioral changes
2. Disturbance of blood circulation
3. Heart damage
4. Effects on eyes and eyesight

5. Reproductive failure
6. Damage to immune systems
7. Stomach and gastrointestinal disorder
8. Damage to liver and kidney functions
9. Hearing defects
10. Disturbance of the hormonal metabolism
11. Dermatological effects
12. Suffocation and lung embolism

2.7.3 Effects of Sulfur on the Environment

Sulfur can be found in the air in many different forms. It can cause irritations of the eyes and the throat with animals when the uptake takes place through inhalation of sulfur in the gaseous phase. Sulfur is applied in industries widely and emitted into the air due to the limited possibilities of destruction of the sulfur bonds that are applied. The damaging effects of sulfur to animals are mostly brain damage, malfunction of the hypothalamus, and damage to the nervous system.

Laboratory tests with animals have indicated that sulfur can cause serious vascular damage in the veins of the brains, heart, and kidneys. These tests have also indicated that certain forms of sulfur can cause fatal damage and congenital effects. Mothers can even carry sulfur poisoning over to their children through their milk. Finally, sulfur can damage the internal enzyme systems of animals.

2.7.4 Sulfur Waste Management

Because of the wide range of uses for sulfur and its compounds, identifying the ultimate disposal of every ton of sulfur consumed is extremely difficult. Quantifying the amount of material that is dispersed through the atmosphere or disposed of in landfills is even more challenging (Ober, 2003). The tremendous variety of industrial processes that use sulfur and sulfuric acid makes it virtually impossible to identify every operation where they are consumed. In addition, the quantity of sulfuric acid required in many of its applications varies considerably according to the quality and type used, the raw materials consumed, and by the exact process employed. An illustration of the flow of sulfur after consumption in the United States, as an example, is shown in Figure 2.13 (Ober, 2003).

The production of phosphate fertilizers results in a phosphor-gypsum material that, in almost all circumstances, is a high-volume waste. Synthetic gypsum is also a byproduct or waste from other industrial processes that use sulfur in the form of H_2SO_4. Some end-uses result in dispersive losses to the environment via various routes, and sulfur as H_2SO_4 is continually recycled through some applications.

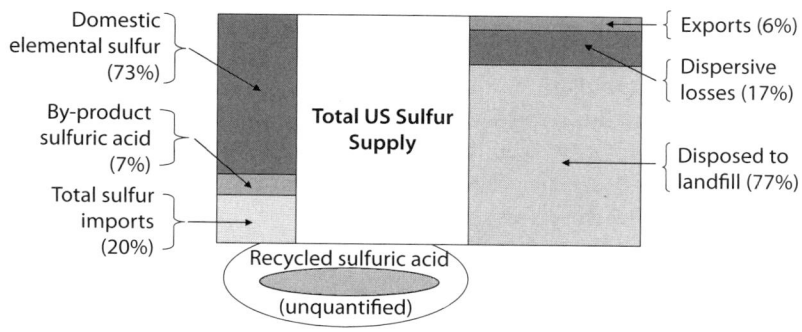

Figure 2.13. Sulfur flow in the United States.

Dispersive Losses

Most agricultural uses of sulfur, where it is applied for crop nutrition or pest and fungus control, disperse excess sulfur through the environment via various routes. In the simplest dispersal process, such sulfur materials as ammonium sulfate or elemental sulfur can be washed away by rain into surface water, contributing no value to the area in which it was applied.

Good agricultural practices are designed to avoid this outcome. Sulfur applied to the cropland mixes into the soil where it reacts with bacteria, forming new sulfur compounds that can be absorbed by plants. If it is absorbed by grass or grain, then it may become part of the food chain and pass through several other organisms before it finally returns to the environment as a waste material. This animal waste could cycle once more through the food chain if it is absorbed again into plant matter after being distributed onto crop land. Portions of sulfur-containing plant residues, which are not consumed as food, decay to enrich soils and undergo other similar processes. Even agricultural sulfur that is used in animal feeds and pest control is deposited eventually in soil where it undergoes the same reactions and transfers as sulfur nutrient materials.

Where sulfur materials are used in construction, the materials undergo slow dispersive losses during use as some portion of it is worn away. For example, as vehicles drive along a highway made of sulfur-extended asphalt, minute particles of the paving material could be loosened from the road and washed by rain water or blown by wind into soil or water. At the same time as traffic creates wear on the road materials, the road surfaces abrade tires that contain small percentages of elemental sulfur that is also dispersed into the environment. Dispersion from these processes is negligible. Roads and tires are engineered to withstand heavy wear and abrasive action as much as possible.

Although H_2SO_4 is constantly cycled through copper-leaching operations, portions of the acid are lost. The H_2SO_4 dissolves the copper oxides and the copper-laden acid drains into collection ponds. The H_2SO_4, however, can react with other minerals and rocks in the leach beds, forming insoluble compounds that remain. The quantity of sulfur compounds dispersed in this way is dependent on the type of rocks and minerals included with the copper oxide ores that are targeted by the leach process.

Disposed to Landfill

Phosphoric acid and phosphate fertilizer production generate huge quantities of phosphor-gypsum. Depending on the source of the phosphate rock that is used, the gypsum may be considered a byproduct of the process, but more frequently, it is treated as waste. Purified phosphor-gypsum may substitute for mined gypsum, but because much of the phosphate rock produced in the United States, especially that from central Florida, contains uranium and radium, the resultant phosphor-gypsum is slightly radioactive. For this reason, the use of phosphor-gypsum in the U.S. is extremely limited. Most phosphor-gypsum produced there is treated as hazardous waste that must be stored in stacks or piles.

Another problem common to phosphor-gypsum stacks is the entrainment of small quantities of acidic liquids in this waste. Neutralization or extensive washing of the material is required before it is stabilized for use in construction. Tremendous quantities of phosphor-gypsum have accumulated during the history of the phosphate fertilizer industry. In the United States, an average of approximately 40 Mt/yr of phosphor-gypsum has been produced since the mid-1980s, nearly all of which has been deposited onto stacks for long-term storage. Some producers, such as the one in Morocco, dispose of the material at sea. Only about 1% of the phosphor-gypsum produced in the United States is used in such applications as agricultural soil conditioners, road base construction, and research activities (U.S. Environmental Protection Agency, 2000).

HF production also results in waste gypsum production. About 4 tons of waste gypsum is produced for every ton of HF produced. Dilute H_2SO_4 resulting from TiO_2 production is sometimes neutralized to form byproduct gypsum. Some TiO_2 producers substitute their byproduct gypsum for natural gypsum in certain applications; others treat it as waste and dispose of it in landfills.

Sulfur used in pulp and paper processing results in waste compounds. Owing to the variety of compounds used, however, it is difficult to identify the ultimate disposition of the sulfur.

Recycling

Because H_2SO_4 is a catalyst in the alkylation process, it should, theoretically, circulate through the process continuously. In practice this is not the case. Over time the acid becomes diluted to the point that it is not effective in this application. It also becomes contaminated with impurities from the petroleum products being processed. When the H_2SO_4 becomes too diluted, portions of it are withdrawn from the process to be replaced with more concentrated acid. The diluted, contaminated material is treated at a regeneration facility for purification and re-concentration. As a result of the H_2SO_4 reacting with impurities, some of it cannot be purified to the point that it can be returned to its original use.

Many processes can be used to re-concentrate or reclaim spent H_2SO_4, depending on the concentration of the acid and the amount and kind of impurities in it. Because so many processes are available, a dominant preferred technology cannot be identified. The goal of all the processes, however, is to remove excess water by evaporation and to dehydrate the acid as well as eliminate impurities. The common denominator at all locations is the need to use materials of construction that can withstand the corrosive effects of hot, dilute H_2SO_4. Depending on the processes used and the uses for which the recycled acid is intended, varying quantities of impurities are removed from the acid.

Like so many other aspects of the sulfur industry, economics are not the driving force behind recycling H_2SO_4. In the past, spent acid was disposed of by dumping it into the oceans or neutralizing it with lime and treating the resulting gypsum material as a waste and then placing it in a landfill. These disposal methods are no longer acceptable in most of the world, necessitating the development of reclamation processes, even when costly (Sander et al., 1984).

The vast majority of sulfur construction materials that are not dispersed into the atmosphere can be recycled more easily than more conventional construction materials. Separation of the aggregate from the binding material is not necessary. Sulfur concrete material that has been crushed to a size to facilitate handling simply can be reheated in the same equipment used for the original installation to melt the sulfur. The recycled material does not suffer any degradation of properties.

2.8 SUMMARY AND CONCLUDING REMARKS

In this chapter we discussed some basic information related to sulfur cycle, supply, trade, uses, and environmental issues. There has been a clear trend of dramatic change in the global sulfur industry over the last decade. Sulfur consumption is dominated by a number of large-scale uses with several complex interactions that complicate the future outlook for demand. Thus, sulfur is used

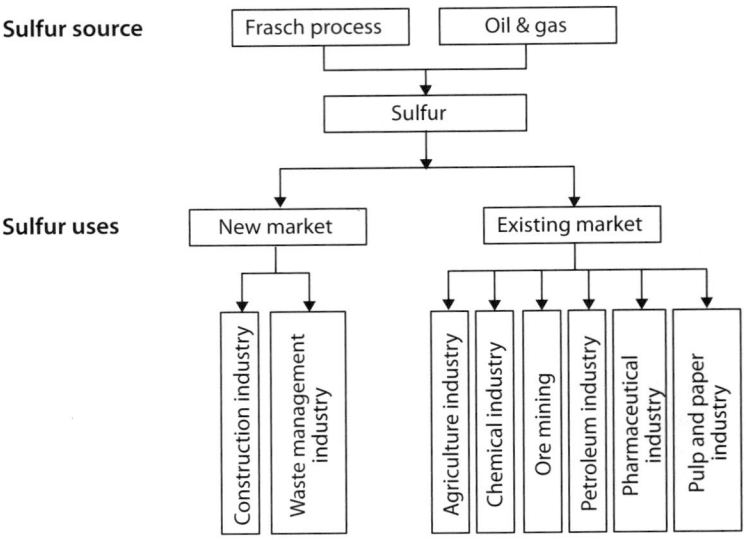

Figure 2.14. Sulfur market dynamics.

to produce sulfuric acid, one of the world's largest volume chemical commodities, but this use competes with sulfuric acid recovered from ore smelting. Uses of sulfuric acid include processing phosphate rock and ore leaching in metallurgical processing (a growing market). There are a number of other uses today for sulfur, including in pigments, pesticides, and rubber vulcanization as well as for an agricultural nutrient.

In general, existing uses for sulfur (Figure 2.14) are relatively mature, and offer limited opportunities to consume significant new supplies. Currently sulfur is in net surplus on a global basis, and environmental regulations mandate ever greater sulfur recovery from petroleum and gas processing. The outlook is clear; there will be substantial and growing surpluses in the global sulfur supply for the foreseeable future. Sulfur prices are likely to be under pressure and producers could face substantial and growing disposal fees.

Sulfur in its elemental form is recognized as an important ingredient in several agronomic applications. As an essential plant nutrient, sulfur has received minimal scientific attention. This is explained by the facts that sulfur was obviously in sufficient supply from the atmosphere, from the soil, and as a byproduct in mineral fertilizers. However, the use of highly concentrated fertilizers containing little or no sulfur has drastically reduced the amount of sulfur supplied to soils. Recent studies have shown that adding sulfur to the soil increased crop yield, drought tolerance, as well as nitrogen efficiency and phosphorus uptake.

The phosphate fertilizer industry is the largest consumer of sulfur, primarily as a consumer of sulfuric acid used in phosphate rock processing. Sulfur itself is also a plant nutrient, mostly supplied as ammonium sulfate and potassium sulfate, and elemental sulfur is a traditional fungicide. Agriculture is thus a significant, but modestly growing outlet for sulfur.

Global sulfur demand has been relatively stagnant at about 57 million Mt/y over the last decade. Based on new regulations limiting sulfur content in diesel and gasoline, the current small global surplus in the sulfur supply is projected to reach between 6 and 12 million Mt by 2011 (The Sulphur Institute), or between 10 to 20% of demand. This projected surplus represents obvious challenges to existing producers, potentially leading to drastically reduced sulfur prices, and even the possibility of costs to producers for disposal of the surplus in some regions. On the other hand, the surplus may also represent opportunities for new uses of sulfur driven by these same reductions in sulfur price that can make new uses more economically feasible and attractive.

Even with relatively small surpluses, the oil and gas industry has already experienced strains on sulfur storage facilities. Sulfur is being stored on site in block, granular or palletized form, or molten (expensively) in rail tank cars because there is insufficient storage capacity to handle sulfur generation at refineries and gas processing plants.

New markets must be found for sulfur to avoid a disposal crises (Figure 2.14). Therefore, this book is designed to identify and evaluate potential new sulfur market applications. The book will be of value to sulfur producers, existing consumers, and government planners with the responsibility for environmental improvement on a national and international scale. In addition, participants and relevant downstream sectors—especially the construction industry, public works, and waste management in which increasing sulfur supplies are likely to demand *new and evolving uses of sulfur*—will find this book to be of great value.

3

SULFUR PROPERTIES

3.1 INTRODUCTION

Sulfur has been known since ancient times; it is the 16th most abundant element in nature. Sulfur was called brimstone—a stone that burns. Sulfur is a complex element that forms many intramolecular and intermolecular allotropes exhibiting a variety of physical properties. The equilibrium composition of liquid sulfur is dependent on temperature and consists of a complex mixture of allotropes. The molecular species that have been observed include S_2, S_3, S_4, S_5, S_6, S_7, S_8, S_{12}, polycatena S_∞ (polymers) and S_π (a mixture of allotropes). The most abundant allotrope is arguably cyclo-octal sulfur (cyclo-S_8) that is the most abundant and stable molecular form. Concentrations of the allotropes control the physical properties of the melt. Discontinuities in density, viscosity, and thermal properties reflect the polymerization process within liquid sulfur; increasing concentrations of S_3, S_4, and S_5 with temperature cause the color change. Variations in the melting point are related to autodissociation of the liquid.

Many solid forms of sulfur have been identified but only orthorhombic (α) and monoclinic (β) sulfur, both composed of cyclo-S_8 sulfur, are stable under terrestrial conditions. Other solid allotropes are composed of various molecular species and may be formed through reactions of sulfur compounds or by quenching the melt.

Physical properties of solid sulfur are dependent on the allotrope and, in some cases, the thermal history. Sulfur's two crystal forms, monoclinic and rhombic, both have a melting temperature just above the boiling point of water at one atmosphere. Under pressure, such as under the earth, water temperature can exceed the melting temperature for sulfur. Since sulfur does not dissolve in water, the liquid sulfur immediately solidifies when it reaches the earth's surface, leaving the distinctive nonmetal, pale yellow, brittle solid.

There is another noncrystalline form of elemental sulfur that can be made by melting crystalline sulfur, but the amorphous allotrope is unstable, reverting to one of the crystalline forms on standing.

Polymerization of liquid sulfur causes discontinuities in viscosity, density, thermal conductivity, and specific heat. Changes in viscosity from several centipoise to almost a thousand poise would have a large effect on flow features and texture.

In this chapter, the nature, properties (physical, chemical, and thermal), and uses of sulfur are discussed.

3.2 OCCURRENCE

Sulfur occurs as a free element near volcanoes and hot springs. Many sulfide minerals are known, and sulfur is widely distributed in minerals such as iron pyrites, galena, sphalerite, cinnabar, stibnite, gypsum, Epsom salts, celestite, and barite. Sulfur occurs in natural gas and crude oil as well as in meteorites. Jupiter's moon Io owes its colors to various forms of sulfur. A dark area on the moon near the crater Aristarchus may be a sulfur deposit.

The abundance of sulfur is shown in Table 3.1 and is given in units of ppb, in terms of weight. Values for abundances are difficult to determine with much certainty, so all values should be treated with some caution, especially so for the less common elements. Local concentrations of any element can vary by orders of magnitude from those given here.

A great deal of elemental sulfur comes from crude oil and natural gas as discussed in Chapter 2. Besides elemental sulfur, sulfur minerals are chiefly found in different forms of metal sulfides and sulfates as illustrated in Table 3.2.

Table 3.1. Sulfur abundance

Abundance	Concentration (ppb by weight)
Universe	500,000
Sun	400,000
Meteorite (Carbonaceous)	41,000,000
Crystal rock	420,000
Sea water	928,000
Stream	4,000
Human	2,000,000

Table 3.2. Sulfur forms

Sulphide pyrite Marcasite (white pyrite)	FeS_2
Pyrrhotite	$FenS_{n+1}$
Cinnabar	HgS
Galena	PbS
Sphalerite	ZnS
Chalcopyrite	$CuFeS_4$
Chalcocite	Cu_2S
Covellite	CuS
Bornite	Cu_5FeS_4
Sulphate	—
Gypsum	$CaSO_4 \cdot 2H_2O$
Anhydrite	$CaSO_4$
Barytes	$BaSO_4$
Celestite	$SrSO_4$

The sulfides are produced by hot, active gases and fluids containing the sulfide ion. Pyrite (FeS_2) is a common mineral, often well-crystallized in characteristic forms (pyritohedrons). It is found, exceptionally, as flat, radial crystal discs near Sparta, Illinois, called *Sparta dollars*.

The sulfur in bituminous coal is often in the form of pyrite. It is hard, heavy (5.02) and has a metallic luster and light-brass color. It frequently occurs in flakes associated with placer gold and is called *fool's gold*. Pyrite is a compound of ferrous iron and the bi-sulfide ion (S_2^-).

Marcasite has the same composition, but a dissimilar crystal structure. It is whiter and has different crystal habits. The regular sulfides, FeS and Fe_2S_3, are not found in nature in pure form. Pyrrhotite is $Fe_{1-x}S$, and pentlandite is a mixed sulphide of nickel and iron, the principal ore of nickel at Sudbury, Ontario, Canada.

Chalcopyrite and bornite have copper with the iron, and arsenopyrite has As replacing one S. The sulfur content ruins pyrite as an ore of iron. SO_2 produced from roasting pyrite often contains arsenic as an impurity, which poisons Pt catalysts in making sulfuric acid.

Galena (PbS), is another familiar sulfide mineral. Not only does it crystallize in the cubic system, its crystals are frequently shiny metallic-looking cubes as shown in Figure 3.1. The metallic luster is due to free electrons in the conduction band since PbS is a semiconductor. It can make a PN-junction with a metal whisker, as in crystal set radios. Galena is lead-gray and heavy (7.4 to 7.6), but not hard and thus scratches easily. It has excellent cubic cleavage and breaks up into

74 Sulfur Concrete for the Construction Industry

| Galena | Cinnabar | Pyrite | Sphalerita |
| (PbS) | (HgS) | (FeS$_2$) | (ZnS) |

Figure 3.1. Various sulfide ores.

smaller cubes when shattered. It is often associated with silver whose sulfide ore is argentite (Ag$_2$S). Argentite can be cut with a knife like lead, which distinguishes it from galena but, like galena, it is heavy and soft. These are the most important ores of lead and silver, respectively.

The great insolubility of sulfides explains their occurrence as minerals. Copper, lead, zinc, silver, mercury, nickel, cobalt, tin, antimony, bismuth and molybdenum are obtained from sulfide ores. Of the common metals only iron, aluminum, and magnesium are obtained from nonsulfide ores.

3.3 PROCESSES

Sweetening

A great deal of sulfur comes from the sweetening of crude oil and natural gas. Much petroleum is sour containing significant amounts of H$_2$S that must be removed before the refined products are sold or the gas is supplied to the pipeline. Since petroleum is liquid or gaseous, this purification is much easier than it is for solid coal.

Roasting

Other sulfur comes from the roasting of sulfide ores of lead, zinc, copper, and pyrite. The recovery of the SO$_2$ in the roasting process, and using it to make sulfuric acid that can either be sold or employed in the ore processing, is not only economical, but beneficial to the surrounding countryside.

Flowering

Flowering of sulfur is the sublimed substance from purification by distillation. It contains minimal oxygen and some sulfuric acid created by slow oxidation, but these are the only impurities. Its colloidal form makes the powder active, and this is the only hazard involved in its use. Flowers of sulfur should be used for chemical purposes, but never as a part of any pyrotechnic mixture. It is also used as an insecticidal dust, producing the SO_2 that is the active substance through slow oxidation.

Rolling

Rolling of sulfur is a method used to grind sulfur into specific particle ranges. Additives such as dispersants, flow aids, and dust suppressants may be added to enhance product performance.

Flouring

Flouring of sulfur is a technique for grounding rhombic sulfur. The powder size in flour is larger than in flowers of sulfur.

3.4 ELEMENTAL SULFUR FORMS

Elemental sulfur has different forms that exhibit a variety of physical properties. The forms are called allotropes and can be divided into two types: (1) intramolecular allotropes, the molecular species resulting from chemical bonding of sulfur atoms and (2) intermolecular allotropes, the lattice structures formed by the arrangement of the molecules within crystals (Meyer, 1965). Complexities in understanding sulfur could be attributed to: (a) most sulfur allotropes are metastable, (b) pure sulfur is difficult to obtain because it reacts readily with other substances, (c) sulfur forms many molecular structures, several of which have similar stabilities, and (d) sulfur has a variety of similarly stable polymorphs.

Sulfur atoms combine to form rings (cyclo-S_n) and chains (catena and polycatena sulfur), allowing several million intramolecular sulfur allotropes to exist, if all possible combinations of the atoms are considered. Theoretically, all S_n molecules with $6 < n < 12$ exist as rings, and, of these, cyclo-S_8, cyclo-S_{12}, and cyclo-S_6 are the most stable. Molecules with more than twelve or less than six atoms can consist of either a ring or a chain configuration. Larger ring structures are unstable and their formation is inhibited or interrupted by the formation of cyclo-S_8 molecules, but they do occur in sulfur liquid.

The chemical difference between rings and chains is that ring molecules have fully paired electrons, whereas chains are considered to be diradicals (Meyer, 1976). This difference is reflected by the chemical reactivity and by the physical properties. The presence and concentration of each allotrope and, therefore, the physical and chemical properties of bulk sulfur, are dependent on the thermal history of the sample. On a broad basis, intramolecular allotropes of sulfur can be divided into four groups: (1) ring molecules with up to 20 atoms that can be isolated as solids, (2) small molecules, (3) large liquid and solid polymers, and (4) ions in solution.

3.4.1 Liquid Sulfur Allotropes

The equilibrium composition of liquid sulfur is dependent on temperature and consists of a complex mixture of allotropes. The molecular species that have been observed include S_2, S_3, S_4, S_5, S_6, S_7, S_8, S_{12}, poly-catena S_∞ (polymers) and S_π (a mixture of allotropes). As noted previously, arguably the most abundant allotrope is cyclo-octal sulfur (cyclo-S_8) that consists of a puckered ring configuration of eight sulfur atoms in which two parallel squares with atoms at the corners are rotated 45° with respect to each other, as shown in Figure 3.2.

This allotrope is relatively stable and also occurs in gaseous and solid states. A part of quickly quenched sulfur can be extracted with CS_2 and precipitated at –78°C. This part of the melt is often called π-sulfur (S_π) and can be further broken down into three fractions, the first having a molecular weight of S_6, the second containing S_8 and having an average composition of $S_{9.2}$, and a third fraction composed of large rings of S_n, for 20 < n < 33. The S_π is formed by autodissociation of the melt and determines the freezing point. At approximately 159.4°C, the character of liquid sulfur changes as a result of polymerization. This temperature is commonly referred to as the λ temperature. The basic theory describing the polymerization of liquid sulfur, whereby cyclo-S_8 changes to chains or catena and poly-catena, has two parts:

$$Cyclo\text{-}S_8 \leftrightarrow Catena\text{-}S_8 \qquad (3.1)$$

$$Catena\text{-}S_8 + Cyclo\text{-}S_8 \rightarrow Catena\text{-}S_{8\cdot 2} \qquad (3.2)$$

The first step involves the conversion of cyclo-S_8 molecules to S_8 chains that initiate polymerization. The chains propagate by the second step through which the S_8 chain combines with another cyclo-S_8 molecule that also converts to a chain structure. The chains continue to grow by adding more S_8 units (Meyer, 1976).

Spectra and other characteristics of liquid sulfur near the critical point indicate that liquid sulfur contains more S_2, S_3 and S_4 than S_5, S_6, S_7, and S_8 (Meyer, 1976).

3.4.2 Solid Sulfur Allotropes

Even though many allotropes of solid sulfur have been defined (Meyer, 1968), only two are thermodynamically stable; orthorhombic (α) sulfur and monoclinic (β) sulfur, both of which are composed of cyclo-S_8 molecules. The other solid

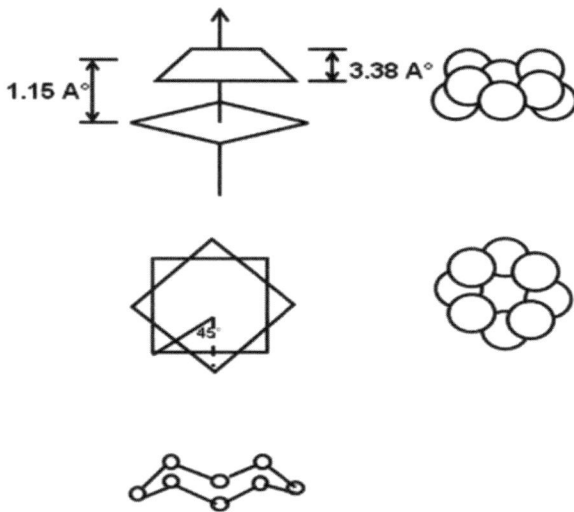

Figure 3.2. Structure of a cyclo-octal sulfur (S$_8$) molecule. The atoms are located at the corners of 2 squares which are rotated 45° with respect to each other (Adapted from Meyer, 1965).

allotropes are generally composed of various intramolecular allotropes and will eventually convert to a more stable form. Some of these metastable allotropes such as cyclo-hexa sulfur (S$_6$), cyclo-dodeca sulfur (S$_{12}$), and other rings are generally formed through reactions of sulfur compounds, but some have been observed in liquid sulfur. Other solid sulfur allotropes are formed by quenching the liquid sulfur and consist of polymers. Orthorhombic sulfur is the most common form of sulfur and is stable up to 95.3°C (Meyer, 1976). The unit cell is composed of 138 atoms (Donohue, 1965). The crystal form exhibits strong anisotropic effects (Meyer, 1977). Many of its properties such as the elastic constant and thermal conductivity are much lower along the c axis than along the a and b axes.

3.5 PROPERTIES OF ELEMENTAL SULFUR

3.5.1 Melting and Freezing Points

Elemental sulfur has a variety of melting and freezing points depending on the solid allotrope that is being melted and the concentrations of allotropes within the melt as shown in Table 3.3. Freezing point depression will occur naturally as a result of autodissociation of the melt to form S$_\pi$, which has a lower freezing point than cyclo-S$_8$. Therefore, the freezing point of the whole mixture is lowered accordingly. The temperature at which the maximum concentration of S$_\pi$ is

Table 3.3. Melting points of sulfur (Meyer, 1976)

Allotrope	Melting points, °C	Remarks
alpha-S	110.06	Natural
	112.80	Single crystal
	115.11	Microcrystal
beta-S	114.50	Natural
	119.60	Ideal: observed
	120.40	Microcrystal
	133.00	Ideal: calculated
gamma-S	106.80	Classic
	108.00	Optical: DTA
	108.60	Microcrystal
insoluble-S	77.00; 90.00; 160.00	Optical: TDA, DTA
	104.00	
S_n	75.00	Optical
	104.00	Classic
S_6	50	Decomposition
S_{12}	148.00	Decomposition
S_{18}	128.00	Decomposition
S_{20}	124.00	Decomposition

reached will be the lowest freezing point and is called the *natural* melting point. The freezing point is also influenced by the pressure and temperature of the melt, indicating slow kinetics and complexities within the liquid as it is cooled. Different reaction paths are followed as sulfur cools that result in various metastable mixtures of different metastable allotropes, probably mostly rings (Meyer, 1976). Single crystals of monoclinic (β) sulfur melt at 119.6°C and are in equilibrium with a liquid of unknown composition. The natural melting point is 114.6°C and calculations indicate that the ideal melting point may be as high as 133°C. Single crystals of orthorhombic sulfur do not readily convert to monoclinic (β) sulfur but tend to melt at 112.8°C. The freezing point depression of orthorhombic sulfur produces a natural melting point of 110.06°C (Tuller, 1954). Microcrystals of both common stable allotropes have higher melting points than larger crystals (Thackray, 1970) such that microcrystals of monoclinic (β) melt at 120.4°C and orthorhombic (α) microcrystals melt at 115.11°C. Melting points of other solid allotropes are given in Table 3.3.

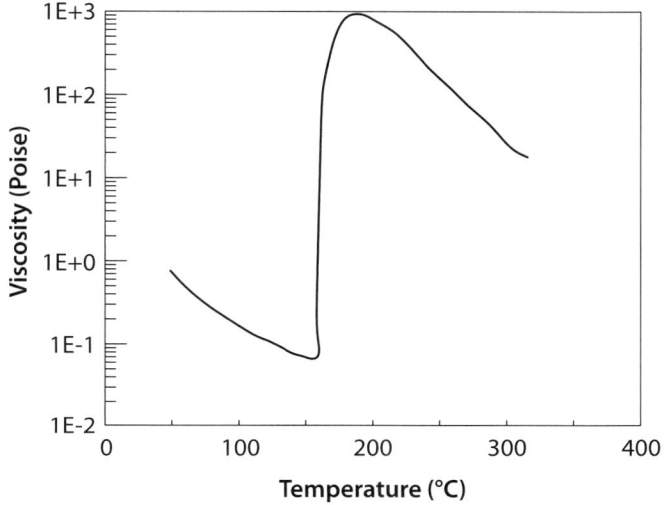

Figure 3.3. Viscosity variations with temperature (adapted from Bacon and Fanelli, 1943).

3.5.2 Viscosity

The viscosity of liquid sulfur decreases with increasing temperature to approximately 7 centipoises at 160°C, at which point the viscosity increases rapidly to a maximum of approximately 932 poise at around 190°C (Figure 3.3). With increasing temperature, the viscosity again decreases. The change in viscosity is related to the concentration and length of poly-catena within the liquid. The jump in viscosity at around 160°C reflects the increase in both the concentration and the length of poly-catena molecules, whereas the decrease with increasing temperature after ~190°C reflects the general decrease in the number of S_8 chains per poly-catena molecule. Original viscosity measurements were made by Bacon and Fanelli (1943).

3.5.3 Density

Meyer (1976) reported the densities of many solid sulfur allotropes as shown in Table 3.4. The density of liquid sulfur decreases with increasing temperature as indicated in Figure 3.4. Polymerization of the melt produces a discontinuity in the density curve at the λ temperature.

3.5.4 Color

The colors of the various sulfur allotropes and of sulfur melt, which may retain its color upon quenching, are important for the study of potential sulfur flows.

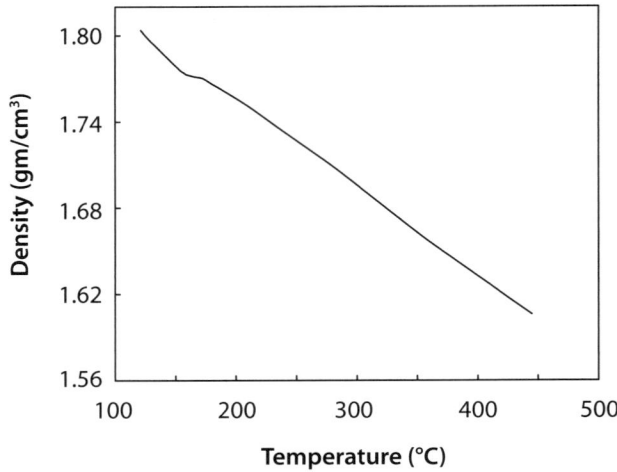

Figure 3.4. Density variations with temperature of liquid sulfur (adapted from Meyer, 1976).

Table 3.4. Density of solid sulfur allotropes (Meyer, 1976)

Allotrope	Symbol	Density (Mg/m³)
Orthorhombic α	S_8	2.069
Monoclinic β	S_8	1.940
Monoclinic γ	S_8	2.190
Cyclo-hexa	S_6	2.209
Cyclo-hepta	S_7	2.090
Cyclo-dodeca	S_{12}	2.036
Cyclo-octadeca	S_{18}	2.090
Cyclo-icosa	S_{20}	2.016
Fibrous	S_∞	2.010

Unfortunately, the color of elemental sulfur is as complex as the chemistry. For solid ring allotropes, the colors are well established as shown in Table 3.5. Pure liquid sulfur is bright, clear yellow at the melting point and consistently changes to deep, opaque red, which is reached at the boiling point (444°C). Dependence of the color on temperature results from the changing chemistry of the melt such that the yellow, from the melting point to above 250°C, is due to thermal broadening of the cyclo-S_8 spectrum, and the overlap of the polymeric sulfur spectrum and the red is due to the absorption spectra of S_3, S_4, and S_5 (Meyer et al., 1971; Meyer, 1976). The commonly observed change to brown and then black as the temperature increases is not observed in pure sulfur (99.99+% sulfur). At high

Table 3.5. Color of solid allotropes (Meyer, 1976)

Allotrope	Symbol	Color
Cyclo-hexa	S_6	Orange-red
Cyclo-hepta	S_7	Light yellow
Orthorhombic α	S_8	Bright yellow
Monoclinic β	S_8	Yellow
Monoclinic γ	S_8	Light yellow
Cyclo-ennea	S_9	Deep yellow
Cyclo-deca	S_{10}	Yellow-green
Cyclo-octadeca	S_{18}	Lemon-yellow

temperatures, sulfur is reactive and the darkening of the fluid to brown and black is considered to be a result of a reaction with impurities, possibly organic (Meyer, 1976).

The color of the quenched solid is dependent on the rate of cooling. If the melt is quenched from near the boiling point to −80°C, the solid will be yellow; whereas if it is quenched in liquid nitrogen (∼−209 to −196°C), a red glass will be produced. This glass will convert to yellow polymeric sulfur at -80°C (∼193 K). Slow kinetics and low conductivity of sulfur allow only a thin film of material to be quenched, and even as the cooling occurs, chemical changes take place within the melt.

Quenching of at least the surface of sulfur flows on Io might be expected because the surface temperature is approximately 130°K. This falls within the temperature range of quenched yellow sulfur (−80°C or 193°K) and red sulfur (∼78°K). Because the surface temperature is below the conversion point of −80°C, red may be the more prominent color; however, more work is needed to determine exactly what color(s) may be expected and the stability of the solid under ionian conditions.

3.5.5 Strength Characteristics

The strength of sulfur is determined by the purity and thermal history of the sample. Dale and Ludwig (1965) reviewed tensile and compressive strength values for sulfur and provided a more comprehensive determination of tensile strength. Compressive strengths of sulfur vary from 12,410.6 kPa (1800 psi) to 22,752.7 kPa (3300 psi). Tensile strengths were determined for a large range of thermal histories and it appears that rate of cooling and the temperature to which the sample was heated are important. A faster rate of cooling and a higher initial temperature of the melt result in higher tensile strengths. Monoclinic (β) sulfur

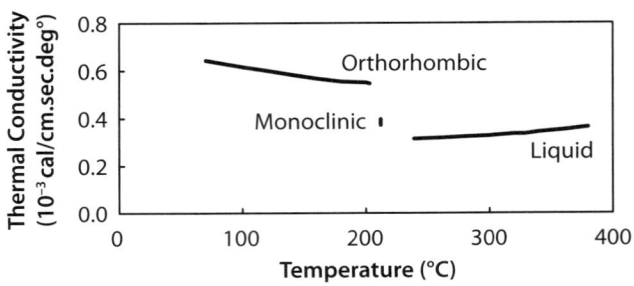

Figure 3.5. Variations of thermal conductivity with temperature for (a) orthorhombic, (b) monoclinic, and (c) liquid sulfur (adapted from Tuller, 1954).

heated to 130°C possessed tensile strengths of 399.9 kPa (58 psi) to 441.3 kPa (64 psi), whereas monoclinic sulfur produced from a melt at 187°C possessed tensile strengths from 427.5 kPa (62 psi) to 1792.6 kPa (260 psi). Orthorhombic (α) sulfur from a melt at 130°C had tensile strengths from 330.9 kPa (48 psi) to 3344.0 kPa (485 psi) and that from 187°C had tensile strengths ranging from 337.8 kPa (49 psi) to 4274.7 kPa (620 psi). Threads of polymeric sulfur had tensile strengths from 17,926.4 kPa (2600 psi) to greater than 96,526.6 kPa (14,000 psi).

3.5.6 Thermal Characteristics

Thermal characteristics of sulfur also exhibit discontinuities caused by polymerization. Thermal conductivity is greater in solid sulfur than liquid sulfur (Tuller, 1954) as shown in Figure 3.5. The thermal conductivity of sulfur is approximately an order of magnitude less than that of most rocks and is comparable to such insulation materials as mica and asbestos. Specific heat values for solid and liquid sulfur, at atmospheric pressure, are the function of temperature (Tuller, 1954). Figure 3.6 illustrates the difference in specific heat between orthorhombic (α) and monoclinic (β) sulfur and the discontinuity in the melt at 160°C. The coefficient of thermal expansion for sulfur reaches a maximum around 152°C and rapidly decreases around 160°C. A second maximum occurs around 288°C (Tuller, 1954). Thermal expansion of fibrous sulfur is 94×10^{-6} cm/deg for the a axis and 72×10^{-6} cm/deg for the b axis (Meyer, 1976).

3.5.7 Allotropic Transformation

Most elements exist in only one form. Different forms of the same element are called allotropes of the element. In sulfur, allotropy arises from two sources: (1) the different modes of bonding atoms into a single molecule and (2) packing of polyatomic sulfur molecules into various crystalline and amorphous forms. Some 30 allotropic forms of sulfur have been reported, but a portion of these

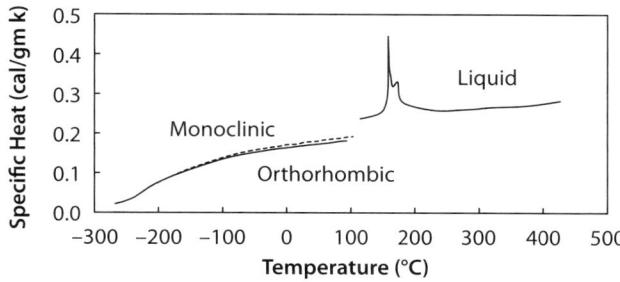

Figure 3.6. Variations of specific heat with temperature for (a) orthorhombic, (b) monoclinic, and (c) liquid sulfur (adapted from Tuller, 1954).

Table 3.6. Allotropic forms of sulfur (Gabel, 1993)

Form	Formula	Structure	Appearance	Phase
Orthorhombic (α)	S_8	Rings	Rhombus	Solid
Monoclinic (β)	S_8	Rings	Needles	Solid
Amorphous		Chains & rings	Tangled rob	Solid-liquid
Lambda (λ)	S_8	Rings	Thin, yellow	Liquid
Mu (μ)	S_8	Chains	Red, black	Liquid
Vapor	S_8	Molecules	Yellow	Gas

represent mixtures. Only eight of the 30 seem to be unique; five contain rings of sulfur atoms and the others contain chains. Allotropes generally differ in physical properties such as color and hardness; they may also differ in molecular structure or chemical activity (Table 3.6) but are generally similar in most chemical properties.

Rhombic and Monoclinic

Rhombic sulfur has orthorhombic crystalline structure (α-sulfur) and is stable below 95.5°C; most sulfur is in this form. The *monoclinic*, or prismatic, form (β-sulfur) has long, needle-like, nearly transparent crystals; it is stable between 95.5°C and its melting point (119°C) as shown in Figure 3.7, which demonstrates the sulfur phase diagram. Rhombic sulfur reverses completely into rhombic sulfur within 20 hours when stored at room temperature as shown in Figure 3.8 (Blight et al., 1978).

The general allotropic class of sulfur is that of the eight-member ring molecules, three crystalline forms of which have been well characterized. One is the orthorhombic (often improperly called rhombic) form, α-sulfur. It is stable at temperatures below 96°C. Another of the crystalline S_8 ring allotropes is the

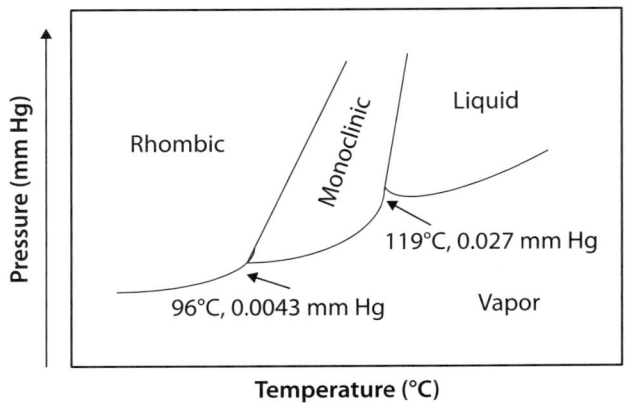

Figure 3.7. Sulfur phase diagram.

monoclinic or β-form, in which two of the axes of the crystal are perpendicular, but the third forms an oblique angle with the first two. There are still some uncertainties concerning its structure; this modification is stable from 96°C to the melting point (118.9°C). A second monoclinic cyclo-octa-sulfur allotrope is the γ-form, unstable at all temperatures and quickly transforming to α-sulfur.

A specimen of crystalline orthorhombic sulfur form is more stable, but it is fairly narrow and conversions between the two forms are slow; α-sulfur has a density of 2.07 and melts at 113°C, β-sulfur has a density of 1.97 and melts at 119°C. Each melts to a straw-yellow liquid of density 1.808 called *lambda* (λ)-sulfur. If the liquid is cooled slowly, needle-like monoclinic crystals form. When the temperature falls below 96°C, these crystals slowly change to orthorhombic, a microcrystal.

The orthorhombic and the needle-like monoclinic forms of sulfur are shown in Figure 3.9. The two forms arrange the S_8 rings differently. Orthorhombic lattices result from stretching a cubic lattice along two of its lattice vectors by two different factors, resulting in a rectangular prism with a rectangular base **a** by **b** and height **c**. Keeping in mind that **a** ≠ **b** ≠ **c**, and the three bases intersect at 90° angles, the three lattice vectors remain mutually orthogonal.

In the monoclinic system, the crystal is described by vectors of unequal length such as in the orthorhombic system. They form a rectangular prism with a parallelogram as base. Hence two pairs of vectors are perpendicular, whereas the third pair makes an angle other than 90° (http://en.wikipedia.org/wiki/Monoclinic).

A typical microstructure picture for elemental sulfur is shown in Figure 3.10 (Mohamed and El Gamal, 2006). It can be observed that elemental sulfur

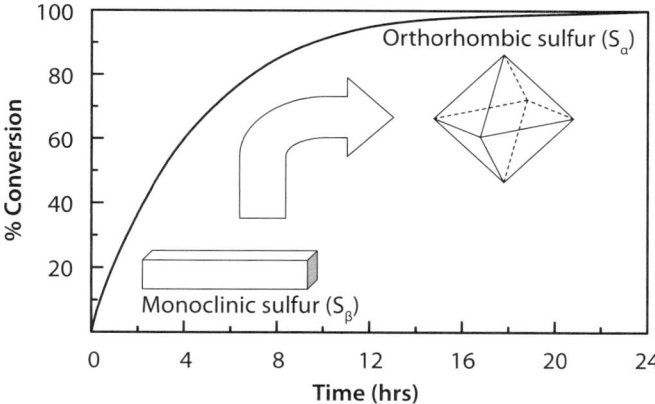

Figure 3.8. Reversion rate of monoclinic sulfur (S_β) to orthorhombic sulfur (S_α) at ambient temperature (adapted from Blight et al., 1978).

crystallizes and forms dense, large alpha sulfur crystals with orthorhombic sulfur morphology.

Amorphous

Amorphous sulfur is a dark, noncrystalline, and gum-like substance. It is often thought to be a super-cooled liquid; it is formed by rapidly cooling molten sulfur, e.g., by pouring it into cold water. It slowly reverts to the rhombic form on

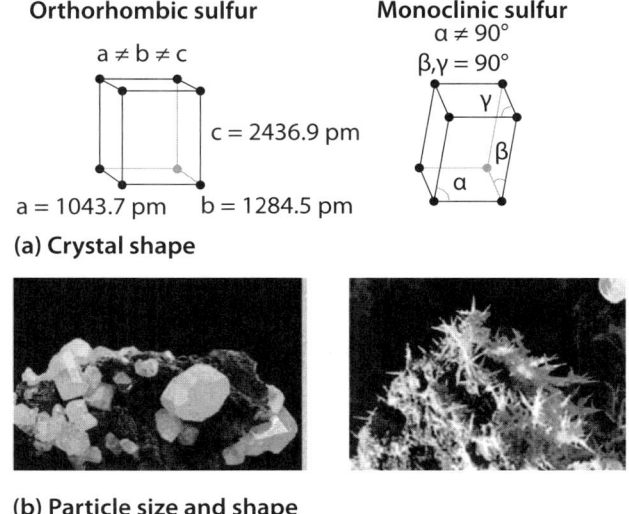

Figure 3.9. Orthorhombic and monoclinic sulfur: (a) Crystal shape; (b) Particle size and shape.

(a) Dense crystalline structure   (b) Large orthorhombic crystals

Figure 3.10. Sulfur crystallization and formation of alpha sulfur crystals (S_α) with orthorhombic sulfur morphology (Mohamed and El Gamal, 2006).

standing. The crystalline forms are readily soluble in carbon disulfide, but the amorphous form is not.

In summary, the general properties of elemental sulfur are summarized in Table 3.7.

3.5.8 Polymerization

Sulfur exhibits a wide variety of allotropes of cyclic molecules (S_x, with x from 6 to 35) and the unusual phenomenon of equilibrium *polymerization* in the liquid state (Meyer, 1976; Steudel, 2003). There is additional interest in understanding this transition because equilibrium polymerization also occurs in organic and biological systems (Greer, 1998). Equilibrium polymerization can be viewed as a particular example of the general tendency in condensed matter for particles to cluster (Dudowicz, 1999). Sulfur is the simplest substance exhibiting this phenomenon and, therefore, should be the most convenient to model. In addition, the polymerization transition in sulfur is an example of a liquid-liquid phase transition. Such transitions are currently of considerable interest (Harrington et al., 1997).

Polymerization in sulfur has been studied for many years, but a coherent picture of this phenomenon is still missing. Sulfur melts at approximately 120°C, forming a light-yellow, low-viscosity liquid mostly consisting of 8-membered-ring molecules. Near 159°C, properties of sulfur change dramatically. Shear viscosity (η), as measured using the Poiseuille technique, increases by four orders of magnitude over a temperature interval 25°C (Bacon and Fanelli, 1943). More recent measurements performed with a sealed pyrex cell, using a falling-ball technique, indicate an even steeper change in η (Ruiz-Garcia et al., 1989).

A specific feature resembling a λ-like singularity has been observed in temperature dependence of the heat capacity (Feher et al., 1958; West, 1959). Changes also occur in the temperature dependence of the density, ρ (Zheng and

Table 3.7. General properties of elemental sulfur

Chemical name: Sulfur
Family name: Element – Sulfur
Chemical formula: S_8
Physical state: Solid
CAS number: 7704-34-9
Appearance: Yellow colored lumps, crystals, powder, or formed shape
Odor: Odorless, or faint odor of rotten eggs if not 100% pure
Purity: 90–100%
Formula: S_8 (Rhombic or monoclinic)
Molecular weight (G): 256.50
Vapor density (Air = 1): 1.1
Vapor pressure: 0 mm HG @ 280°F
Solubility in water: Insoluble
Specific gravity: 2.07 @ 70°F
Boiling point: 832°F (444°C)
Freezing/melting point: 230–246°F (110–119°C)
Bulk density: Lumps 75–115 lbs./ft³ Powder 33–80 lbs./ft³
Flashpoint: 405°F (207.2°C)
Flammable limits: LEL: 3.3 UEL: 46.0
Auto-ignition temperature: 478–511°F (248–266°C)

Greer, 1992) and in the sound velocity, c (Timrot et al., 1984; Kozhevnikov et al., 2001). The color of sulfur becomes darker, and the refractive index exhibits a minimum (Tamura et al., 1986). A minimum is also observed in the dielectric permittivity (Greer, 1986). Electron spins resonance (ESR) spectra indicate a sharp increase in the number of unpaired electrons. The ESR signal is interpreted as being due to the appearance of linear polymers. An estimated degree of polymerization (average number of monomers in a polymeric chain) near the transition temperature is on the order of 10^6 (Koningsberger, 1971). Finally, molten sulfur quenched from temperatures below 159°C is easily soluble in carbon disulfide, whereas sulfur quenched from higher temperatures yields an insoluble fiber-like residue whose mass fraction, Φ (also called extent of polymerization), grows with the quenching temperature and reaches saturation at about 300°C (Steudel, 2003; Koh and Klement, 1970). The data for η, ESR spectra, and Φ suggest that liquid sulfur above 159°C is a solution of linear chain polymers in a mono-meric solvent (Kozhevnikov et al., 2001).

The equilibrium polymerization of sulfur is investigated by Monte Carlo simulations. The potential energy model is based on density functional results

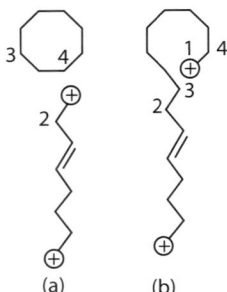

Figure 3.11. Schematic view of the bond interchanges mechanism. The large circles with a cross represent unsaturated atoms (adapted from Ballone and Jones, 2003).

for the cohesive energy, structural, and vibration properties as well as reactivity of sulfur rings and chains (Ballone and Jones, 2003). Thermally activated bond-breaking processes lead to an equilibrium population of unsaturated atoms that can change the local pattern of covalent bonds and allow the system to approach equilibrium. The concentration of unsaturated atoms and the kinetics of bond interchanges are determined by the energy required to break a covalent bond (Ballone and Jones, 2003). Broken bonds open the way to propagation that takes place via a bond-switching mechanism involving one unsaturated atom (Figure 3.11), and the corresponding energy barrier depends on the interaction of the radical atom with all the others. If this interaction is the same as for regular atoms, bond switching would not be observed during runs of practicable length. However, the radical atoms can approach regular atoms much closer than a pair of regular atoms and the energy barrier is relatively low.

Sulfur undergoes a liquid–liquid transition usually interpreted as the ring opening polymerization of elemental sulfur S_8. The increase in temperature is accompanied by an increase in motion, and the bond within the ring become strained and finally broken. As the covalent bond breaks equally in half, a di-radical is formed. Ring opening gives rise to triplet diradical chains. Polymerization occurs to form long chains as shown in Figure 3.12 (Deanna Dunlavy, 1998; Mohamed and El Gamal, 2006).

3.6 CHEMICAL PROPERTIES

Sulfur is a yellow solid; it is a typical nonmetal and is found in Group VII of the periodic table. The notable characteristics are that of a soft and light substance and bright yellow in color. Although hydrogen sulfide (H_2S) has the distinct smell of rotten eggs, it should be noted that elemental sulfur is odorless. It burns with a blue flame that emits sulfur dioxide, notable for its peculiar suffocating odor.

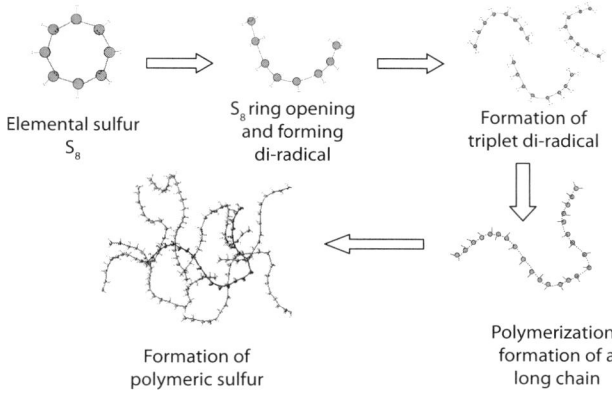

Figure 3.12. Sulfur polymerization mechanisms (adapted from Deanna Dunlavy 1998) http://www.thecatalyst.org/experiments/Dunlavy/Dunlavy.html

Sulfur is insoluble in water but soluble in carbon disulfide and other nonpolar solvents. Sulfur forms stable compounds with all other elements except the noble gases.

3.6.1 Electronic Structure

The electronic structure of sulfur has an important effect on the forms in which it exists. If a neutral atom gains two electrons, it completes the stable argonic structure. Sulfur is the chemical element in the periodic table that has the symbol S. Its atomic number is 16 and its atomic weight is 32.07; its electron configuration is $1s^2 2s^2 2p^6 3s^2 3p^4$, with six valence electrons outside a neon core. The first ionization potential of the atom is 10.357 V, the second is 23.405 V, and so the atom strongly holds its electrons.

Although sulfur is in the same periodic table group as oxygen, there are more differences in the chemical characteristics of these elements than there are similarities. Thus, whereas oxygen always displays a valence of two, sulfur displays valences of four (in sulfur dioxide [SO_2]), six (in thionyl chloride [$SOCl_2$]) and eight (in sulfur hexafluoride [SF_8]).

The sulfur ion S^- is stable in aqueous solution in which the polar water reduces the energy penalty of the charge. Two sulfur atoms can make a *coordinate covalent* single bond, and this S_2 molecule is found in the vapor at high temperatures. Sulfur does not form S_2 with a double bond as oxygen does. If two electrons are added to this molecule, the result is the disulfide ion (S_2^-) which is also stable in aqueous solution. Both these ions form salts with metals that are disposed to

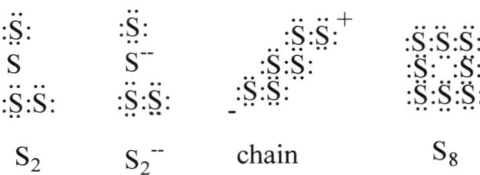

Figure 3.13. Lewis electron diagram for sulfur.

furnish the two electrons. Lewis electron diagrams are shown for sulfur in Figure 3.13 (Lewis, 1916).

Sulfur atoms can aggregate further and form long chains, each one sharing one of its electron pairs with the preceding atoms. The end atom is always in an awkward position, but can acquire two electrons for an argonic structure. In Figure 3.13, the electron distribution makes one end polar and the other negative so that combining ends would attract one another. If one electron is transferred from one end to the other, the ends are alike but they still act as free radicals. Silicon forms chains with intermediate oxygen atoms, –Si–O–Si–O–, and carbon forms chains if hydrogen satisfies the extra valences (–(CH$_2$)n-). However, left to their own devices, each of these atoms prefers a three-dimensional structure. Sulfur naturally forms chains if left to itself. One result of this is the polysulfide ions S_x^- that are easily made. The bonds angles in these structures are expected to be 90°, since p orbital are involved. There is free rotation about a bond.

It is clear that a sulfur chain could bite its own tail and become a ring. This would solve the problems of the terminal atoms and produce a stable structure. The smallest ring that has no strains consists of 8 atoms. It forms a crown-shaped molecule with 4 atoms on top and 4 atoms on the bottom, rotated relatively by 45°. The Lewis diagram suggests this, schematically. There is a little steric interaction between nonbonded atoms so that the bond angle is stretched to 102°, and the bond length is 0.208 nm. This shows that the size of the covalent sulfur atom is close to 0.1 nm. There is also a less-stable ring of 6 atoms. In any sample of sulfur, thermal agitation is constantly breaking and reassembling chains, so they gradually assume the most stable state for the particular temperature. The yellow color of sulfur is probably due to a small number of thermally-generated unterminated chains.

3.6.2 Oxidation States

Sulfur exhibits oxidation numbers of –2, 0, +2, +4 and +6 (Figure 3.14). Oxidation number –2 is seen in the sulfide ion (S$^-$) and a typical compound is hydrogen sulfide or hydro-sulfuric acid (H$_2$S).

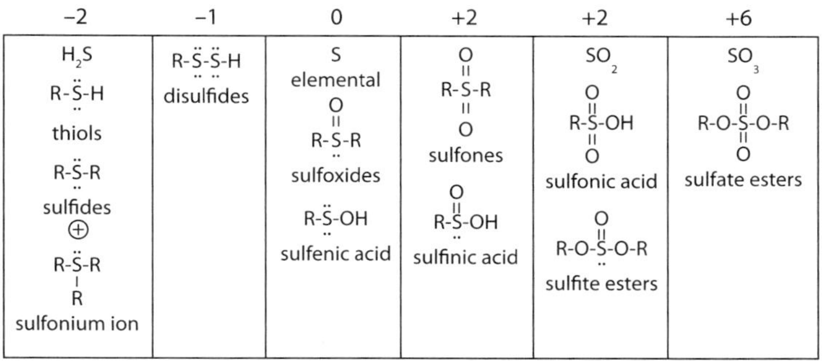

−2	−1	0	+2	+2	+6
H$_2$S	R-S̈-S̈-H	S elemental	O‖R-S-R	SO$_2$	SO$_3$
R-S̈-H thiols	disulfides	O‖R-S-R sulfoxides	O‖O sulfones	O‖R-S̈-OH‖O sulfonic acid	O‖R-O-S-O-R‖O sulfate esters
R-S̈-R sulfides		R-S̈-OH sulfenic acid	O‖R-S̈-OH sulfinic acid	O‖R-O-S-O-R sulfite esters	
R-S̈-R $\oplus$ R sulfonium ion					

Figure 3.14. Sulfur oxidation states in organic compounds.

Unlike H$_2$O, this is a gas at room temperature, melting at −82.9°C and boiling at −61.8°C because hydrogen bonding is absent. In aqueous solution, its ionization constant to H$^+$ and HS$^-$ is 5.7×10^{-8}, and further to S$^-$ is 1.2×10^{-15}. It is a stronger acid than water, but it is still weak. H$_2$S is poisonous (more poisonous than HCN), but fresh air will overcome small concentrations.

The uses of sulfur can be classified under the oxidation number as sulfides (−2), elemental sulfur (0), SO$_2$ and sulfites (+4), and SO$_3$ and sulfates (+6). Sulfuric acid (+6) has already been mentioned as the most important product of the chemical industry—the King of Chemicals.

Oxidation number 0 occurs in S$_8$ and in the polysulfides as well. However, it does not correspond to a distinct group of compounds. +2 is represented by the unusual compound SO that is not stable like the oxygen analogue, the extremely stable CO, because S does not like to make double bonds of this type. When sulfur is burned in air, the dioxide SO$_2$ is formed in which the oxidation number appears to be +4. The structure of this molecule is surprising because the S kernel is surrounded by 12 electrons rather than the usual 8 of the argonic structure. Pauling calls this a *trans-argonic* structure. It is formed by promoting one electron to a d level so that five bonding molecular orbital can be based on the configuration sp^{3}d. This gives a lower energy than the alternatives. In fact, SO$_2$ is quite stable. The bond angle is 119.5° and the bond length is 0.143 nm. Carbon dioxide, for comparison, is linear. SO$_2$ is made by burning sulfur in air

$$S + O_2 \rightarrow SO_2 \tag{3.3}$$

or by roasting sulfide ores such as pyrite:

$$4FeS_2 + 11O_2 \rightarrow 2Fe_2O_3 + 8SO_2 \tag{3.4}$$

It is an irritating gas and is used for fumigation.

Sulfur dioxide is an excellent and inexpensive refrigerant, and was once widely used in home refrigerators before it was replaced by the freons. The only problem with SO_2 is its irritating nature. The freons are nontoxic and noninflammable, but are now forbidden because of the damage that the chlorine does to the ozone layer. Perhaps sulfur dioxide will come back as a common refrigerant. Its critical temperature is 143.12°C, and its critical pressure is 77.65 atm so that it is easy to liquefy at room temperature by simple compression. The liquid, of density 1.4601 g/cc, boils at −10°C at one atmosphere with a latent heat of 172.3 Btu/lb. SO_2 melts at −75.2°C. Its ratio of specific heats is 1.256, and the molecular weight is 64.06, from which the speed of sound can be calculated.

Try drawing Lewis-structures for the sulfur atoms in these compounds. If you restrict your formulas to valence shell electron octets, most of the higher oxidation states will have formal charge separation. The formulas written here neutralize this charge separation by double bonding that expands the valence octet of sulfur. Indeed, the S=O double bonds do not consist of the customary σ and π-orbitals found in carbon double bonds. As a third row element, sulfur has five empty 2d-orbitals that may be used for *p-d* bonding in a fashion similar to *p-p* (π) bonding. In this way, sulfur may expand an argon-like valence shell octet by two (e.g., sulf-oxides) or four (e.g., sulfones) electrons.

Sulf-oxides have a fixed pyramidal shape (the sulfur nonbonding electron pair occupies one corner of a tetrahedron with sulfur at the center). Consequently, sulf-oxides that have two different alkyl or aryl substituents are chiral. Enantiomeric sulf-oxides are stable and may be isolated.

Thiols also differ dramatically from alcohols in their oxidation chemistry. Oxidation of 1° and 2°-alcohols to aldehydes and ketones changes the oxidation state of carbon but not oxygen. Oxidation of thiols and other sulfur compounds changes the oxidation state of sulfur rather than carbon.

In the following examples, representative sulfur oxidations are demonstrated. In Example 1, mild oxidation converts thiols to di-sufides. An equivalent oxidation of alcohols to peroxides is not normally observed. The reasons for this different behavior are not hard to identify. The S–S single bond is nearly twice as strong as the O–O bond in peroxides, and the O–H bond is stronger than the S–H bond by more than 25 kcal/mole. Thus, thermodynamics favors di-sulfide formation over peroxide. Mild oxidation of di-sufides with chlorine gives alkyl-sulfenyl chlorides, but more vigorous oxidation forms sulfonic acids (Example 2). Finally, oxidation of sulfides with hydrogen peroxide (or peracids) leads first to sulf-oxides and then to sulfones (Example 3):

Example 1 $2\ R\text{-}S\text{-}H \xrightleftharpoons[\text{[H]}]{\text{[O]}} R\text{-}S\text{-}S\text{-}R$ $\begin{cases} \text{[O] = mild oxidizing agents} \\ \text{[H] = mild reducing agents} \end{cases}$

Example 2 $2\ R\text{-}S\text{-}Cl \xleftarrow{\underset{-20°C}{Cl_2}} R\text{-}S\text{-}S\text{-}R \xrightarrow{\underset{\text{or } HNO_3}{HCO_3H}} 2\ R\text{-}SO_3H$

Example 3 $R\text{-}S\text{-}R + H_2O_2 \xrightarrow{25°C} R\text{-}\overset{O}{\underset{..}{\overset{\|}{S}}}\text{-}R + H_2O_2 \xrightarrow{100°C} R\text{-}\overset{O}{\underset{\overset{\|}{O}}{\overset{\|}{S}}}\text{-}R$

3.6.3 Chemical Reactions of Elemental Sulfur

Chemical reactions, also called chemical changes, are not limited to happening in a chemistry laboratory. Here are some examples of chemical reactions of elemental sulfur with the corresponding chemical equations (Brown and Lemay, 1981; Shakhashiri, 1983).

Reaction of Sulfur with Air

Sulfur burns in air to form the gaseous dioxide sulfur (IV) oxide (SO_2):

$$S_8(s) + 8O_2(g) \rightarrow 8SO_2(g) \tag{3.5}$$

Reaction of Sulfur with Water

Sulfur does not react with water under normal conditions. Many researchers have examined the reaction of sulfur with water at elevated temperatures and pressures in solving geothermal problems or for technological purposes such as the hydrometallurgical processing of sulfur ores (Oana and Ishikawa, 1966; Dadze and Sorokin, 1993; Tohji, 2002). A paper on the effect of reaction temperature and time presented at the 14th International Conference on the Properties of Water and Steam in Kyoto (Lin et al., 2004) indicated that when the reaction temperature is higher than 119.6°C, which is the melting point of elemental sulfur, the melting sulfur forms an independent liquid phase surrounded by aqueous. Therefore, the solubility of sulfur into the aqueous phase may be controlled by the diffusion of reactants between the liquid sulfur phase and aqueous phase or surface reactions. The result suggests that the external diffusion is the rate determining step in the dissolution reaction of sulfur phase. On heating a 0.5M NaOH solution containing elemental sulfur in a Teflon autoclave from 90 to 250°C, the color of resultant solution is changed. At temperatures less than 150°C, all the resultant solution samples are yellow which suggests that polysulfide HS_n^- exist in the solutions (where $n = 1, 2, 3, 4$, etc.). With increasing temperature and time, the sample solutions become colorless and imply the degradation of polysulfides, that is to say, polysulfide HS_n^- are thermally unstable.

Reaction of Sulfur with the Halogens

Sulfur reacts with all the halogens upon heating. Molten sulfur reacts with chlorine to form disulfur dichloride (S_2Cl_2). The odor is awful. With excess chlorine and in the presence of a catalyst, such as $FeCl_3$, SnI_4, etc., it is possible to make a mixture containing an equilibrium mixture of red sulfur (II) chloride (SCl_2) and disulfur dichloride (S_2Cl_2).

$$S_8 + 4Cl_2 \rightarrow 4S_2Cl_2(l) \text{ [orange]} \quad (3.6)$$

$$S_2Cl_2(l) + Cl_2 \leftrightarrow 2SCl_2(l) \text{ [dark red]} \quad (3.7)$$

Reaction of Sulfur with Acids

Sulfur does not react with dilute nonoxidizing acids.

Reaction of Sulfur with Bases

Sulfur reacts with hot aqueous potassium hydroxide (KOH) to form sulfide and thio-sulfate species:

$$S_8(s) + 6KOH(aq) \rightarrow 2K_2S_3 + K_2S_2O_3 + 3H_2O(l) \quad (3.8)$$

Reactions of Sulfur with Iron and Zinc

Iron (II) sulfide and zinc sulfide are quite stable compounds and their formation from the metal and sulfur is highly exothermic. However, iron and sulfur and zinc and sulfur can be mixed at room temperature and no reaction takes place. When heated to the temperature of a red-hot metal rod, they react rapidly—in the case of zinc, almost explosively. In addition to ZnS (s) some ZnO (s), some SO_2 (g) are produced by the high temperature of the second reaction:

$$8\ Fe(s) + S_8(s) \rightarrow 8\ FeS(s) \quad (3.9)$$

$$8\ Zn(s) + S_8(s) \rightarrow 8\ ZnS(s) \quad (3.10)$$

Reactions of Sulfur with Silver

Silver reacts with sulfur in the air to make silver sulfide, which forms the black material we call tarnish—*a silver spoon tarnishes*:

$$2\ Ag + S \rightarrow Ag_2S \quad (3.11)$$

3.6.4 Chemical Reactions of Sulfur with Olefins

It is also known that di-olefins will plasticize sulfur to give a fairly rigid, resinous, and noncrystalline product. Di-olefins and tri-olefins known to effect such plasticization include dicyclopentadiene, myrcene, alloocimene, limonene, and di-pentene. Some di-olefins will not react with the molten sulfur or the degree of reaction is so small that there is essentially no apparent change in the sulfur properties. In some cases, compounds that might normally react give no discernible product due to being too volatile to stay in contact with the sulfur under the conditions used for reaction. As indicated, where reaction does occur, the product is a rigid, noncrystalline product and because of such properties the utility of the composition is limited.

The plasticization of sulfur with mixtures is known. For example, U.S. Pat. No. 3,459,717 discloses the use of at least one ethylenic hydrocarbon together with a phosphorus containing di-ester to react with molten sulfur. Also, Polish Pat. No. 83,123 is reported by Chemical Abstracts (C.A. 86:75381, 1977) to disclose plasticization of sulfur with a solution of an aromatic olefin such as styrene in a paraffin oil, polyethylene wax, benzene, or xylene. It has been found that reaction of certain di- or tri-olefins with sulfur may be achieved to obtain a flexible product that is thereby of greater utility. Thus, the flexible plasticized sulfur compositions may be used in formulations for road surfaces and road markings (e.g., paint striping), in adhesive and caulking formulations to impart flexibility.

3.6.5 Chemical Reactions of Sulfur Compounds

Sulfur analogs of alcohols are called thiols or mercaptans, and ether analogs are called sulfides. The chemical behavior of thiols and sulfides contrasts with that of alcohols and ethers in some important ways. Because hydrogen sulfide (H_2S) is a much stronger acid than water (by more than ten million fold), we expect and find that thiols are stronger acids than equivalent alcohols and phenols. Thiolate conjugate bases are easily formed and have proven to be excellent nucleophiles in SN_2 reactions of alkyl halides and tosylates:

$$R-S(-) \ Na(+) + (CH_3)_2CH-Br \rightarrow (CH_3)_2CH-S-R + Na(+) \ Br(-) \quad (3.12)$$

Although the basicity of ethers is roughly a hundred times greater than that of equivalent sulfides, the nucleo-philicity of sulfur is much greater than that of oxygen, leading to a number of interesting and useful electro-philic substitutions of sulfur that are not normally observed for oxygen.

3.6.6 Biological Reactions of Sulfur Compounds

There is a complex sulfur cycle in the biosphere involving several oxidized and reduced inorganic and organic sulfur species. Inorganic sulfur chemistry is quantitatively most important in marine sediments, whereas organic sulfur chemistry is most important in freshwater sediments and terrestrial soils. As much as 90% of the sulfur in soils is organic; 50% as sulfate esters (C-O-S), 20% as S-amino acids in proteins, and 20% as a wide variety of other organo-sulfur compounds (Hagedorn, 2003). Most of the oxidation and reduction reactions of inorganic and organic sulfur compounds are mediated by several species of indigenous bacteria and fungi. Multispecies consortia of sulfur bacteria and fungi control the sulfur cycle in most soils, sediments, and water bodies (Hamilton, 1998).

Several species of bacteria and fungi can oxidize or reduce various inorganic and organic sulfur compounds under aerobic and anaerobic conditions, deriving the energy necessary for metabolic activities (Hagedorn, 2003). A few of the reactions involving elemental sulfur are summarized below.

Several species of *Thiobacillus* can oxidize elemental sulfur to sulfate under aerobic and anaerobic conditions at pH ranging from 1.9 to 8.5. Energy is liberated at each step in the reaction sequence as follows:

$$\underset{\text{Sulfur}}{S^0} \overset{(-15 \text{ kcal})}{\Rightarrow} \underset{\text{Thiosulfate}}{S_2O_2^{-2}} \overset{(-5 \text{ kcal})}{\Rightarrow} \underset{\text{Tetrathinate}}{S_4O_6^{-}} \overset{(-100 \text{kcal})}{\Rightarrow} \underset{\text{Sulfate}}{S_2O_4^{-2}} \quad (3.13)$$

Aerobic oxidation of elemental sulfur in soil by *Thiobacillus thiooxidants* produces sulfuric acid that acidifies the soil:

$$2S^0 + 3O_2 + 2H_2O \Rightarrow 2H_2SO_4 \quad (3.14)$$

If limestone ($CaCO_3$) is abundant in the soil, the sulfuric acid reacts with it to produce gypsum ($CaSO_4.2H_2O$). Elemental sulfur is used frequently as a soil amendment in agriculture and horticulture as discussed in Chapter 2. Sulfur is used to acidify alkaline or neutral soils to the optimum pH for cultivation of acidophilic crops. Ionic calcium derived from acid dissolution of calcium carbonate exchanges with sodium ions attached to clay colloids; the sodium reacts with sulfate to form soluble Na_2SO_4 that is washed from the root zone with percolating water. These reactions, plus the formation of gypsum in the soil, tend to break up clays, improving soil texture for agriculture.

Sulfate reduction is the quantitatively most important pathway of microbial oxidation of organic matter in sub-oxic layers of sediments in soils. Various oxidized sulfur species, including SO and organo-sulfur compounds, are reduced by anaerobic bacteria and fungi to sulfide that usually volatilizes to the atmosphere or precipitates as various metal sulfides. Microbial reduction of elemental sulfur in anoxic sediments to sulfide is as follows:

$$4S^0 + 4H_2O \Rightarrow 3H_2S + SO_4^{-2} + 2H^+ \quad (3.15)$$

The oxygen in the sulfate produced in this anaerobic reaction is from water. Most of the sulfide in sub-oxic layers of freshwater sediments and soils is derived from the microbial breakdown of organo-sulfur compounds (Dunnette et al., 1985).

If reactive Fe(III) and Fe(II) compounds are present, they react with H_2S to form pyrite (FeS_2) that is extremely insoluble and stable. Elemental sulfur may react directly with acid-soluble iron mono-sulfides, such as mackinawite (FeS), or iron oxides, such as goethite (FeOOH), to form pyrite (Gagnon et al., 1996):

$$FeS + S^0 \Rightarrow FeS_2 \quad (3.16)$$

or

$$FeOOH + \frac{3}{2}H_2S + \frac{1}{16}S_8^0 \Rightarrow FeS_2 + H_2O \quad (3.17)$$

The reaction expressed by Eq. 3.16 has been observed at a temperature around 100°C and above. At lower temperatures, down to 60°C, the reaction is slower, and results in greater percentages of marcasite, the dimorph of pyrite.

The main fate of sulfur in marine sediments is precipitation as pyrite. Salt-marsh sediments examined by Luther et al. (1991) contained 23 to 452 micromole of sulfur per gram of dry wt of solid pyrite and 8 to 82 micromole of sulfur per gram of elemental sulfur. Freshwater sediments and soils usually contain much lower concentrations of iron, and sulfur is converted primarily to various organo-sulfur compounds (Luther and Church, 1992).

Elemental sulfur can be synthesized from sulfide, thio-sulfate, and sulfur by various soil and sediment bacteria. For example, Thamdrup et al. (1994) reported the reversible formation of elemental sulfur and sulfite from thio-sulfate in Dutch salt-marsh sediments according to the reaction:

$$S_2O_3^{-2} \Leftrightarrow S^0 + SO_3^{-2} \quad (3.18)$$

Sulfur can be oxidized aerobically to elemental sulfur in freshwater and marine sediments by species of sulfur oxidizing bacteria, *Thiothrix* and *Beggiatoa* (Whitcomb et al., 1989; Hagedorn, 2003). The predominant sulfur oxidizing bacteria in soils belong to the genus *Thiobacillus*. *Thiobacillus* can couple sulfide oxidation to synthesis of carbohydrate by the reaction:

$$CO_2 + 2H_2S \Rightarrow CH_2O + 2S + H_2O \quad (3.19)$$

As described above, *Thiobacillus* can then oxidize elemental sulfur to sulfate, usually in the form of sulfuric acid (H_2SO_4). These reactions can occur in both oxic and anoxic environments. Under aerobic conditions, oxygen is the electron acceptor; under anaerobic conditions, another electron acceptor such as nitrate is used. Typically, sulfur oxidizing bacteria will preferentially oxidize sulfide until it is depleted and then shift to oxidizing elemental sulfur. This is because the energy yield from sulfide oxidation is greater than that from elemental sulfur oxidation.

Sulfate-reducing bacteria in freshwater and marine sediments rapidly reduce sulfate to sulfide if biodegradable organic matter is present. *Desulfobacter* spp. and *Desulfobacterium* spp. are important sulfate reducers in marine sediments and *Desulfovibrio desulfuricans, Desulfobulbus propionicus,* and *Desulfococcus multivirans* often are abundant in freshwater sediments and moist soils (King et al., 2000).

Sulfate reducing bacteria reduce sulfate in a multistep enzymatic reaction sequence through sulfite (SO_3^{-2}) to hydrogen sulfide (H_2S); these reactions are coupled to oxidation of organic matter in sediments (Hagedorn, 2003). Sulfide may subsequently be oxidized to elemental sulfur as described previously.

There is a lack of information in the literature on the impacts of *Thiobacillus* bacteria on modified sulfur cement and concrete products and their durability. This is an important area that should be investigated. If applications of sulfur concrete grow, the question of long-term durability and mechanical integrity becomes more significant.

Under anoxic conditions, elemental sulfur may react with sulfide to produce a wide variety of polysulfides (S_n^{-2}) (Luther et al., 1991). Polysulfides are important in complexing with several metal sulfides, increasing their concentrations in apparent solution in sediment pore water. Sulfur may also react with sulfite (SO_3^{-2}) to form thio-sulfate ($S_2O_3^{-2}$).

It is highly probable that aqueous polysulfides are involved in pyrite formation. Sulfur is known also to dissolve quite rapidly in alkaline sulphide solutions to give polysulphides, and pyrite itself is an iron polysulphide. The probable reaction for pyrite formation in aqueous solution is thus of the form:

$$FeS + S_n^{-2} \Rightarrow FeS_2 \qquad (3.20)$$

with the immediate oxidants being polysulphide anions.

The meta-stable ferrous sulphides are produced through the reaction between detrital iron minerals, particularly ferric oxy-hydroxides, and bacteriogenic sulphide. The iron sulphide-precipitating environment is limited by the physico-chemical tolerance of the bacteria and the stability of the iron sulphides towards dissolution and oxidation. Although dissimilatory sulphate-reducing bacteria of the genera *Desulfovibrio* and *Desulfotomaculum* are sensitive to oxygen, extremes of pH and temperature, their size, and production rate mean that they may exist in reduced micro-environments with different physico-chemical characters to the sediment as a whole. The stability limits of the meta-stable iron sulphides

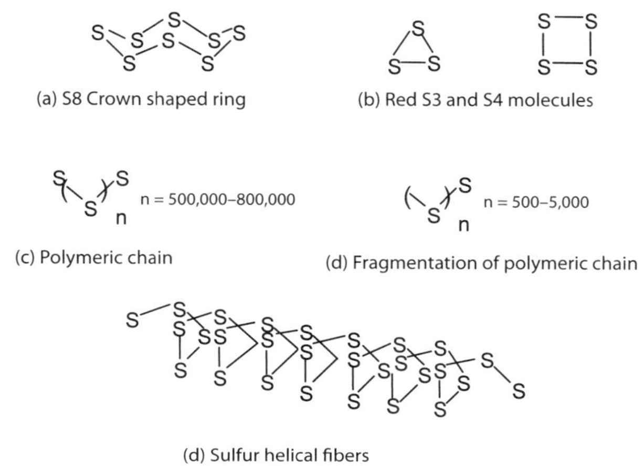

(a) S8 Crown shaped ring

(b) Red S3 and S4 molecules

(c) Polymeric chain, n = 500,000–800,000

(d) Fragmentation of polymeric chain, n = 500–5,000

(d) Sulfur helical fibers

Figure 3.15. Sulfur forms with temperature.

toward dissolution and oxidation coincide to a large extent with pH and pε tolerances of the bacteria. It is necessary to have total dissolved sulphide activities of more than 10^{-6}, in order to precipitate these meta-stable ferrous sulphides from solution containing a total dissolved iron activity of 10^{-7} or less at pH 9.

The concentration of pyrite is limited by the availability of utilizable organic matter, sulphate, iron, and oxygen. In areas with a high rate of organic matter sedimentation, aerobic microorganisms rapidly remove the bulk of the dissolved oxygen, at or near the sediment-water interface. The rate of biologic oxygen uptake in sediment is also temperature dependent. The most reducing environments are, therefore, those containing fine-grained, organic rich sediments associated with warmer waters. Sulphate-reducing bacteria are heterotrophic and the amount of sulphate-reduction is approximately quantitatively proportional to the concentration of organic matter. Organic rich sediments not only remove oxygen most rapidly, but also produce the greatest quantities of sulphides.

3.7 THERMAL PROPERTIES

Common yellow sulfur consists of S_8 molecules in which eight sulfur atoms are connected in a crown-shaped ring (Figure 3.15a). Upon heating, sulfur melts to provide a yellow liquid that flows readily. As the temperature increases, the color of sulfur changes to red and eventually darkens further. The color is caused by the presence of a small amount of red S_3 and S_4 molecules (Figure 3.15b). At 160 to 195°C the color of sulfur becomes dark red and its viscosity sharply increases (by a factor of 10,000).

The reason for such a high viscosity is that sulfur rings open and combine to form long polymeric chains with more than 500,000 to 800,000 sulfur atoms per chain (Figure 3.15c). Such long polymer chains become entangled, resulting in a dramatic increase in viscosity. As a result of the increase in viscosity, the molten sulfur stops flowing. This phenomenon can be easily observed when melting sulfur in a test tube. Thus, molten sulfur, which flows readily at a lower temperature, will stop flowing and the test tube can be turned upside down without spilling any sulfur.

Above 200°C, as the polymer chains begin to fragment into smaller pieces (Figure 3.15d), the viscosity again decreases. The sulfur, now dark brown to black, begins to flow again. Sulfur chains form helical fibers (Figure 3.15e) that can be stretched, giving sulfur rubbery or *plastic* properties. At 445°C, the black liquid boils to emit a pale yellow vapor that is composed of S_n molecules (n = 2-10, mostly 8) (Cotton el al., 1999).

The thermal conductivity of α-sulfur is 0.277 J/m-s-K, and of β-sulfur, 0.156 J/m-s-K. These are low figures for materials of equivalent density (122 lb/cu ft and 129 lb/cu ft, respectively) and especially compared with metals. The specific heat of α-sulfur is 5.40 cal/mol, and β-sulfur is 5.65 cal/mol.

Thermal analysis techniques such as thermal gravimetric analysis (TGA) and differential scanning calorimetric (DSC) have been used effectively for determining the temperature history of sulfur (Syroezhko et al., 2003). The TGA measures the weight loss of a material from a simple process such as drying, or from more complex chemical reactions that liberate gases and structural decomposition. Thermo Gravimetric Analyzer Perkin Elmer TGA7 was used to measure weight changes in sample materials as a function of the temperature. A furnace heats the sample while a sensitive balance monitors loss or gain of sample weight due to chemical changes and decomposition. TGA coupled with mass spectrometry and FT-infrared spectroscopy provides elemental analysis of decomposition products. Weight, temperature, and furnace calibrations have been carried out for the usable range of the TGA (100 to 600°C) at a scan rate of 20°C/min. The results shown in Figure 3.16 suggest that the maximal thermal effect was observed at 180 to 450°C, which was accompanied by active liberation of hydrogen sulfide (Mohamed and El Gamal, 2006).

The sulfur heat capacity is measured using differential scanning calorimeter (DSC) through phase transitions on heating. Ten mg of the tested sample was heated up to 150°C with a heating rate of 5°C/min. The sample was allowed to be self-cooled to room temperature for 24 hrs, then the sample was reheated up to 150°C. The DSC results of sulfur are shown in Figure 3.17. Melting of the alpha and beta was detected in the first run, whereas only alpha melting was detected in the second run.

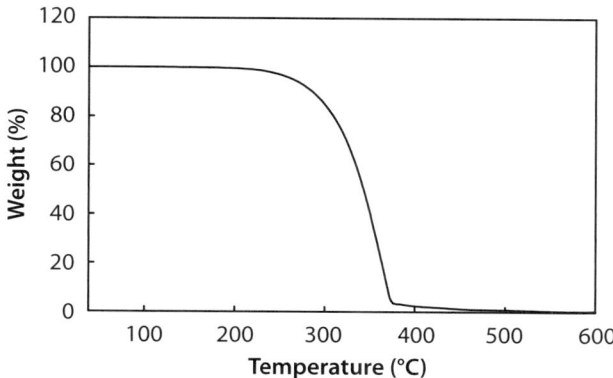

Figure 3.16. Thermo gravimetric analysis for pure sulfur with a heating rate of 20°C/min (Mohamed and El Gamal, 2006).

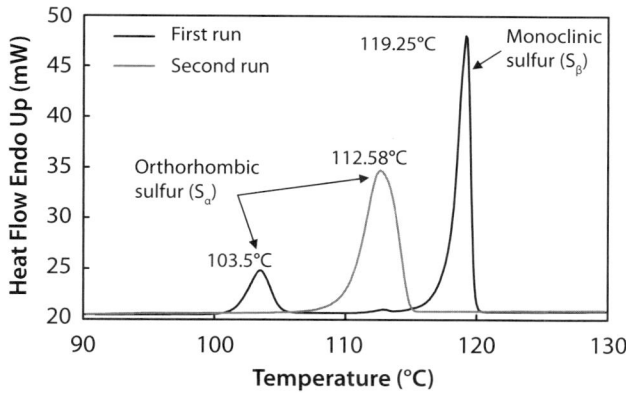

Figure 3.17. Differential scanning calorimetric curves for sulfur (Mohamed and El Gamal, 2006).

At 40°C, sulfur has a relatively high linear coefficient of thermal expansion, 7.4×10^{-5} /C and a low thermal conductivity, 6.1 gcal/cm²/sec ($\Delta t = 1$ C). When a material containing adjacent macrocrystals of sulfur is subjected to changing temperatures, there will be a constant movement between these macrocrystals as one example or contracts relative to its neighbor. This movement will gradually break the bonds with other cross-laid crystals, causing microfractures and eventual formation of cleavage planes (Vroom, 1998).

Typical temperatures of interest are:
  a. *Melting Point:* The temperature at which the vapor pressure of the solid and the liquid are the same and the pressure totals one

atmosphere. Typical melting point value is 388.36°K or 115.21°C (239.38°F).

b. *Boiling Point:* The temperature at which the vapor pressure of a liquid is one atmosphere. Typical boiling point is 717.87°K or 444.72°C (832.5°F).

c. *Critical Temperature:* The temperature above which it is no longer possible to liquefy the substance in question by increasing the pressure. Typical critical temperature is 1314° K or 1041°C (1906°F).

3.8 ELECTRICAL PROPERTIES

Sulfur is the best electrical insulating material known with a resistivity of about 2×10^{23} $\mu\Omega$-cm. It is not easy to measure resistivities this large. The reason for this large resistivity is probably the electron traps produced by thermal breaking of S_8 rings, which means that electron-attracting terminal atoms are always present. This may also be the explanation for the yellow color of sulfur. Balls of sulfur were the preferred material for static electricity generators. Otto von Guericke is generally cited as having constructed the first frictional electrical machine consisting of a ball of sulfur in a large glass globe.

3.9 ISOTOPES

Sulfur has 18 isotopes of which four are stable: S-32 (95.02%), S-33 (0.75%), S-34 (4.21%), and S-36 (0.02%). Other than S-35, the radioactive isotopes of sulfur are all short-lived. Sulfur-35 is formed from cosmic ray spallation of argon-40 in the atmosphere. It has a half-life of 87 days. Table 3.8 indicates key sulfur isotopes.

When sulfide minerals are precipitated, isotopic equilibration among solids and liquid may cause small differences in the dS-34 values of co-genetic minerals. The differences between minerals can be used to estimate the temperature of equilibration. The dC-13 and dS-34 of co-existing carbonates and sulfides can be used to determine the pH and oxygen fugacity of the ore-bearing fluid during ore formation.

In most forest ecosystems, sulfate is derived mostly from the atmosphere; weathering of ore minerals and evaporates also contribute some sulfur. Sulfur with a distinctive isotopic composition has been used to identify pollution sources, and enriched sulfur has been added as a tracer in hydologic studies. Differences in the natural abundances can also be used in systems in which there is sufficient variation in the S-34 of ecosystem components. Rocky Mountain lakes thought to be dominated by atmospheric sources of sulfate have been found

Table 3.8. Key sulfur isotopes

Nuclide	^{32}S	^{33}S	^{34}S	^{35}S	^{36}S
Atomic mass	31.972	32.971	33.968	34.969	35.967
Natural abundance (%)	95.02	0.75	4.21	0	0.02
Half-life	Stable	Stable	Stable	87.9 days	Stable

to have different dS-34 values from lakes believed to be dominated by watershed sources of sulfate.

3.10 POTENTIAL ECOLOGICAL EFFECTS OF ELEMENTAL SULFUR

Elemental sulfur is a natural ingredient of soils, sediments, and water bodies. It is readily transformed to many other inorganic and organic sulfur compounds, primarily by bacteria. Currently, elemental sulfur is registered by the U.S. Environmental Protection Agency (EPA) as an insecticide, fungicide, and rodenticide for use on several hundred food and feed crops, as well as ornamental, turf, and residential sites (U.S. EPA, 1991). It is an active ingredient in nearly 300 registered pesticide products. It is also applied directly to soil as a soil acidifier.

Sulfur has been shown to be practically nontoxic to bobwhite quail; the 8-day dietary median lethal concentration is greater than 5260 ppm (Howard, 1991). Ecological field studies have confirmed the low toxicity of agricultural sulfur to bobwhite quail (U.S. National Library of Medicine, 1995). Elemental sulfur also has a low toxicity to two species of fish, water fleas (Daphnia), mysid shrimp, and honey bees (U.S. EPA, 1991). Thus, there is minimal hazard associated with the broadcast of elemental sulfur as a pesticide or soil amendment.

Elemental sulfur also has a low acute and chronic toxicity to humans and other mammals. It has a low oral toxicity and does not ordinarily irritate the skin. However, sulfur dust can cause eye irritation and is an inhalation hazard. Chronic inhalation of sulfur dust may cause inflammation of the nasal mucosa and tracheobronchitus. Chronic inhalation may eventually lead to bronchiopulmonary diseases that may, after several years of continued exposure, be complicated by emphysema. There is no evidence that sulfur is mutagenic or carcinogenic.

Recently, Sevenson et al. (1996) reported that elemental sulfur in sediments is highly toxic in the Microtox assay. Because elemental sulfur is nearly insoluble, they brought it into solution by extracting sediments with organic solvents. The median effects concentration of the extracts in the Microtox assay ranged from 15 to 19 ppb elemental sulfur. Juvenile freshwater fish were more tolerant, with median lethal concentrations of 58 to 670 ppb. The toxicity of the extracts disappeared within 1.5 hrs of preparation, indicating that the toxic sulfur species were extremely labile. The authors concluded that the toxic sulfur species may have been formed during the solvent extraction of the sediment and that *in-situ* elemental sulfur in soils and sediments may be much less toxic.

It is highly doubtful that the sulfur in sulfur cement, concrete, and asphalt would be toxic to plants and animals including humans. Any elemental sulfur leaching from sulfur concrete structures and asphalt pavements is likely to be oxidized rapidly to sulfate in the soil. The sulfate may combine with soil water to form sulfuric acid that will acidify the soil. Soil acidification may facilitate mobilization of some metals from the soil and may change the texture and pH of the soil rendering it less suitable for some plants and animals that live in soil. However, at the concentrations likely to be released from sulfur-containing concrete and pavement structures, sulfur in unlikely to be toxic to plants and animals.

3.11 PRODUCTS

3.11.1 Product Groups

Sulfur products fall into three major categories, according to use:
1. *Agricultural sulfur* formulated for use as pesticides, plant nutrients, and soil amendments
2. *Industrial sulfur* used primarily in nonrubber applications
3. *Rubber-maker's sulfur* is primarily used in the production of rubber-related products

3.11.2 Practical Applications

Sulfur is an element used for everything from adhesives to matches. Its most common use is as a vulcanizing agent in the manufacture of rubber products, including tires. Table 3.9 illustrates the types of sulfur bought most often for each use or application (Georgia Gulf Sulfur Corporation).

Agricultural Sulfur

Agricultural sulfur products are formulated for use as nutrients, soil amendments, and pesticides. They are generally finely divided, averaging 93% through a #325 US sieve and formulated into wettable powders or dusts, or homogenized into flowable products with an average particle size of 2.5 microns. Their main uses are as fungicides, insecticides, and miticides. Another common use is as a soil additive to correct alkalinity or sulfur deficiency.

    a. *Wettable sulfur*: Wettable powders are formulated by blending dispersants and surfactants together and then milling to a very fine

Table 3.9. Types of sulfur bought most often for each use or application (http://www.georgiagulfsulfur.com/processing.htm)

Sulfur Products — group, type, and pecific brand names are listed across top
Practical uses and applications, down left-hand side, are indicated with a √ for each product.

	Rubbermaker		Industrial			Agricultural			
	Rms	Triangle®	Arrow Roll®	Emulsified	Formed	Wettable	Dusting	Flowable	Degradable
Animal Feed		√							
Explosives		√							
Fertilizer	√	√		√	√				√
Inorganic chemicals	√	√			√				
Matches		√							
Organic chemicals		√			√				
Pesticides						√	√	√	
Petroleum refining		√	√		√				
Poultry						√		√	
Rubber	√								
Salt blocks, animal feed		√							
Soil Amendments						√		√	√
Steel			√		√				
Sugar refining			√		√				

particle size. They can be applied as a spray or dust. Dry-milled sulfur has been used in pesticide applications for over 100 years.

b. *Flowable sulfur*: Dispersion- or flowable-type products are generally used on vine crops such as grapes, tomatoes, and peanuts. In use since the early 1970s, this product is homogenized to a particle size of 2.5 microns and suspended in water using dispersants and thickeners. Benefits include both ease of measurement and spray application as well as the absence of irritating dust. The micronized particle size increases surface area, enhancing efficacy. Flowable sulfur is primarily used as a fungicide. This micronized product can also be used as a soil amendment if immediate pH correction is required.

c. *Dusting sulfur*: Formulated at 98%, this product is primarily used as a fungicide.

d. *Degradable sulfur*: Formulated at 90%, this product is primarily used as a plant nutrient. Formed as a pastille or granule, degradable sulfur is also available in various sizes to conform to specific blend requirements.

Industrial Sulfur

Industrial sulfurs are 99.5% minimum purity and are processed into various physical shapes to provide a full range of particle sizes. This market includes pulp and paper, metals reclaiming, mining, steel, and oil refining, and a multitude of other uses. Sulfur is also used in the public utilities sector as a scale inhibitor. Industrial sulfur is available as flakes, milled, formed pastilles, or formed briquettes. Flake sulfur can be screened to a variety of specifications:

a. *Commercial grades*: Ground sulfurs milled to various specifications.

b. *Arrow Roll® refined sulfur*: refined sulfur molded into half cylinder-shaped bricks averaging 0.75 to 1.5 lb each. Arrow Roll® sulfur is primarily used in oil refining, steel production, and secondary smelting.

c. *Animal feed sulfur*: The mineral sulfur has been used for some time with other minerals in nutrient supplements for animal feeds and in salt blocks. This market is served with milled and crimped sulfur.

d. *Granular/pastille*: Produced using rotoform technology to produce 4mm size split pea-shaped particles.

e. *Emulsified*: Emulsified sulfur is produced by homogenization in water to form a 70% by weight dispersion. The particle size averages 2.5 microns.

f. *Flake*: Molten sulfur is flaked on water-chilled drums and is screened to various particle sizes.

Rubber-Maker's Sulfur

Sulfur has been used as a rubber chemical since Charles Goodyear discovered its vulcanizing properties in the mid 1800s. Sulfur gives the compound hardness, allowing it to be molded but still retaining its elasticity. Pencil erasers, rubber bumpers on automobiles, and latex gloves all use the same type of product but in different quantities and heat variations.

Rubber-maker's sulfur products vary widely in formulation and use. Sizes milled range from 44 to 74 microns. Conditioning agents are added to improve flowability, handling, and dispersion characteristics of finely ground sulfur. Oil is often added as a dust suppressant, reducing the risk of a sulfur dust explosion. Flow agents prevent caking, and are especially useful when using automatic feed equipment for batching.

The following sulfur options are called grades and offer a wide choice of purity, fineness, and conditioning agents for rubber processing:

a. *Grinding and screening*: There are various grades of rubber-maker's sulfur that differ primarily in fineness. Choice of screen size is usually made by the user based on his own operating experience and needs. Coarsely ground grades are considered somewhat less dusty whereas finely ground grades have dispersion advantages.
b. *Conditioning agents:* In general, conditioning agents improve flow and dispersion characteristics of finely ground sulfur. Conditioned sulfur may contain up to 2.5% of individual or combinations of flow agents.
c. *Oil treatment*: Sulfur can be treated with a small amount of oil, generally between 0.5% to 1.0%, to reduce dusting. If requested, most of the grades described can be supplied with an appropriate oil treatment.

3.12 SUMMARY AND CONCLUDING REMARKS

Sulfur occurs as the free element near volcanoes and hot springs. Many sulfide minerals are known, and sulfur is widely distributed in various types of minerals. Elemental sulfur can assume different forms exhibiting a variety of physical properties. The forms are called allotropes and can be divided into two types: (1) intramolecular allotropes, the different molecular species resulting from chemical bonding of sulfur atoms and (2) intermolecular allotropes, the different lattice structures formed by arrangement of the molecules within crystals. Complexities in understanding sulfur occur because: (a) most sulfur allotropes are metastable, (b) pure sulfur is difficult to obtain because it reacts readily with other substances, (c) sulfur forms many different molecular structures, several of which have similar stabilities, and (d) sulfur has a variety of similarly stable polymorphs.

Orthorhombic (α) and monoclinic (β) sulfur, both composed of cyclo-S_8 sulfur, are stable under terrestrial conditions. Physical properties of solid sulfur are dependent on the allotrope and, in some cases, the thermal history. Sulfur's two crystal forms, monoclinic and rhombic, both have a melting temperature just above the boiling point of water at one atmosphere. Under pressure, water temperature can exceed the melting temperature for sulfur. Since sulfur does not dissolve in water, the liquid sulfur immediately solidifies as it reaches the earth's surface leaving the distinctive nonmetal, pale yellow, brittle solid. Elemental sulfur has another noncrystalline form that can be made by melting crystalline sulfur, but the amorphous allotrope is unstable and reverts to one of the crystalline forms on standing.

Sulfur atoms combine to form rings and chains, allowing several million intramolecular sulfur allotropes to exist, if all possible combinations of the atoms are considered. The chemical difference between rings and chains is that ring molecules have fully paired electrons, whereas chains are considered to be diradicals. This difference is reflected by the chemical reactivity and by the physical properties. The presence and concentration of each allotrope and, therefore, the physical and chemical properties of bulk sulfur, are dependent on the thermal history of the sample. Polymerization of liquid sulfur causes discontinuities in viscosity, density, thermal conductivity, and specific heat.

The thermal conductivity of sulfur is approximately an order of magnitude less than that of most rocks and is comparable to such insulation materials as mica and asbestos. Specific heat values for solid and liquid sulfur, at atmospheric pressure, are a function of temperature. Sulfur is the best electrical insulating material known, with a resistivity of approximately 2×10^{23} $\mu\Omega$-cm.

Sulfur is an element used for everything from adhesives to matches. Its most common use is as a vulcanizing agent in the manufacture of rubber products, including tires. Agricultural sulfur products are formulated for use as nutrients, soil amendments, and pesticides. Industrial sulfurs are used in pulp and paper, metals reclaiming, mining, steel, and oil refining, and a multitude of other uses.

New uses of sulfur in construction, asphalt pavement, and sewerage management industries are explored and discussed in Chapters 4 to 8.

ELEMENTAL SULFUR CONCRETE

4.1 INTRODUCTION

Sulfur is readily available in many countries as a byproduct of the oil and gas industry. The relatively low cost and unique properties of sulfur have led to its utilization as a construction material particularly to replace or extend the use of Portland cement concrete or asphalt-extended cement. The use of elemental sulfur with aggregates to make sulfur concrete has been attempted for many years. Sulfur can be used in mortars, concretes, and asphaltic paving materials.

The term sulfur concrete is applied to mixtures of sulfur, as the binder, and a variety of aggregates. Sulfur concretes were manufactured using relatively simple equipment. As with normal concrete, it is possible to use naturally occurring heavy aggregates, e.g., sand, gravel, stone chips or ballast, and naturally occurring light aggregates. Sulfur concretes are manufactured by allowing mixtures of the aggregate and molten sulfur to cool, whereupon the mixtures solidify to give products of hardness comparable to concrete.

Sulfur concrete has unique properties, including rapid strength development, high ultimate strength, low permeability, and superior resistance to strong acids and saline solutions. However, the main shortcoming of this sulfur concrete (in which the used sulfur was in its elemental form without any modifications) is the high shrinkage volume that occurs when the formed article cools to ambient temperature. Such high shrinkage frequently results in distortion as well as inaccurate final dimensions, which is unacceptable in blocks used in a mortarless building system that must be producible with accurate predetermined dimensions.

This chapter discusses what constitutes elemental sulfur concrete and the fundamental aspects of manufacturing it. What are the factors that contribute to its strength development? What are the impacts of the environmental variables on elemental sulfur concrete strength?

4.2 HISTORY OF SULFUR CONCRETE DEVELOPMENT

The utilization of sulfur as a molten bonding agent dates back to prehistoric times (Sheppard, 1975). During the 17th century, sulfur was used to anchor metal in stone, and similar practices are currently being used in Latin America (Rybczynski et al., 1974). During World War I, however, there was a strong demand for sulfur, and several deposits were opened in North America for exploitation. This operation doubled the annual production of sulfur in the United States and had, as a consequence, a surplus stock of sulfur that enhanced the interest in developing new applications for this element.

Researchers were fascinated with sulfur concrete because of its high strength, corrosion resistance, impermeability, and fast setting characteristics of laboratory products that were obtained by mixing heated aggregates with molten sulfur and allowing the mixture to cool at room temperature. The potential use of surplus sulfur in the manufacture of construction materials was reported by Bacon and Davis (1921); they found that a mixture of 60% sand and 40% elemental sulfur produced an acid-resistant material with excellent strength. The same product was further studied by Duecker (1934) who found an increase in volume on thermal cycling with a loss in flexural strength. The testing methods for sulfur materials reported by McKinney (1940) have been adopted and documented in ASTM specifications for chemical-resistant sulfur mortar (ASTM C287-98, 2003).

In the late 1960s, Dale and Ludwig (1966, 1968) pioneered the work on sulfur aggregate systems, pointing out the need for well-graded aggregates to obtain optimum strength. This work was followed by the investigation of Crow and Bates (1970) for the development of high strength elemental sulfur basalt concrete. After that, more researchers undertook studies related to the potential use of sulfur in construction activities that led the division of the industrial and engineering chemistry of the American Chemical Society to organize a symposium in 1977 and document the research findings in a special volume entitled "New Uses of Sulfur-II", Advances in Chemistry Series, Volume 165, published by the American Chemical Society (Bourne, 1978).

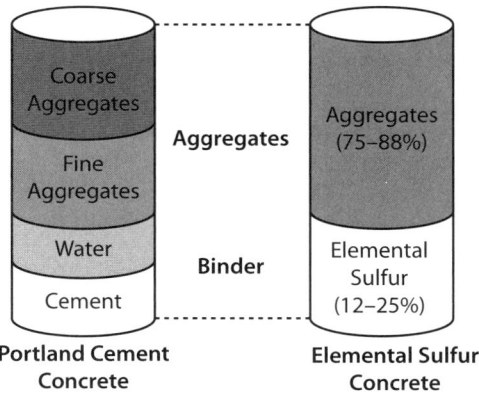

Figure 4.1. Comparison of the composition of Portland cement concrete and sulfur concrete.

4.3 TERMINOLOGY

In searching the literature related to this topic, we have noticed that various terms are being used without a clear distinction. To prevent confusion, terms used in this chapter are defined first. We begin with terms used in the construction industry. Cement-based concrete or Portland cement concrete is a mixture of specific portions of cement, water, and aggregates—sand and crushed stone (Figure 4.1). Therefore, the word *concrete* is used to refer to a product in which aggregates are bound with hydraulic products such as Portland cement. In the sulfur business, sulfur is being used in its molten state as a binding agent for aggregates, hence, it replaces the cement and water components of a regular cement-based concrete mix. In the case of sulfur products, heat is used to render the sulfur molten and, upon cooling, a material of concrete-like hardness is obtained. Therefore, *sulfur concrete* is used to refer to a product in which aggregates are bound with molten sulfur.

Furthermore, because the sulfur being used is in its elemental state and lacks chemical additives, the term *elemental sulfur concrete* is used. Therefore, throughout this book, *elemental sulfur concrete* is used to identify products that are formed from elemental sulfur and aggregates without the addition of any chemical additives (Figure 4.2).

4.4 COMPRESSIVE STRENGTH

Elemental sulfur concrete has properties that are superior to Portland cement concrete for certain applications. Its rapid strength development makes it good

Figure 4.2. Sulfur terminologies used for elemental sulfur cement production.

choice for both low- and high-temperature concreting. Portland cement bonded materials attain their maximum strength only after several weeks of moist curing as shown in Figure 4.3. The figure shows the comparison between strength development of elemental sulfur concrete and Portland cement concrete. Cement concrete requires approximately 28 days for hydration to achieve 90% of its final strength if certain moisture and temperature conditions are observed. Elemental sulfur concrete, on the other hand, reaches this final strength after only a few hours, and moisture and temperature do not influence the development of strength (Gregor and Hackl, 1977). The compressive strength of elemental sulfur concrete ranges from 40 to 50 MPa.

The following sections discuss the basic factors that control strength development and reduction in both Portland cement concrete and elemental sulfur concrete.

4.4.1 Strength Development for Portland Cement Concrete

When Portland cement reacts with water, heat evolves. The physical changes in the paste are detected by the increased stiffness. The physical and chemical processes occurring are therefore of practical importance because they lead to a decrease in workability.

The raw materials used in the manufacture of Portland cement are mainly constituted by four mineral oxides: lime, silica, alumina, and iron oxide. These compounds interact with each other as well as with the water in the *kiln*. Four components are usually regarded as the major constituents of cement: *tri-calcium silicate* (C_3S); *di-calcium silicate* (C_2S); *tri-calcium aluminate* (C_3A); and *tetra-calcium alumino-ferrite* (C_4AF). In the presence of water, the silicates and aluminates form products of hydration that in time, produce stiff and hard mass, i.e., the hydrated cement paste. These products of hydration have a low solubility in water.

The main products of hydration of Portland cement are the *calcium silica hydrate* ($C_3S_2H_3$), *tri-calcium aluminate hydrate* (C_3AH_6), and *calcium alumino-ferrite hydrate* (C_6AFH_{12}). These phases contribute significantly to the development of mechanical strength. They are generated as a consequence of hydration of the silicate, aluminate, and alumino-ferrite phases according to the following reactions:

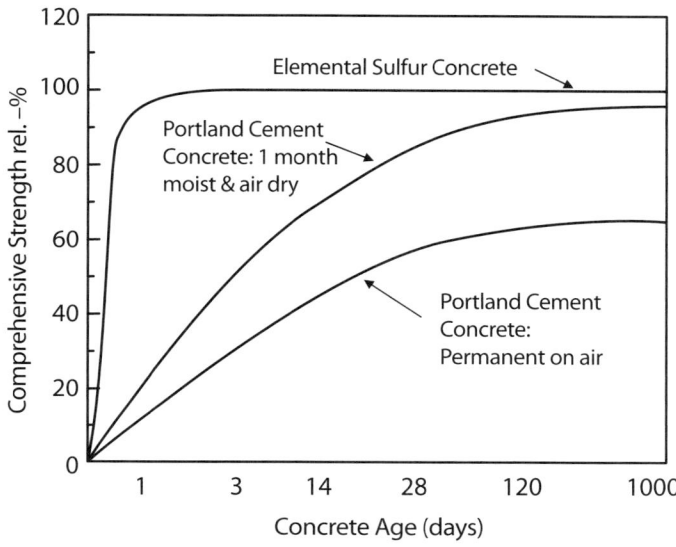

Figure 4.3. Strength development of sulfur concrete and cement concrete as function of concrete age and moisture supply (adapted from Gregor and Hackl, 1977).

$$\begin{array}{lllll}
2(3CaO.SiO_2) & + & 6H_2O & \rightarrow & 3CaO.SiO_2.3H_2O + 3Ca(OH)_2 \\
\text{Tri-calcium silicate} & + & \text{Water} & \rightarrow & \text{Tombermonte gel + Calcium hydroxide} \\
2C_3S & + & 6H & \rightarrow & C_3S_2H_3 \quad + \quad 3CH
\end{array}$$
(4.1)

$$\begin{array}{lllll}
2(2CaO.SiO_2) & + & 4H_2O & \rightarrow & 3CaO.SiO_2.3H_2O + 3Ca(OH)_2 \\
\text{Di-calcium silicate} & + & \text{Water} & \rightarrow & \text{Tombermonte gel + Calcium hydroxide} \\
2C_3S & + & 4H & \rightarrow & C_3S_2H_3 \quad + \quad 3CH
\end{array}$$
(4.2)

$$\begin{array}{lllll}
3CaO.Al_2O_3 & + & 6H_2O & \rightarrow & 3CaO.Al_2O_3.6H_2O \\
\text{Tri-calcium aluminate} & + & \text{Water} & \rightarrow & \text{Trialcium aluminate hydrate} \\
C_3A & + & 6H & \rightarrow & C_3AH_6
\end{array}$$
(4.3)

$$\begin{array}{lllll}
4CaO.Al_2O_3.FE_2O_3 & + 10H_2O + 2Ca(OH)_2 \rightarrow & 6CaO.Al_2O_3.FE_2O_3.12H_2O \\
\text{Tri-calcium alumino-ferrite} & + \text{Water} + \text{Lime} \rightarrow & \text{Calcium alumino-ferrite hydrate} \\
C_4AF & + 10H + C_2OH \rightarrow & C_6AFH_{12}
\end{array}$$
(4.4)

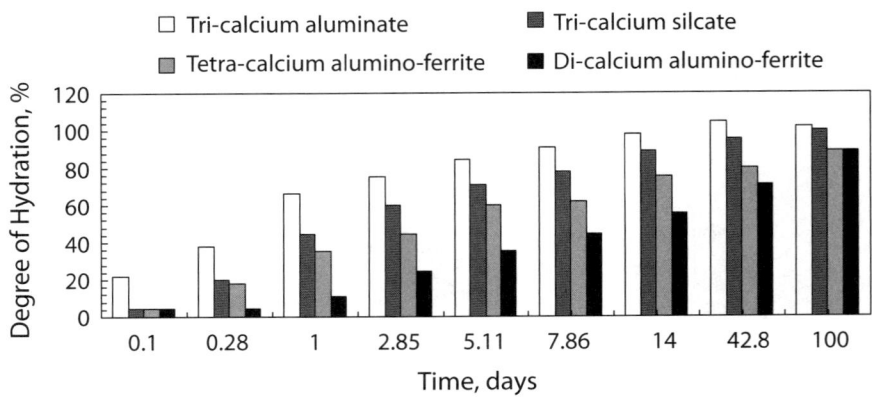

Figure 4.4. Degree of hydration of the principal phases in a sample of ordinary Portland cement.

The usual method of measuring the rate of chemical reaction involves the determination of the changes in the concentrations of reactants or products with time. It is possible, as shown in Figure 4.4, to follow the changes of the hydration products (tri-calcium silicate (C_3S); di-calcium silicate (C_2S); tri-calcium aluminate (C_3A) and tetra-calcium alumino-ferrite (C_4AF), in cement hydration.

Calcium silicate hydrate ($C_3S_2H_3$) is poorly crystalline with only a few broad bands in the x-ray diffraction pattern. This fact and its uncertain composition have resulted in the general use of the notation C-S-H to represent it. The hyphens indicate that the composition is indefinite. Because C-S-H has a high specific area, it is considered to be a gel. The formation and growth of C-S-H can be observed from the increase in the peak intensity of the major (d = 4.15 Angstrom) and secondary (d = 2.67 Angstrom) basal spacing. Tri-calcium aluminate hydrate (C_3AH_6) is formed as platelets with hexagonal symmetry and morphology resembling that of calcium silicate hydrate.

4.4.2 Strength Reduction for Portland Cement Concrete

Portland cement strength could be reduced due to deterioration as a reaction to various chemical, physical, mechanical, and biological processes. Frequently, more than one alteration mechanism takes place simultaneously or chemical processes or biological processes have physical or mechanical effects. This, in turn, makes it difficult to establish a classification of the deterioration mechanisms. Therefore, we point out some of the processes of deterioration commonly observed in Portland cement concrete.

Dissolution and Leaching

Pure waters (a result of the condensation of fog or water vapor) and soft waters (rain water or melted snow and ice) contain minimal amounts of calcium. When these waters get in contact with the hardened Portland cement concrete, they spread through the porous system of the material and dissolve the hydrated phases that are rich in calcium. In Portland cement concrete, $Ca(OH)_2$ is the most soluble constituent, and it is, therefore, the most easily leached phase.

The dissolution increases the porosity of the system and, consequently, the permeability. This, in turn, will reduce the overall mechanical strength and increase the susceptibility for attack by other aggressive chemical agents.

Leaching processes of calcium salts can also have other undesirable effects from an aesthetic point of view. Frequently, the leachate ($Ca(HCO_3)_2$) precipitates on the surface of the material or even on adjacent materials giving way to white efflorescence of $CaCO_3$.

Interaction with Atmospheric Pollutants

Atmospheric pollutants can exist in three forms: gas, solid particles, and dissolved substances. The most important pollutants are SO_2, NO_x, and CO_2. Because of the low pH of acid rain, hardened concrete allows the formation of calcium nitrates, sulfates, and bicarbonates. Some of these salts are highly soluble and easily leached.

The degree of interaction between atmospheric SO_2 and concrete is controlled by the known processes of dry deposition and wet deposition (acid rain). In the case of the dry deposition of SO_2, the rate of reaction increases according to the following order: $SO_2 < SO_2+O_3 < SO_2+H_2O < SO_2+O_3+ H2O$.

The chemical reaction of SO_2 with cement components in the presence of water and an oxidant is 35 times faster than in the absence of those elements. The rate of conversion of SO_2 to sulfate increases when the water/cement ratio decreases. In the case of acid rain, the increase of water/cement ratio involves the increase of gypsum formation on the surface of the affected specimens. The produced gypsum is a consequence of the oxidation/hydration of SO_2 and the later interaction of the sulfuric acid with the $CaCO_3$ in the substrate. The dry deposition of SO_2 generates a low proportion of SO_4^{2-}; meanwhile, when the humidity conditions are appropriated, SO_4^{2-} is formed in large amounts.

Recently, experimental investigations have suggested the possible formation of thaumasite and ettringite in concrete when they are exposed to an atmosphere containing SO_2. First gypsum is produced and then it interacts with calcium carbonate and C-S-H gel giving way to thaumasite.

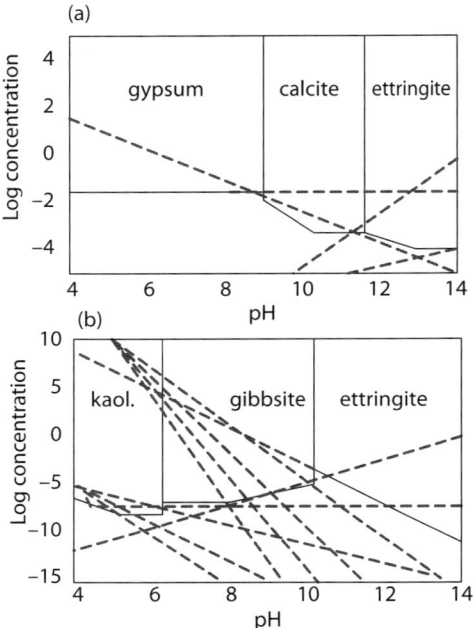

Figure 4.5. Solubility diagrams of: (a) calcium with log [Al (OH)$_4^-$] = -2M, and log [SO$_4^{-2}$] = -3M and (b) aluminum with log [Ca^{+2}] = -2M, and log [SO$_4^{-2}$] = -3M (Mohamed et al., 1995).

Expansive Reactions

Hydration of CaO and MgO

Due to incomplete clinkerization, crystallized oxides do exist in cement. The reaction of these oxides with water produces portlandite and brucite which are expansive minerals and can cause cracking problems.

Ettringite Formation

In the presence of calcium sulfate, the product of hydration described by Eq. 4.5 is an aluminosulfate (C$_3$A.CaSO$_4$.32H$_2$O) known as *ettringite*, a mineral with prismatic crystals and a hexagonal cross section.

$$\begin{array}{lcccl}
3CaO.Al_2O_3 & + & 26H_2O & + & CaSO_4.2H_2O \rightarrow 3CaO.Al_2O_3.3CaSO_4.32H_2O \\
\textit{Tri-calcium aluminate} & + & \textit{Water} & + & \textit{Gypsum} \quad \rightarrow \quad \textit{Ettringite} \\
C_3A & + & 26H & + & C_2S_3H_6 \quad \rightarrow \quad C_6AS_3H_{32} \quad (4.5)
\end{array}$$

Ettringite crystallization involves a high increment of volume. Most often, this takes place during the first stages of hydration before the hardening of the cement slurry. In these initial stages, the system is in a plastic state, and it is able to absorb that increment of volume. Much of the water in the ettringite structure is loosely held and associated with the sulfate ions. Water is held in channels between columnar units that are parallel to the needle axis. This structure is identified by an empirical formula $[Ca_3Al(OH)_6 12H_2O]^{3+}$ (Moore and Taylor, 1968).

Ettringite dissolves in water to form calcite, gypsum, and alumina gel in the absence of lime and sulfate as shown in Figure 4.5 (Mohamed et al., 1995). Stability fields of aluminum- and calcium-bearing products are also shown in Figure 4.5. For calcium-bearing products, the activity ratios of gypsum, calcite, ettringite, portlandite, monosulphate, C_3AH_6, C_2AH_8, and C_4AH_{13} are shown in Figure 4.5a. For aluminum-bearing minerals, the activity ratios of kaolinite, gibbsite, ettringite, C_3AH_6, C_2AH_8, C_4AH_{13}, and mono-sulphate are shown in Figure 4.5b. The figures clearly identify the thermodynamic stability of ettringite.

However, if ettringite is formed when the system is hardened, then an expansive phenomenon with cracking and loss of mass can take place (Mehta, 1969, 1973; Mehta and Wang, 1982, Mohamed, 2000). The damage produced due to ettringite formation depends on various factors such as cement composition, cement/water ratio, amount of available sulfates, and aluminum present for reactions (Hossein et al., 1999; Mohamed, 2003; Mohamed et al., 2003). The recommended way to increase resistance to sulfate attack is the reduction of the porosity of the system, but it is also important to modify the chemical composition of the cement in order to reduce the C_3A content.

Thaumasite Formation

Thaumasite formation can be the result of different mechanisms. It can be the consequence of the evolution of ettringite by incorporating Si^{4+} in its structure, substituting Al^{3+} ions within the columns $Ca_6[Al(OH)_6]_2$, and the interstitial replacement of $(SO_4)_3.2H_2O$ by $(SO_4)_2.(CO_3)_2$. Also, it can be the result of the interaction of sulfates and carbonates with C-S-H gel. Both mechanisms might simultaneously develop within the same period.

The destructive effect of thaumasite formation is due to expansion mechanism, which is similar to those described for ettringite formation. Also, thaumasite formation involves the destruction of C-S-H gel and, therefore, strength reduction.

Freeze-Thaw Cycles

The increase of specific volume (9%) that takes place in the water when passing from liquid to ice generates pressures in the pore walls of the mixture that can produce cracking and breaking.

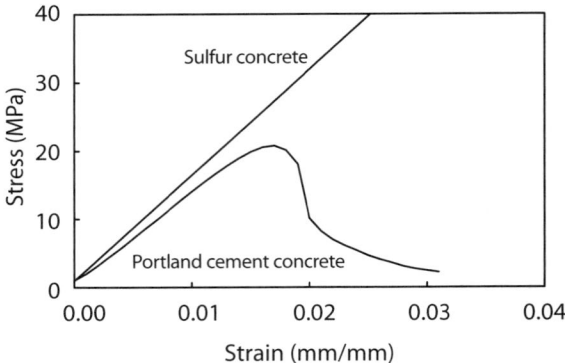

Figure 4.6. Stress-strain relationship for sulfur concrete and normal Portland cement concrete (adapted from Mohamed and El Gamal, 2006).

Biological Effects

The biodegradation is the process of the destruction of a material due to active microorganisms or the products of their metabolisms. The attack of natural materials through biochemical and biological processes, as well as the influence that microorganisms have in the formation of ground, is well known. The process of colonization of a matrix is favored by its physical characteristics such as porosity, composition, and grain surface roughness. These characteristics favor the retention of water in the material and the consequent growth of algae and bacteria. This could be avoided by biocide treatments.

4.4.3 Strength Development for Elemental Sulfur Concrete

Compared to Portland cement concrete, elemental sulfur concrete products offer advantages because they attain their final compressive strength and flexural tensile strength a short time after solidification. Elemental sulfur concrete samples, consisting of elemental sulfur, sulfur, fly ash, and sand, were prepared according to the procedure described in (ACI 548.2R-93). Figure 4.6 shows a typical stress-strain curve for Portland cement concrete and elemental sulfur concrete. The compressive strength of Portland cement concrete increases as strain increases, to approximately 0.017 and reaches a maximum value of 20 MPa. Then, it decreases. However, for elemental sulfur concrete, the compressive strength continues to increase with the strain in excess of 0.025 reaching a value of approximately 40 MPa after which the test was stopped because the stress level reached the maximum capacity of the testing machine (Mohamed and El Gamal, 2006).

To evaluate the strength development for the elemental sulfur concrete mixture, we need to identify the possible formation of mineralogical composition via

Elemental Sulfur Concrete 119

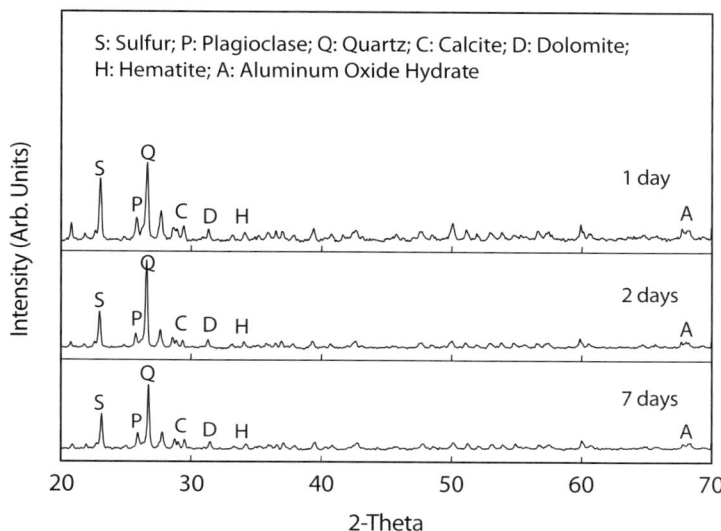

Figure 4.7. X-ray diffraction analysis of elemental sulfur concrete with time (Mohamed and El Gamal, 2006).

x-ray diffraction analysis. Changes in the mineral composition of sulfur concrete were performed using x-ray Philips PW/1840, with Ni filter, Cu-K α radiation (λ = 1.542 Å) at 40 KV, 30 mA, and scanning speed 0.02°/S. The diffraction peaks between 2 θ = 2° and 2 θ = 80° were recorded and the newly formed minerals due to the chemical reaction were determined manually by comparison with standard reference patterns and measurements (ASTM cards). The results of tested elemental sulfur concrete samples left in dry air at 40°C for one, two, and seven days are shown in Figure 4.7.

The mineralogical structure of elemental sulfur concrete samples for one day is made up of the major components; sulfur and quartz (SiO_2). The minor constituents include the following in decreasing percentages; plagioclase ($CaAlSi_3O_8$), aluminium oxide hydrate ($5Al_2O_3!H_2O$), calcite ($CaCO_3$), hematite (Fe_2O_3), and dolomite ($CaMg(CO_3)_2$). The large mechanical strength for elemental sulfur concrete is attributed to the reactive alumino-silicate and calcium alumino-silicate components that are routinely present in their oxide nomenclatures such as silicon dioxide, aluminum oxide, and calcium oxide. Fly ashes tend to contribute to concrete strength when the alumino-silicate components react with calcium oxides to produce additional cementatious materials.

It is worth noting that the amount of water molecules associated with the formed hydrated compounds is unknown because the process does not utilize water at all. This leads one to pose the following question. Where is the water

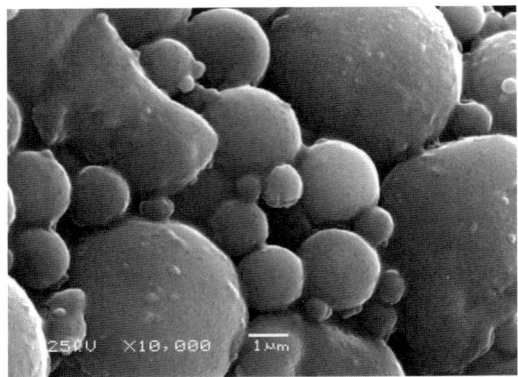

(a) SEM Micrograph

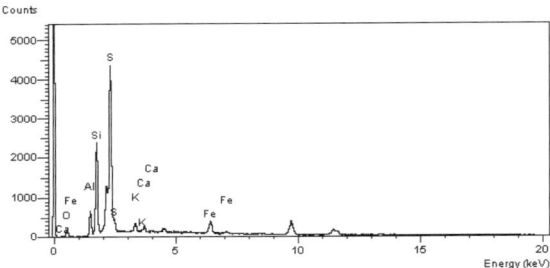

Atomic	%
O	22.9
Al	6.30
Si	23.5
S	23.5
K	2.50
Ca	1.60
Fe	4.90

(b) EDX spectrum

Figure 4.8. (a) SEM micrograph showing variety of particle shapes, sizes, and voids for elemental sulfur concrete and (b) EDX spectrum showing the residue main chemical element for sulfur concrete (Mohamed and El Gamal, 2006).

coming from? Is it from the humidity of the atmosphere or from the bonded water in the recycled aggregates? These issues need further investigation. The mineral composition of elemental sulfur concrete samples, which were treated to the same temperature for two and seven days, did not show any changes from that shown for one day, indicating no new mineral formation and that all the strength was acquired at early stages.

Mechanical strength is directly related to the defects in mortar microstructure which are generally characterized by SEM. The microscopic studies were determined from a section cut at a distance of 0.5 cm from the surface of elemental sulfur concrete samples. The microstructure characterization, performed by scanning electron microscopy (SEM), provides important data on mortar studies. Figure 4.8a is an image for elemental sulfur concrete in which some common

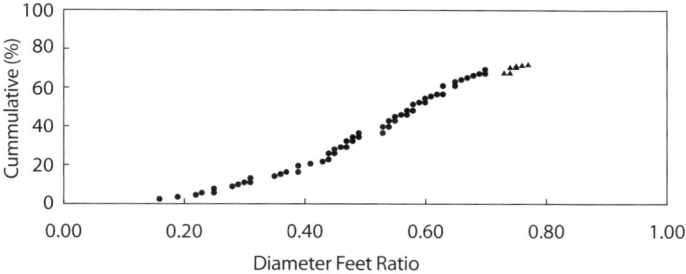

Figure 4.9. Cumulative percentage of pores sizes for elemental sulfur concrete using image analysis program (Mohamed and El Gamal, 2006).

crystallization features were observed. The microstructure shows a considerable degree of packing with some pores. The chemical analysis performed by energy dispersive spectroscopy (EDS) for sulfur concrete shown in Figure 4.8b indicates that the samples are composed mainly of sulfur and silicon containing compounds. For sulfur concrete, sharp and intense signals with low atomic sulfur percentage were found, indicating high crystalline structure.

It is obvious that the voids entrained within the elemental sulfur concrete are small and discontinuous. The voids serve as sites for stress relief and for improving the durability of the material (McBee et al., 1983).

Furthermore, to evaluate the size of the voids from the results of SEM, we have utilized image analysis software. The results are shown in Figure 4.9 in terms of the cumulative percentage of voids for elemental sulfur concrete. The results indicate that the voids within the matrix are uniformly distributed. It is worth noting that when diameter Feret ratio reaches one, the pore sizes are spherical.

From these experimental results, it is clear that the strength gain in elemental sulfur concrete is solely physical in nature without any chemical reactions. The basic aggregate components are coated with molten sulfur and, upon cooling, a solidified matrix is developed. The strength is fully developed within the first three days with 90% of it developed the first day (Figure 4.9).

4.5 MATERIAL COMPOSITION

4.5.1 Elemental Sulfur

Sulfur is an element with atomic no. 16, atomic mass 32, melting point 119°C, boiling point 444.6°C, and is found as an organic element in mines, but the most important source of sulfur is petrochemical. In most of the Middle Eastern countries, sulfur is produced through the oil and gas desulphurization process. The produced sulfur has purity in excess of 96% (*The Merck Index*, 7th ed.). (More information about sulfur sources and properties can be found in Chapter 3.) The elemental sulfur used in the preparation of sulfur concrete may be commercial grade, crystalline, or amorphous. Either primary or recovered sulfur can be used (McBee et al., 1977); the form and purity of the sulfur is of relatively minor importance, providing that the impurities do not represent more than about 4% clay or other water expansive material. It was reported by Mohammadi (2006) that impurity does not have any effect on sulfur strength. Since the sulfur is employed in molten form, particle size is generally insignificant. The sulfur may be either solid or liquid (molten) form. This suggests that *off-spec* sulfur that may be unmarketable for other uses because of color, carbon, or ash content, may be suitable for elemental sulfur concrete production.

Assessment of the Sulfur Binder Demand

The quantity of elemental sulfur needed can be calculated by ascertaining experimentally the vibration density of the aggregates used in any given case. From the vibration density given on the vibrating table, the void space of the grain heap is calculated using the following:

$$PV = 100 - \frac{S_R}{R_D} \cdot 100 \qquad (4.6)$$

Where PV is total pore volume of the vibrated mineral mass (vol. %), S_R is the vibration density of the vibrating mineral mass (Mg/m³) and R_D is the raw density of the aggregate (Mg/m³).

Usually 10 to 25 wt% in elemental sulfur is added to the pore volume obtained to glue the individual grains in the elemental sulfur concrete mixture (Gregor and Hackl, 1978).

4.5.2 Aggregates

Aggregates may be of various types, including gravel, crushed rock, natural or manufactured sand, slag, volcanic ash and lava, crushed clay brick, and most mining wastes. Whereas small amounts of salt, organic matter, or other impurities may be tolerated, the aggregates should be essentially free (< 1 percent) of unfired clay or other swelling materials such as shale, lignite, and cellulosed fiber. Sea sand and common dune sand are generally usable without washing. Although certain dolomites aggregates occurring in Middle Eastern counties have been found unsuitable, the acceptability of a relatively wide range of fine-grained sands is significant.

Molten sulfur sand grouts have been used for many years in the construction of acid vats because of their excellent acid resistance. In the mid 1960s, an extensive characterization of sulfur concretes employing fine and coarse aggre-

Table 4.1. Physical properties of aggregates (Gregor and Hackl, 1978)

Aggregate type	Specific gravity	Compressive strength (MPa)
Gravel	2.64	216
Granulit	2.65	236
Basalt	2.90	305

Table 4.2. Quantitative analysis of aggregates (Gregor and Hackl, 1978)

Element (%)	Gravel	Granulit	Basalt
SiO_2	33.3	70.7	42.7
Al_2O_3	—	15.5	16.6
Fe_2O_3	—	3.4	10.9
K_2O	—	3.8	2.5
Na_2O	—	4.9	5.0
CaO	Rest	—	—
MgO	—	—	—
TiO_2	—	—	—
Loss at red heat	26.6	—	—

gates was undertaken for the U.S. Air Force by the Southwest Research Institute (Dale and Ludwing 1966, 1967, 1968). Typical compressive strength of 33 to 47 MPa was attained in 6 to 12 hours after solidification with limestone aggregate. The major problem with grout or concrete employing elemental sulfur as the binder was poor resistance to thermal cycling and subsequent loss of strength.

Grain Shape

The influences of grain distribution, grain shape, and the basic character of aggregates on the strength of elemental sulfur concrete have been studied by Gregor and Hackl (1978). As for the grain form (Table 4.1), gravel as spherical material and crushed stone (granulit, basalt) as sharp materials were used as aggregates. In mineral composition (Table 4.2), granulit is an acidic rock, whereas basalt consists mainly of basic mineral components. The aggregates used were screened to obtain defined grain size distribution (Table 4.3). The largest grain was limited to 8 mm. For testing purposes cubes with edge lengths of 71 mm and 100 mm, respectively, and prisms 40 × 40 × 160 mm were produced by casting the hot sulfur aggregate mixture and using vibration compaction.

When the influence of the grain shape on the strength of elemental sulfur concrete is examined, complete analogies to Portland cement concrete are found

Table 4.3. Sieve analysis of aggregates (Gregor and Hackl, 1978)

Sieve line	Square sieve opening (mm)						
	0.09	0.2	0.5	1	2	4	8
	Cumulative percentage passing each sieve (wt %)						
1	–	7.25	18.50	30.00	46.50	68.50	100
2	3.75	3.50	18.50	30.00	46.50	68.50	100
3	9.00	14.50	25.70	37.80	54.20	75.79	100
4	10.61	15.81	25.00	35.36	50.00	70.71	100
5	16.62	22.87	32.99	43.54	57.43	75.79	100
6	26.02	33.07	43.53	53.59	65.98	81.23	100

(Gregor and Hackl, 1978). Because of its spherical form, gravel can be compacted more easily and therefore has lower grain heap porosity. Because of its smooth surface and its spherical form, the binder demand is lower.

Because the grain shape of crushed stone aggregates deviates considerably from the spherical form and the aggregates have a rough surface texture, it is more difficult to compact them and they have a greater void space. In the case of the same sieve line and similar consistency, the quantity of binder that they require, is higher by 10 to 20 vol. %. Crushed stone concrete has a remarkable advantage. Because of good compaction and linking, the sharp grains form a closer matted structure and thus have a better inner binding for the same reasons that they adhere better in the concrete than do spherical and smooth pebbles. When the grain distribution was the same, the sulfur concrete had a compressive strength that was 32% higher than with crushed stone aggregate.

Grain Size Distribution

To manage with low elemental sulfur content and to utilize the compressive strength of the sulfur most effectively, it seemed necessary to build up a maximum bearing mineral structure with minimal void space (Gregor and Hackl, 1978). The liquid sulfur was to glue the individual grains to each other and to fill the grain heap pores completely. Therefore, the binder, and thereby the sulfur demand, is not only determined by the void space to be filled but also by the sum of the grain surfaces. Following Austrian norm (ONORM, 1969), for a favorable grain size distribution of the aggregates to produce cement concrete, a defined sieve line area is recommended for aggregate sizes from 0 to 8 mm in diameter (Figure 4.10).

In the first tests to produce elemental sulfur concrete, the aggregates were built up according to their grain size distribution following the sieve lines 1 and 2 and were, therefore, within the favorable sieve line range. However, mixtures

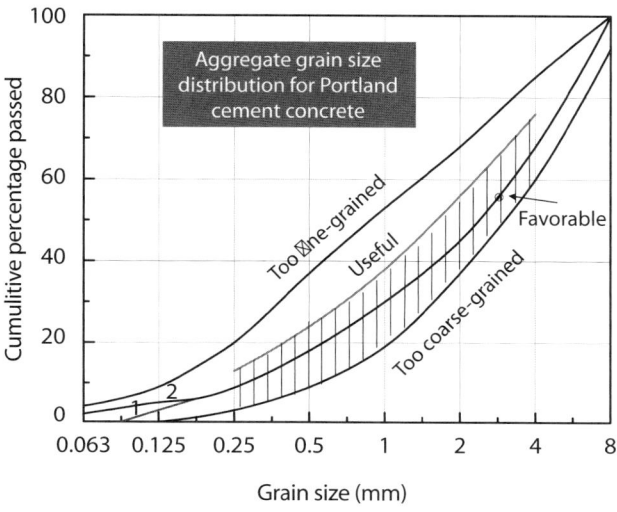

Figure 4.10. Grain size distributions for aggregate sizes from 0 to 8 mm in diameter to produce Portland cement concrete (ONORM, 1969).

whose aggregates were composed according to sieve lines 1 and 2 had extremely poor workability because there was a strong tendency to separate and no homogenous products could be produced. The acceptable grain size distributions was chosen by Fuller (Grun, 1937). Fuller's parabola follows:

$$A = 100 \cdot (\frac{d}{D})^n \quad n = 0.5 \tag{4.7}$$

Where A is the percentage of a grain class of diameter d (from 0 to d); D is the largest grain diameter of the grain mixture; d is any grain diameter between 0 and D. The Fuller parabola for aggregate sizes from 0 to 8mm in diameter is represented by the sieve line 4 in Figure 4.11. The elemental sulfur concrete mixture whose aggregates corresponded to the grain size distribution of the Fuller curve no longer separated and was plastic in consistency. The value for compressive and flexural strength rose.

Density and Sulfur Content

Hummel (1960) showed that the Fuller parabola for $n = 0.5$ does not yet furnish the optimal packing densities. Gregor and Hackl (1978) used $n = 0.4$ for spherical aggregates and $n \approx 0.3$ for crushed aggregates to achieve maximum density for the aggregate as shown in Figure 4.12.

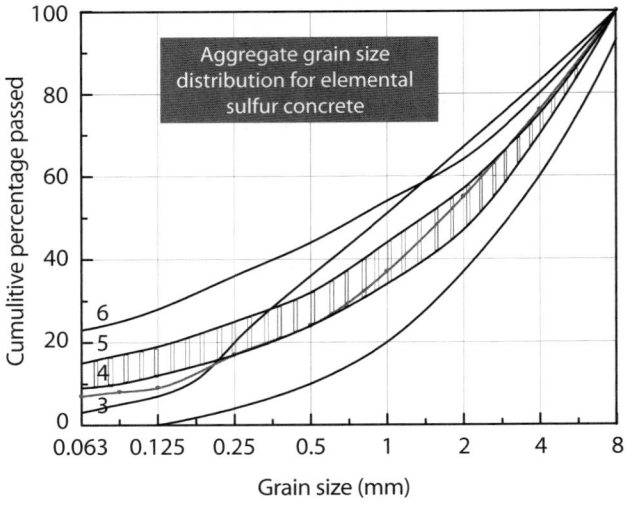

Figure 4.11. Grain size distributions for aggregate sizes from 0 to 8 mm in diameter to produce elemental sulfur concrete (adapted from Gregor and Hackl, 1978).

For sieve line 5 in which aggregates were built with $n = 0.4$, the void spaces were reduced because of the increase of the filler content (16.62%) with grain size diameters less than 0.09mm. This, in turn, contributes to an increase of the specific surface area. Such a mixture was stiff and difficult to compact. Higher

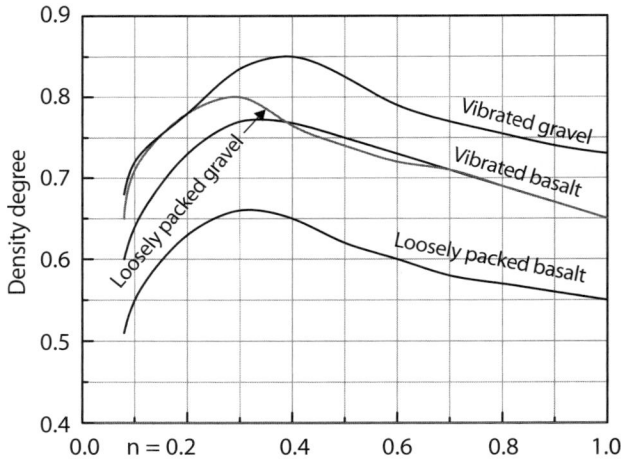

Figure 4.12. Density degrees of gravel sands and basalt grain mixture 0/30 mm grain composition (adapted from Gregor and Hackl, 1978).

compressive and flexural strength values could be reached only by increasing the sulfur and thus the binder content (Figure 4.13).

Aggregate compositions based on sieve line 6, $n = 0.3$, had lower void spaces than sieve line 5. This is understandable because of the increase of the filler content (26%) with grain sizes less than 0.09 mm. Such an increase of fine content contributes to the proportional increase of the sulfur (binder) content (Figure 4.13).

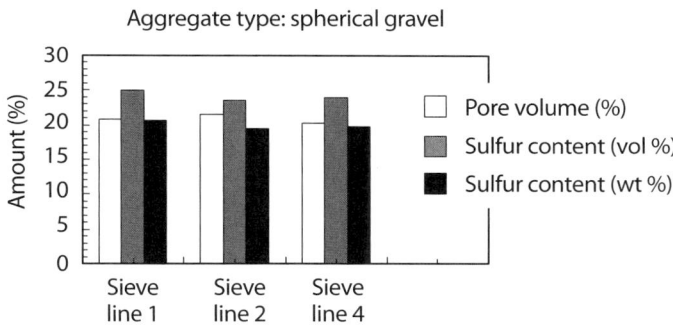

Figure 4.13a. Variations of pore volumes and sulfur contents with spherical gravel and its grain size distributions (data from Gregor and Hackl, 1978).

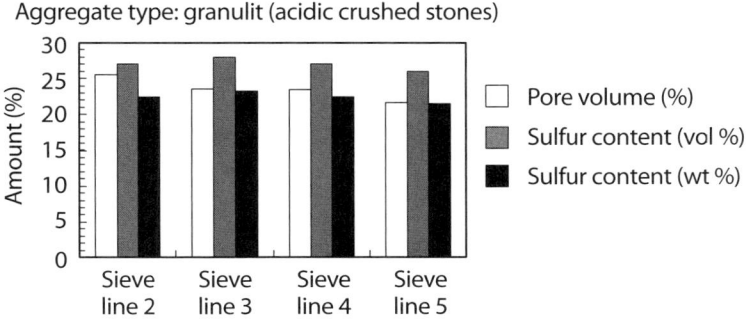

Figure 4.13b. Variations of pore volumes and sulfur contents with acidic crushed stones and its grain size distributions (data from Gregor and Hackl, 1978).

128 Sulfur Concrete for the Construction Industry

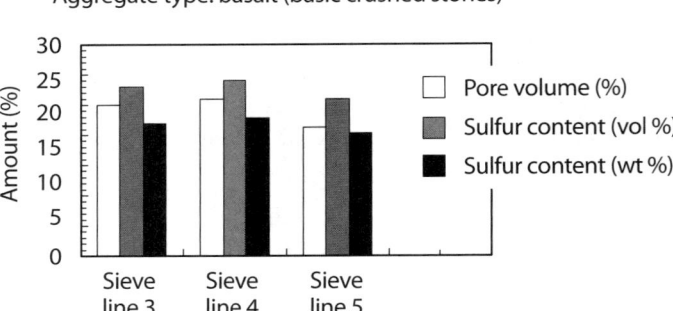

Figure 4.13c. Variations of pore volumes and sulfur contents with basic crushed stones and its grain size distributions (data from Gregor and Hackl, 1978).

Strength

The highest strength values were obtained with grain mixtures which were built up according to sieve line 5. The most favorable sieve line area for the production of elemental sulfur concrete lies within the area limited by the sieve lines 4 and 5. By continuously building up these grain compositions, high packing densities with low grain heap porosity result. The highest compressive and flexural strength values obtained is shown in Figure 4.14. Favorable grain distributions were found to be important to the strength of the elemental sulfur concrete.

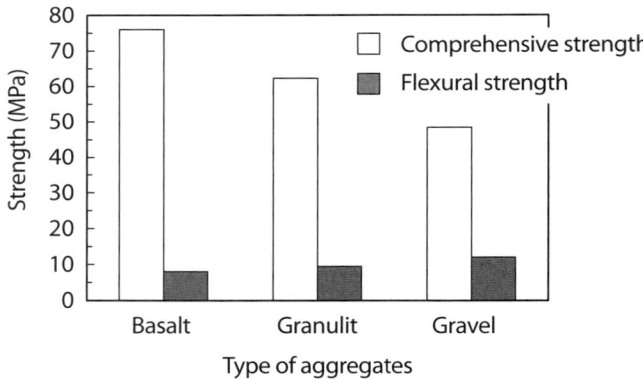

Figure 4.14. Effect of the aggregate type on the compressive and flexural strengths of elemental sulfur concrete (data from Gregor and Hackl, 1978).

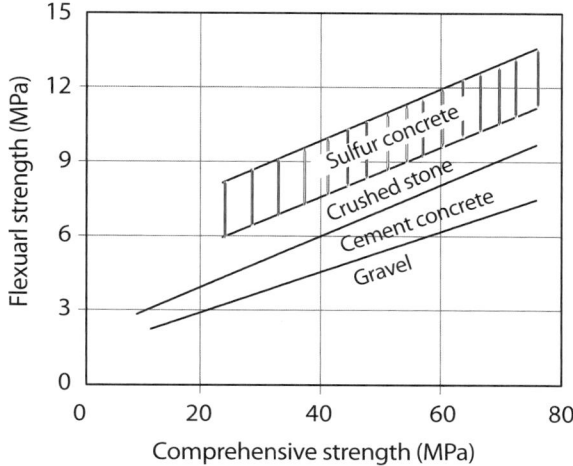

Figure 4.15. Comparison of the ratio between the flexural strength and compressive strength of cement concrete and sulfur concrete (adapted from Gregor and Hackl, 1978).

The strength of sulfur concrete is often greater than that of ordinary Portland cement concrete. Compressive strength of sulfur concrete prepared from elemental sulfur and different aggregates (limestone, sandstone, and basalt) obtained at 2 days ranged from 50-58 MPa (Abdel Jawad and Qudah, 1994). Whereas the compressive strength of Portland cement concrete obtained at 28 days ranged from 24-35 MPa (McBee and Weber, 1990).

If one compares the ratio of flexural to compressive strength for elemental sulfur concrete and Portland cement concrete, further interesting aspects appear (Figure 4.15). If the ratio of flexural strength to compressive strength of Portland cement concrete is 1:10 to 1:8 according to the type of aggregate used, then the ratio for elemental sulfur concrete is 1:6. Low strength ratios are important in case of heavy loading (Gregor and Hackl, 1978).

Optimum Grain Size Distribution and Sulfur Content

The highest strength values were obtained with grain mixtures that were built up according to sieve line 5. The most favorable sieve line area for the production of elemental sulfur concrete lies within the area limited by the sieve lines 4 and 5 shown in Figure 4.11 (Gregor and Hackl, 1978). Within this area, the amount of fine material ranges between 10.61 to 16.61% and the amount of sulfur needed as a binder ranges from 19 to 23 wt %, which facilitates the successful mixing of the elemental sulfur concrete.

The preceding underscores the importance of having fine materials within the elemental sulfur concrete mix not only to provide a dense matrix with high strength properties but also to control the sulfur crystallization process discussed in Chapter 3. By using grain size distributions falling within lines 4 and 5, a dense network of nucleation points is created within the elemental sulfur concrete mixture. In this way, sulfur crystal growth would be controlled and a compact structure with small crystals would be obtained in which adverse effects such as volume changes, cracking under thermal cycles, and loss of strength could be avoided. This instability comes from the fact that sulfur, after solidifying at 119°C from a liquid state, undergoes a solid transformation at 95.4°C between two crystal structures: monoclinic sulfur (S_β) that is stable over that temperature and orthorhombic sulfur (S_α) that is stable below 95.4°C and at room temperature (more details are provided in Chapter 3). This transformation implies a densification, and thus reduction of volume that can produce an internal tensional state in elemental sulfur concrete matrix. Sulfur is then subject to cracking and disintegration under thermal cycles or other instabilities.

4.5.3 ACI Guide for Material Selection

The following discussion is based on ACI 548.2R-93 (re-approved 1998) "Guide for Mixing and Placing Sulfur Concrete in Construction." Selection of quality aggregates appropriate for each application is essential for sulfur concrete materials. Aggregates should meet ASTM C33 specification with respect to durability, cleanliness, and limits of deleterious substances. The aggregates should be resistant to chemical attack by the environment in which elemental sulfur concrete is to be used; for example, quartz aggregates are suitable for use in both acidic and salt environments, whereas limestone aggregates are suitable for use in salt environments but not for acidic environments. Crushed aggregates are preferable to rounded ones because they produce higher strength materials.

Aggregate Gradation

Good distribution of all sized material creates a desirable dense-graded aggregate. Three size fractions of aggregate are generally used in preparing the dense-graded product; coarse aggregate, fine aggregate, and mineral filler (minus 200 mesh, 0.075 mm). The consequence of a less densely graded aggregate is a greater void space fraction in the sulfur concrete. In sulfur concrete mix proportioning, a maximum density gradation is used to minimize the aggregate void space and thus the amount of sulfur used as a binding material. Figure 4.16 shows the recommended ranges of grain size distributions for 25 and 9.5 mm of aggregate sizes based on ACI 548.2R-93. Data for other aggregate sizes as recommended by ASTM D3515 are shown in Table 4.4. It is clear that the grain size distribution

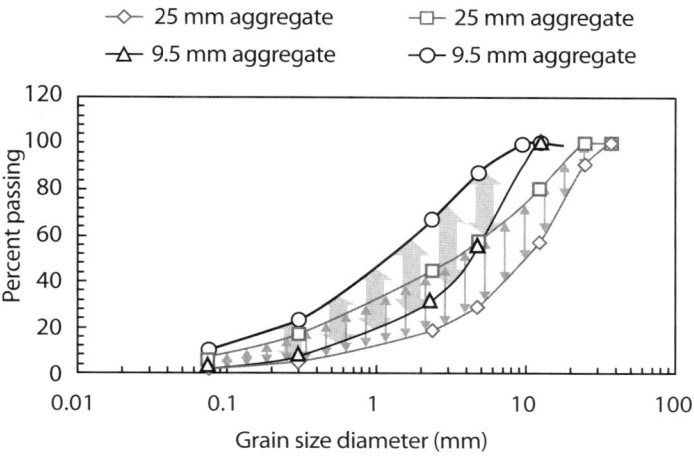

Figure 4.16. Ranges of recommended grain size distributions for 25 and 9.5 mm of aggregate sizes (data obtained from ACI 548.2R-93).

of some of these aggregates is characterized as *gap-graded* hence requiring more sulfur filler to eliminate the void space (Okumura, 1998).

Corrosion Resistance

Aggregates for sulfur concrete in acidic environments should not show any effervescence when tested in acid of the expected concentration and at the expected temperature of exposure. Aggregates for sulfur concrete to be used in acidic environments should also have less than 2 percent loss in weight when leached for 24 hr in acid of the expected concentration and at a temperature of 60 ± 3°C. Aggregates for sulfur concrete in salt environments show no reaction or deterioration after leaching 24 hr in the expected salt solution at a temperature of 60 ± 3°C.

Moisture Absorption

Porous aggregates should not be used. Instead, the aggregates should be highly impervious and highly resistant to freeze-thaw stresses. Maximum moisture absorption should be less than 1 percent for coarse aggregates and less than 2 percent for fine aggregates, as determined by using ASTM C127 and C128 procedures, respectively.

Table 4.4. Dense graded aggregate gradation limits (ASTM D3515)

Sieve size (mm)	25 mm aggregate, % passing	19 mm aggregate, % passing	13 mm aggregate, % passing	9.5 mm aggregate, % passing
37.50	100			
25.00	90-100	100		
19.00		90-100	100	
12.50	56-80		90-100	100
9.500		56-80		90-100
4.750	29-59	35-65	44-74	55-85
2.380	19-45	23-49	28-58	32-67
0.300	5-17	5-10	5-21	7-23
0.075	1-7	2-8	2-10	2-10

Table 4.5. Range of sulfur content with aggregate sizes

Maximum aggregate size (mm)	25.4	19.1	12.7	9.5
Percent passing sieve size 0.075 mm	1-7	2-8	2-10	2-10
Sulfur (% wt.)	12-15	13-16	14-17	16-19

Optimum Mix Design

The objective of mixture design is to determine the sulfur content in combination with specific aggregates that provide the most desirable balance between mechanical strength, high specific gravity, low absorption, and good workability. Mixture proportions are the amounts of aggregates and sulfur required to obtain the high quality sulfur concrete. Aggregate meeting the gradation requirements of ASTM D3515 will provide more workable sulfur concrete than those complying with ASTM C33 specifications. Each mixture design should be specific to a certain use. The mixture design described here is specific for sulfur concrete materials suitable for the construction of floors, foundations, tiles, sumps, side walls, and electrolyte cells for use in acid and salt environments.

Table 4.5 shows the range of sulfur levels for maximum sizes of dense-graded aggregates. The matrix was designed for moisture absorption less than 0.1 percent by mass. Figure 4.17 shows the changes of compressive strength with sulfur and aggregate percent for dense-graded quartz aggregates of a diameter less than 9mm. The range of sulfur loadings should be adjusted according to the type, size, and gradation of the aggregates. Table 4.5 and Figure 4.17 could be used to select approximate sulfur limits for mix design purposes. Actual proportions may fall slightly outside of this range for certain material combinations.

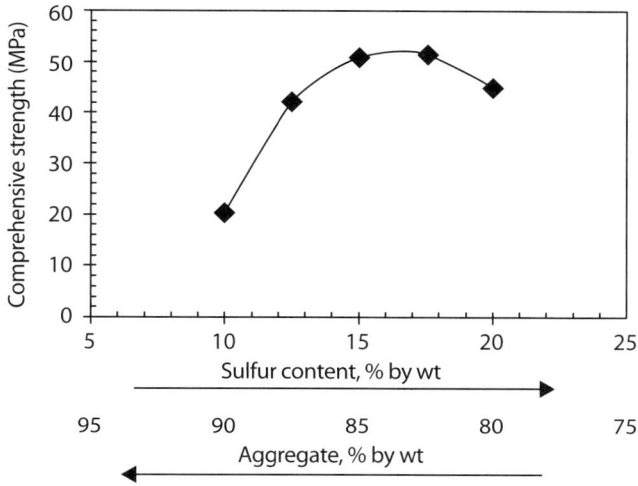

Figure 4.17. Changes of compressive strength with sulfur and aggregate percent for dense-graded quartz aggregates of diameter less than 9mm (data obtained from ACI 548.2R-93).

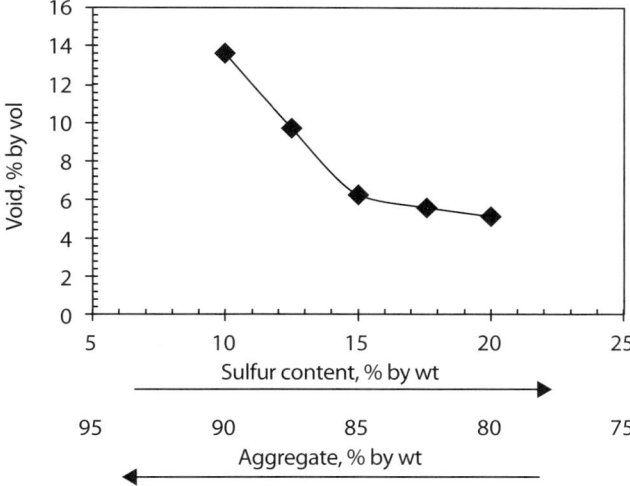

Figure 4.18. Changes of voids with sulfur and aggregate percent for dense-graded quartz aggregates of diameter less than 9mm (data obtained from ACI 548.2R-93).

Voids

Figure 4.18 shows the changes of voids with sulfur and aggregate percent for dense-graded quartz aggregates of a diameter less than 9mm (ACI 548.2R-93). Voids in sulfur concrete are important for the following reasons:

- (a.) Serve as stress relief sites, improving the durability of the material (McBee et al., 1983)
- (b.) Reduce the quantity of sulfur required to coat the mineral aggregates, thereby minimizing the sulfur-related shrinkage.

Void contents may be determined in two ways:

- (a.) By determining the specific gravity of sulfur concrete using ASTM C642. Voids are determined by calculation of the void content from the measured specific gravity and the theoretical specific gravity of the aggregate-sulfur mixture. A distinction between discrete and continuous voids should be stated as provided in ASTM C642.
- (b.) By microscopic determination on a section cut from a sulfur concrete specimen in accordance with ASTM C457 using the linear traverse method.

Absorption

Absorption is determined by calculating the amount of water being absorbed after 24 hr at room temperature. Sulfur concrete specimens are weighed, immersed in water at room temperature for 24 hr, surface dried, and reweighed. Moisture absorption is calculated as follows:

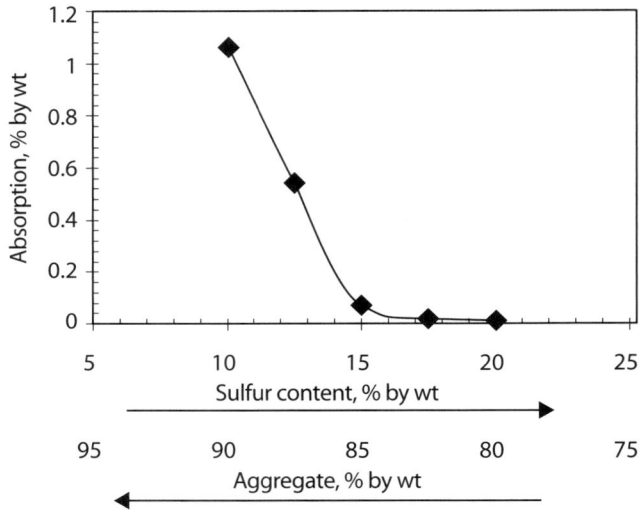

Figure 4.19. Changes of absorption with sulfur and aggregate percent for dense-graded quartz aggregates of diameter less than 9mm (ACI 548.2R-93).

$$A = \frac{C-B}{B} \times 100 \qquad (4.8)$$

Where A is the percent by the weight moisture absorption; B is the weight of the dry 76 by 153 mm cylindrical specimen; and C is the weight of the surface-dried specimen after immersion.

Figure 4.19 illustrates the changes of absorption with the sulfur and aggregate percent for dense-graded quartz aggregates of a diameter less than 9mm (ACI 548.2R-93).

4.6 DURABILITY

Although sulfur concrete materials could be prepared by hot mixing elemental sulfur and aggregate, the durability of the resulting product was a problem. Elemental sulfur concretes failed when exposed to repeated cycles of freezing and thawing, humid conditions, or immersion in water. When unmodified sulfur and aggregate are hot mixed, cast, and cooled to prepare sulfur concrete products, the sulfur binder, on cooling from the liquid state, first crystallizes as monoclinic sulfur (S_β) at 114°C with a volume decrease of 7%. On further cooling to below 96°C, the S_β starts to transform to orthorhombic sulfur (S_α), which is the stable form of sulfur at ambient temperatures (Lin et al., 1995). This transformation is rapid, generally occurring in less than 24 hr. Since S_α form is more dense than S_β, high stress is induced in the material by solid sulfur shrinkage. Thus, the sulfur binder can become highly stressed and can fail prematurely.

Over recent years, major drawbacks to the potential use of elemental sulfur as a construction material have been highlighted by many researchers (Gillott et al., 1983; Czarnecki and Gillott, 1989, 1990; Abdel Jawad and Qudah, 1994).

4.6.1 Effect of Sulfur Loading, Aggregate Type, and Different Admixtures

Czarnecki and Gillott (1990) studied the durability of elemental sulfur concrete made with different aggregates shown in Table 4.6 (syenite, granite, diorite, basalt, and greywacke) and various loadings of sulfur binder (20%, and 23%). Results showed significant expansion for some of the elemental sulfur concretes exposed to water. The amount of expansion depends on the type of rock used as aggregates and the loading of the sulfur binder. For sulfur binder loading of 20%, the highest expansion was observed when syenite, granite, and greywacke were used as aggregates and, after 90 days of immersion in water, expansion values reached approximately 0.28, 0.14, and 0.05%, respectively. Sulfur concrete made with basalt and diorite aggregates showed an expansion level of about 0.04%.

Table 4.6. Aggregate mineralogical composition and properties

Rock type	Mineral composition	Clay fraction content (%)	Density (Mg/m³)	Specific surface area (m²/g)
Granite	Quartz, plagioclase, biotite, chlorite	2.3	2.694	1.13
Syenite	Plagioclase, albite, microcline, pleochroic hornblende, quartz, iron, pyroxene	3.2	2.679	1.54
Diorite	Plagioclase, biotite, hornblende	2.1	2.886	1.17
Basalt	Feldspar, pyroxene, augite, iron, calcite, sericite	2.5	2.968	1.39
Greywacke	Plagioclase, quartz, calcite, sericite, mica	4.2	2.945	1.50
Marble	Calcite, quartz, mica	2.4	2.697	0.67

When the loading of sulfur was increased to 23%, moisture expansion decreased but remained unacceptable for concrete made with syenite, granite, and basalt aggregates. Concrete made with marble aggregate showed minimal change in the dimensions upon immersion in water. For concrete made with syenite, granite, and basalt aggregates and immersed in water for 90 days, the expansion remained unacceptably large, reaching values approximately 0.09, 0.05, and 0.05%, respectively.

Studies have shown that weathered rocks, and particularly those with a larger specific surface area, may cause moisture durability problems if used as aggregate in sulfur concrete; expansion may be reduced by the use of proper admixtures.

Attempts were made to overcome this expansion through the pretreatment of the aggregates with admixtures such as glycerin, crude oil, and silane (Czarnecki and Gillott, 1990):

- Pretreatment of the aggregate with 5% of glycerin decreased the moisture expansion of the concrete made with syenite, granite, diorite, and basalt aggregates, but it had only a marginal effect when greywacke was used. Expansion remained unacceptably high for the concrete made with syenite (0.17%), granite (0.05%), and greywacke (0.04%) aggregates.
- When crude oil with the amount of 5% by weight was incorporated in the mix, moisture expansion increased when syenite, granite, basalt, and greywacke aggregates were used. Changes were minimal with the diorite and marble aggregates. Expansion increases have resulted in strength decrease and the increase of flexibility and strain capacity of the sulfur concrete.

Table 4.7. Compressive strength variations with mineral composition of elemental sulfur concrete (data from Vroom, 1981)

No.	Aggregate type	Aggregate %	Sulfur %	Fly ash %	Density	Strength MPa
1	Crushed brick	55.0	27.4	17.6	2.18	69.89
2	Expanded clay	41.0	34.0	25.0	1.79	64.31
3	Crushed quartzite	65.6	18.8	15.6	2.40	63.96
4	Crushed granite	66.2	21.2	12.6	2.37	56.93
5	Crushed Portland cement	64.8	23.0	12.2	2.24	54.10
6	Crushed limestone	60.8	21.8	17.4	2.32	53.34
7	Crushed barite	79.6	12.8	7.6	3.25	52.33
8	Siliceous tailings	61.2	34.8	4.0	2.20	44.30
9	Expanded shale	35.8	35.8	28.4	1.68	33.53

- Pretreatment of the aggregates with 1.5% by weight silane generally decreased the moisture expansion of the sulfur concrete. These admixtures were capable of altering the critical surface tension that is associated with the wettability potential of substrates. When a liquid has a surface tension below the critical surface tension of a substrate, it will wet the surface (contact angle = 0). The critical surface tension is unique for any solid.
- Concrete that contained glycerin and silane admixtures showed reduced moisture expansion.

The experimental results shown in Table 4.7 (Vroom, 1981) were prepared by the addition of: (a) the required sulfur (less than contained in the pre-reacted material), (b) the pre-reacted material, and (c) fly ash to achieve the desired consistency of the sulfur cement. Aggregate was added to provide the sulfur concrete.

The components were mixed at about 130°C in a heated 1/3 cu ft concrete mixer for approximately 15 minutes before pouring into the molds. Compaction was obtained through vibration or tapping of the molds. The compressive strength varies from 69.89 to 33.53 MPa, depending on the type and amount of aggregates, as well as the sulfur and fly ash contents as shown in Figure 4.20. The use of these elemental sulfur concretes would be limited to reasonably isothermal applications such as underground or underwater structures subjected to mild thermal cycling.

4.6.2 Effect of Water and Temperature

The combined effects of water and temperature on the strength of elemental sulfur concrete have been extensively researched. The microstructure of sulfur

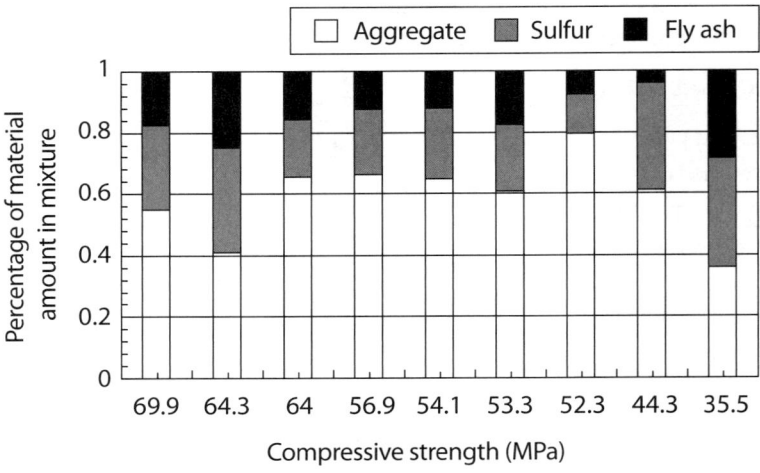

Figure 4.20. Compressive strength variations with mineral composition of elemental sulfur concrete (data from Vroom, 1981).

Table 4.8. Aggregate compositions and properties

Aggregate type	Mineral composition	Specific gravity	Unit weight (Mg/m³)	Absorption (%)
Limestone	Calcite, smectite, illite	2.45	1.845	5.31
Sandstone	Quartz, calcite	2.45	1.821	1.43
Basalt	Pyroxene, calcic-plagioclase, kaolinite	2.83	2.320	2.47

concrete, using different types of aggregates shown in Table 4.8 (limestone, sandstone, and basalt) has been studied (Abdel Jawad and Qudah, 1994). It was found that sulfur concrete developed its maximum compressive strength within a few hours after casting. No significant strength gain with time was observed when specimens were kept at ambient air temperature to solidify. Visible cracks and surface cavities have been noticed on the surfaces of both concretes made with limestone and basalt aggregates (Figure 4.21).

No cracks were observed in case of sandstone surface concrete immersed in water. Cracks that developed in the case of utilizing limestone and basalt aggregates may be due to moisture uptake by the clay minerals, illite and smectite in limestone concrete, and kaolinite in basalt concrete. Sandstone aggregate concrete showed no sign of visible cracks because it was free of clay minerals. The clay minerals resulted in improper bonding between the binder (sulfur) and the aggregates. The swelling and water absorption characteristics of clay would

Figure 4.21. Limestone elemental sulfur concrete sample after 14 days of water immersion (adapted from Abdel Jawad and Qudah, 1994).

generate expansive forces leading to disintegration of the concrete whether the clay was on the surface or inside the aggregate particles (Shrive et al., 1977). The formation of scattered surface cavities and micro-cracks would facilitate the penetration of water into the specimen.

The development of surface disintegration and cracking was attributed to the noncompatible volume changes (due to differences in thermal expansion coefficient) of the sulfur matrix and aggregates. Also, when sulfur concrete cools from the liquid to the solid state, the reduction in the volume of sulfur is much larger than the volume changes of the aggregate particles resulting in the developments of micro-cracks. When the concrete is immersed in water, these cracks allow the

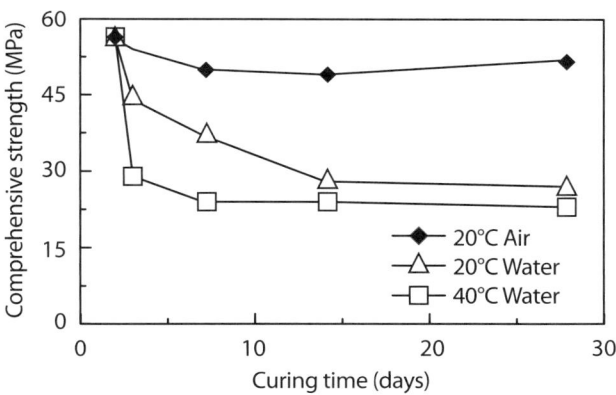

Figure 4.22. Compressive strength variations with time for limestone sulfur concrete at different curing conditions (adapted from Abdel Jawad and Qudah (1994).

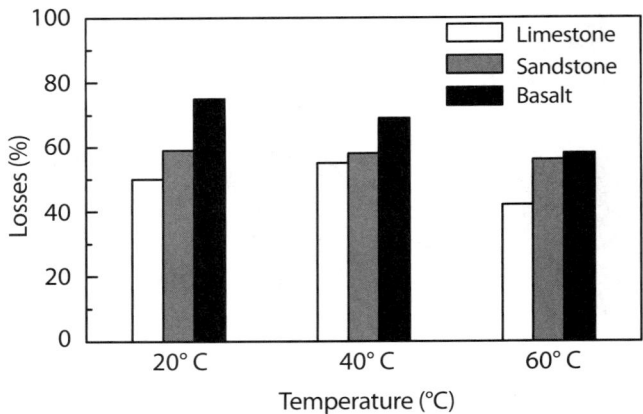

Figure 4.23. Compressive strength losses after 28 days of water immersion (adapted from Abdel Jawad and Qudah (1994).

water to easily penetrate into the sulfur-aggregate interface, resulting in separation between the two components.

Abdel Jawad and Qudah (1994) found that a sharp reduction in strength had occurred for all tested materials when the elemental sulfur concrete specimens were immersed in water. Most of the reduction took place during the first three days of immersion, especially at a curing temperature of 40°C. Figure 4.22 shows that limestone sulfur concrete specimens immersed continuously in water at varying temperatures developed lower strength than the samples exposed to normal laboratory air conditions. No clear trend was noticed on the effect of varying temperature on strength reduction over time. The reduction in strength ranges from 42% to 75%, depending on the type of aggregate and water temperature (Figure 4.23).

It is obvious that the highest strength loss was shown by the basalt concrete, followed by sandstone and limestone concrete, respectively. For nearly all types of concrete, the immersion in water at 20°C results in the highest losses, whereas the smallest losses took place at 60°C. Specimens immersed in water at 40°C resulted in the intermediate value.

4.7 SUMMARY AND CONCLUDING REMARKS

As discussed in Chapter 3, elemental sulfur undergoes a complex transition in two steps between allotropic forms, from liquid sulfur above the melting point (119.2°C) to solid sulfur at room temperature (below 95.5°C). Upon solidification, sulfur undergoes 7% contraction in volume. Using elemental sulfur as a binder with mineral aggregates leads to subpressure in pores and on the surfaces from this contraction. The tensile capacity of sulfur, which is only 0.3 to 0.4 MPa (Ecker and Steidl, 1986), is not capable of enduring the strain, and microcracking in inevitable (McBee, 1993). This exposes the elemental sulfur concrete material to moisture penetration.

After solidification, sulfur takes a monoclinic (amorphous) β-phase that on further cooling, transforms into the stable orthorhombic form, or α-phase, at 95.5°C. This transition is rather rapid (less than 24 hr) and leads to a further 6% vol. decrease. It will cause a strain on the binder and cracking within the material, whether or not volume compensation has been made at solidification when further cracking occurs.

Historically, elemental sulfur concretes failed (in the mechanical sense, due to disintegration) when it was exposed to humid conditions, repeated cycles of freezing and thawing, and immersion in water.

In principle, there are two ways to relieve the material from imposed stress due to contraction; either by modifying the sulfur binder in such a way that it remains in the β-phase for a longer duration or by accepting the transition into the α-phase but preventing the formation of microsulfur crystals that would cause contraction. Sulfur modification methodologies are discussed in Chapter 5.

A durable sulfur concrete material does not only require a stable binder but also such a composition of aggregates and binder that the full composite remains stable and durable (limited absorption) under fluctuating temperature and moisture conditions.

Aggregates play a key role in the mix design for a durable sulfur concrete product. Moisture absorption can be limited by the use of a dense-graded mineral aggregates and the proper composition design with binder, mixing, and consolidation. The selection of quality aggregates that will be appropriate for each particular application is necessary for a sulfur concrete material. To meet the requirement of durability, cleanliness and limits of harmful substances, the composite aggregates must meet the ASTM C 33 specifications according to the ACI Committee 548. To determine an aggregate's suitability for a particular use, it is recommended that preliminary testing be carried out for verification.

The properties for corrosion resistant aggregates must be clean, hard, tough, strong, durable and free of swelling constituents. They must also resist chemical attacks and moisture absorption from exposure to acid and salt solutions. Moisture absorption and dissolution losses should not exceed 1% in a 24-hr period.

The clay contained within the solidified sulfur concrete is believed to have an absorptive capacity that will allow water to permeate through the material. When clay absorbs water, expansion occurs and results in a deterioration of the product. Therefore, clay containing aggregates should not be used in producing sulfur concrete unless it has been treated to limit the swelling capacity.

The mechanisms involved in preparing sulfur concrete are different from those of Portland cement concrete. New gradation designs were developed based on the technology for asphalt concrete. The intention was to develop aggregate mixtures with maximum density and minimum voids in the mineral aggregate. Adopting this design means that less sulfur is needed to fill the voids of the mixture. Pickard (1984) indicated that the optimum range of sulfur concrete should, therefore, be just slightly less than the amount necessary to fill the aggregate to 100% saturation, yet high enough to keep the final void content less than 8%. This, in most cases, results in higher strength materials, because improved aggregate contact means less shrinkage after solidification (McBee et al., 1983).

The choice of mineral filler is important because it forms with the binder paste that coats and binds the coarse and fine aggregate particles strong and dense product. Soderberg (1983) defines the function of filler as:

a. Controls the viscosity of the fluid sulfur-filler paste, workability, and bleeding of the hot plastic concrete
b. Provides nucleation sites for crystal formation and growth in the paste and minimizes the growth of large needle-like crystals
c. Fills the voids in the mineral aggregate that would otherwise be filled with sulfur
d. Reduces the hardening shrinkage and the coefficient of thermal expansion
e. Acts as a reinforcing agent in the matrix, increasing the strength of the formation

Therefore, the filler must be reasonably dense-graded and finely divided to provide a large number of particles per unit weight, especially to meet the provision of nucleation sites.

5

SULFUR CEMENT

5.1 INTRODUCTION

The need to plasticize sulfur for use as cement was recognized during the 1930s. The researchers showed that combing elemental sulfur and aggregates produced a high strength, acid-resistant product. However, as discussed in Chapter 4, the durability of this type of material was poor and most materials failed in less than one month, especially in moist environments and when subjected to thermal cycling. Improvements in sulfur-based cement formulations have been achieved by adding a modifier component to the cement formulations. Sulfur polymer cement is a thermoplastic polymer consisting of ~95 wt% elemental sulfur and ~5 wt% organic modifiers to enhance long-term durability. Sulfur polymer cement was originally developed by the United States Bureau of Mines in 1972 to use byproduct or waste sulfur as an alternative to hydraulic cement for construction applications and to significantly improve the stability of the product. Many types of organics have been employed in an effort to accomplish production of a durable cement, but only a few are used commercially.

This chapter discusses the basic ingredients of modified sulfur-based cement formulations and their physicochemical properties.

5.2 DEVELOPMENT BACKGROUND

The use of sulfur in the preparation of construction materials had been proposed following World War I when an acid-resistant mortar compound of 40% sulfur binder and 60% sand was prepared. However, upon thermal cycling such mortars exhibit a loss in flexural strength, resulting in failure of the mortars. However, after solidification, the sulfur in these concretes undergoes allotropic transformation wherein the sulfur reverts to the more dense orthorhombic form that results in a product that is highly stressed and, therefore, vulnerable to failure by cracking. Improvements in sulfur-based cement formulations have been achieved by adding a modifier component to the cement formulations.

In 1972, Dr. Alan Vroom, in cooperation with the National Research Council of Canada and McGill University, commenced a research program aimed at overcoming the problem of durability in sulfur concrete that other researches had experienced (Beaudoin and Sereda, 1974). The first sulfur concrete manufacturing plant began production of precast products in Calgary, Alberta, Canada, in 1975 (Sulfur Concrete Go Commercial, 1976).

Diehl (1976) has shown improved sulfur concrete formulations by the addition of small quantities of dicyclopentadiene as a modifier to the sulfur. Such modified cement formulations exhibit improved compressive strength characteristics.

McBee et al. (1976) have shown a variety of sulfur cement formulations such as sulfur concretes, sand-sulfur-asphalt paving, and others wherein the sulfur binder component is modified by the presence of small quantities of dicyclopentadiene.

Sullivan et al. (1975a, b; Sullivan and McBee, 1976) have also described various sulfur cement formulations in which the sulfur binder is modified with dicyclopentadiene (DCPD). Sullivan et al. (*Advances in Chemistry* No. 140) have also described sulfur cement formulations in which DCPD, dipentene, methylcyclopentadiene, styrene, and an olefinic liquid hydrocarbon were investigated as modifiers. Although the modification of sulfur cement formulations with various unsaturated hydrocarbon materials results in cement formulations of improved characteristics, nevertheless a need continues to exist for modified sulfur cement formulations of improved freeze-thaw stability and strength characteristics.

The key to the durability of sulfur concrete produced by the preceding process lies in the stability of microcrystalline sulfur (Gannon et al., 1983) for a successful 20-year performance record in a variety of aggressive environments. The sulfur concrete development in Canada is recognized as a valuable construction material in many geographic areas.

The South Western Research Institute, under government control, has constructed several test sections utilizing plasticized sulfur as binder (Plato. J. S., 1980).

Since 1984, the American Concrete Institute, through its Committee 548D (Sulfur Concrete), has been developing guidelines for the use of sulfur concrete (ACI 548.2R-1993). The same committee is currently finalizing a state of the art report on precast sulfur concrete.

The ASTM Committee for Chemical Resistance Materials has been activity developing standards for sulfur concrete and has published the following standard specifications and test methods on the subject:

- ASTM C1312–97 "Standard practice for making and conditioning chemical-resistant sulfur polymer cement concrete test specimens in the laboratory."
- ASTM C1159–98 "Standard specification for sulfur polymer cement and sulfur modifier for use in chemical-resistant, rigid sulfur concrete."

In 2006, the authors of this book have shown improved sulfur cement formulation by the addition of small quantities of bitumen as a modifier to the elemental sulfur. Such modified cement formulation exhibits improved durability performance.

5.3 TERMINOLOGY

To prevent confusion, terms found in various publications on the topic are further defined. The ASTM C1159-98 defines *sulfur polymer cement* as a polymer product consisting of small amounts of chemical modifying additives dispersed in sulfur. The term SPC is generally used as an indicator for sulfur polymer cement. Once it is loaded with aggregates or waste products, it becomes sulfur concrete and the term SC is used as an indicator. This nomenclature follows the same principle as that for Portland cement. The paste of Portland cement is represented by PC, whereas the concrete obtained by the addition of sand and aggregates to PC is designated by PCC, or Portland cement concrete. Other publications used the terminology modified sulfur cement instead of SPC; however, both names refer to the same material.

In the preceding terminologies, the word cement is included without clear indication of the type of additives that will produce this cementious behavior. Generally, chemical additives are added to modify elemental sulfur, nearly all of which fall under the heading of polymeric polysulfide or, alternatively, substances that may react with elemental sulfur to give *in situ* formation of polymeric polysulfide. Therefore, terms such as polymerized sulfur or modified sulfur should be used instead.

For inclusion of the word cement in the definition, the mixture should include, in addition to the chemical additives, physical stabilizers or mineral fillers such as fly ash, furnace slag, cement kiln dust, talc, mica, silica, graphite, carbon black, pumice, insoluble salts (e.g., barium carbonate, barium sulfate, calcium carbonate, calcium sulfate, magnesium carbonate, etc.), magnesium oxide,

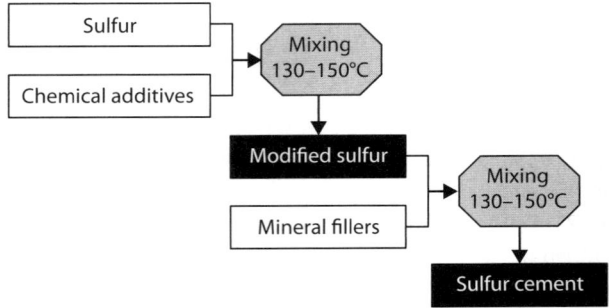

Figure 5.1. Sulfur terminologies used for sulfur cement production.

and mixtures. Such fillers typically have a particle size less than 100 mesh (U.S. Standard Testing Sieves) and preferably less than 200 mesh. Such fillers generally act as thickening agents and generally improve the hardness or strength of the sulfur cement product.

Therefore, in this book, the term *modified sulfur* is used when chemical additives are added to the elemental sulfur, and *sulfur cement* is used when mineral fillers are added to the modified sulfur (Figure 5.1). In addition, *sulfur concrete* is used when aggregates are added to sulfur cement.

5.4 MODIFIED SULFUR

5.4.1 Mechanism

As discussed in Chapter 3, if pure elemental sulfur is heated to 140°C and then cooled to ambient temperature, monoclinic sulfur is instantaneously formed followed by a reversion to orthorhombic sulfur that is nearly complete in about 20 hr. Also, as discussed in Chapter 4, elemental sulfur concrete products exhibit large expansion and contraction when subjected to thermal cycles, hence variations in the hydromechanical-thermal properties are expected. Therefore, it is necessary to modify the elemental sulfur with chemical additives to overcome such problems.

Many investigators have attempted to plasticize or stabilize sulfur in its polymeric form. Pryor (1962) pointed out that in sulfur-olefin reactions, only the carbon-sulfur (C-S) bond is formed and no new carbon-carbon (C-C) bond is formed. Therefore, in such a co-polymerization reaction, one expects to get a strictly alternating structure. The research showed that elemental sulfur reacts with olefins at 90 to 160°C in the liquid phase to form several types of polysulfide products. As a result of these interactions, new molecules of cyclic polysulfide

Sulfur Cement 147

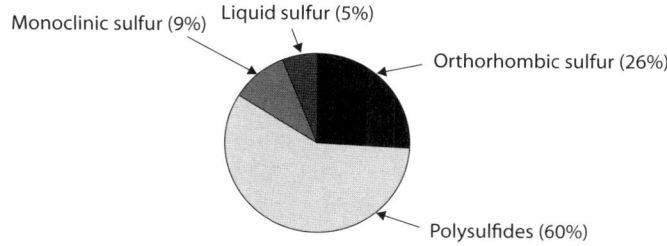

Figure 5.2. Sulfur components after elemental sulfur reaction with styrene at 140 for 3 hr (data from Blight et al., 1978).

distinguished by the number of sulfur atoms are obtained. There are bases to assume that in the process of these reactions two weak bonds sulfur-sulfur (S-S) are broken off; π-bond of unsaturated compound and a new bond (S-S) as well as two rather strong C-S bonds are formed (Rylova, 2002).

Many chemical additives have been proposed to modify elemental sulfur. Added chemical substances may react with elemental sulfur to give *polymeric polysulfide*. The additional sulfur in the matrix is not polymerized or held in an amorphous state by the polymer. It is allowed to crystallize to the only physical form that is stable below 90°C (194°F), i.e., orthorhombic sulfur. For example, if styrene is added to elemental sulfur (25% w/w of sulfur) and heated at 140°C for 3 hours, the resulting material will be a mixture of polysulfides and unreacted elemental sulfur as shown in Figure 5.2.

The polymer does, however, prevent the growth of macro-sulfur crystals. When one part of polymer is dissolved in ten parts, by weight, of liquid sulfur, and the solution is allowed to cool, the sulfur crystals are generally smaller than one micron. This is important, as discussed in Chapter 3, because sulfur has a relatively high linear coefficient of thermal expansion, 7.4×10^{-5}/C and a low thermal conductivity, 6.1 gcal/cm²/sec ($\Delta t = 1°C$), both at 40° C (104°F) (Texas Gulf Sulfur Co., 1961; Vroom, 1998). When a material containing adjacent macrocrystals of sulfur is subjected to changing temperatures, there will be a constant movement between these macrocrystals as one expands or contracts relative to its neighbor. This movement will gradually break bonds with other cross-laid crystals, causing microfractures and a severe weakening of the structure as well as an eventual formation of cleavage planes.

A similar mechanism may be responsible for the aging phenomenon in certain types of commercially formed solid elemental sulfur. An independent scanning electron microscopic study, published by ASTM, has revealed graphically how the polymer controls the crystallization of sulfur, preventing the growth of macrocrystals and thereby imparting durability (Gannon et al., 1983).

5.4.2 Types

It is expected that, depending on the type of the chemical modifiers, the percent of polysulfides and unreacted sulfur varies. Blight et al. (1978) reviewed the various substances used as additives and compared quantitatively their effectiveness in retarding crystallization of elemental sulfur. The additives studied were a range of olefins and certain Thiokols (polymeric polysulfides). The effect of each modifier is shown in Figure 5.3. All additives were added to elemental sulfur at 25% by weight and heated at 140°C for 3 hr. These additives have a significant effect in reducing the formation of orthorhombic sulfur, the olefins (except styrene) being particularly effective with no orthorhombic sulfur formed, even after 18 months.

The preceding underscores the main aim of adding chemical additives to elemental sulfur, which is to inhibit the transformation to orthorhombic sulfur (α-sulfur). As a result of the chemical reaction with the substance, sulfur should remain in the monoclinic state after cooling.

Sulfur cement can be unaltered sulfur and/or plasticized sulfur. The term *plasticized sulfur* refers to the reaction product of sulfur with a *plasticizer* and/or mixtures of sulfur and plasticizers and/or the reaction product of sulfur with a plasticizer. When a plasticizer is used, the amount of the plasticizer(s) will vary with the particular plasticizer and the properties desired in the cement. The cement can contain 0.1 to 10% of the plasticizer and typically will contain approximately 2 to 7%, preferably 2.5 to 5% by weight, based on the total weight of both free and combined (or reacted) sulfur in the composition (herein referred to as total sulfur).

The term *sulfur plasticizer* or plasticizer refers to materials or mixtures of materials that, when added to sulfur, lower its melting point and increase its crystallization time (Nimer et al., 1985). One convenient way to measure the rate of crystallization is as follows: the test material (0.040 g) is melted on a microscope slide at 130°C and is covered with a square microscope slide cover slip. The slide is transferred to a hot plate and is kept at a temperature of 70°C ± 2°C, as measured on the glass slide using a surface pyrometer. One corner of the melt is seeded with a crystal of test material. The time required for complete crystallization is measured. Plasticized sulfur, then, is sulfur containing an additive that increases the crystallization time within experimental error, i.e., the average crystallization time of the plasticized sulfur is greater than the average crystallization time of the elemental sulfur feedstock.

Inorganic plasticizers include the sulfide of iron, arsenic, and phosphorus. Generally, the preferred plasticizers are organic compounds that react with sulfur to give sulfur-containing materials.

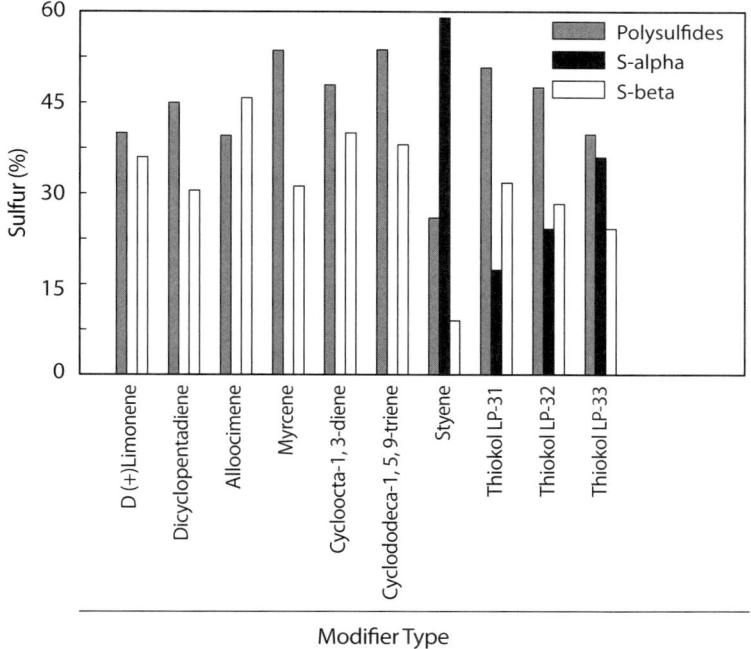

Figure 5.3. Effects of various modifiers on the formation of beta sulfur crystals and reduction formation of alpha sulfur crystals. Each of these modifiers (25% w/w sulfur) was mixed with elemental sulfur and heated at 140°C for 3 hr (data obtained from Blight et al., 1978).

Suitable sulfur chemicals that can be used include aliphatic polysulfides, aromatic polysulfides, styrene, DCPD, dioctylphthalate, acrylic acid, epoxidized soybean oil, triglycerides, tall oil fatty acids, and compatible mixtures (Nimer et al., 1985).

One class of preferred plasticizers is the aliphatic polysulfides, particularly those that will not form cross-linking. Thus, butadiene is not a preferred constituent to form the aliphatic polysulfide, because it may form cross-linking sulfur bonds; whereas DCPD is a preferred compound for forming the aliphatic polysulfide useful as the sulfur plasticizer.

Another class of preferred plasticizers for use in the composition of aromatic polysulfides formed by reacting 1 mol of an aromatic carbo-cyclic or heterocyclic compound, substituted by at least one functional group of the class –OH or –NHR in which R is H or lower alkyl with at least 2 mol of sulfur.

Suitable organic compounds of this type include phenol, aniline, N-methyl aniline, 3-hydroxy thiophene, 4-hydroxy pyridine, p-aminophenol, hydroquinone, resorcinol, meta-cresol, thymol, 4, 4-dihydroxy biphenyl, 2,2-di(p-hydroxy-phenol) propane, di(p-hydroxy-phenyl) methane, etc., p-phenylene diamine, methylene di-aniline. Phenol is an especially preferred aromatic compound to form the aromatic polysulfide.

The aromatic polysulfides are generally prepared by heating sulfur and the aromatic compound at a temperature in the range of 120 to 170°C for 1 to 12 hr, usually in the presence of a base catalyst such as sodium hydroxide. The polysulfide product made in this way has a mol ratio of aromatic compound: sulfur of the 1:2 to 1:10, preferably from 1:3 to 1:7. Upon completion of the reaction, the caustic catalyst is neutralized with an acid such as phosphoric or sulfuric acid. Organic acids may also be used for this purpose. The resulting aromatic polysulfide may be used immediately, or it may be cooled and stored for future use.

Another type of aliphatic polysulfide useful as a plasticizer is the linear aliphatic polysulfides. Although these polysulfides may be used alone as the sulfur plasticizer, it is preferred to use them in combination with either (a) DCPD or (b) the aromatic polysulfides, especially with the phenol-sulfur adduct. In this connection, the preferred plasticizer mixtures contain from 5% to 60% by weight linear aliphatic polysulfide, based on total plasticizer, preferably 20 to 50% by weight.

These aliphatic polysulfides can have branching indicated as follows (Nimer et al., 1985):

$$\begin{array}{c} C \\ \| \\ B \end{array} = C + S \rightarrow S_x - \begin{array}{c} C \\ | \\ B \end{array} - C - S_x -$$

Wherein x is an integer of from 2 to 6 and wherein B is H, alkyl, aryl, halogen, nitrile, ester or amide group. Thus, in this connection, the aliphatic polysulfide is preferably a linear polysulfide. The chain with the sulfur preferably is linear, but it can have side groups as indicated by B. Also, this side group B may be aromatic. Thus, styrene can be used to form a phenyl-substituted linear aliphatic polysulfide. The preferred aliphatic polysulfides of this type are both linear and nonbranched.

Unbranched linear aliphatic polysulfides include Thiokol LP-3 that contains an ether linkage and has the recurring unit (Nimer et al., 1985): — $S_xCH_2CH_2OCH_2OCH_2CH_2S_x$—, wherein x has an average value of approximately 12. The ether constituent of this aliphatic polysulfide is relatively inert to reaction. Other suitable aliphatic polysulfides have the following recurring units:

a. —S_x—(—CH_2—)$_y$—S_x— from reaction of alpha, omega-di-haloalkanes and sodium polysulfide
b. —S_x—(—CH_2CH_2—S—CH_2CH_2—)$_y$—S_x— from reaction of alpha, omegadihalosulfides, and sodium polysulfide

c. $-S_x-(-CH_2CH_2-O-CH_2CH_2-)_y-S_x-$ from reaction of alpha, omegadihaloesters, and sodium polysulfide

Where x is an integer of 2 to 5 and y is an integer of 2 to 10.

In some instances, it is preferred to use mixtures of materials having different reactivity with sulfur as the plasticizer. For example, good results can be obtained using a mixture of cyclopentadiene and/or DCPD with oligomers of cyclopentadiene as discussed later in this chapter.

Other class of additives is olefinic hydrocarbon polymers such as RP220 and RP020 (Exxon Chemical), Escopol (Vroom, 1977, 1981 & 1992; Nnabuife, 1987), bitumen (Mohamed and El Gamal, 2006), 5-ethylidene-2-norbornene (ENB) and 5-vinyl-2-norbornenen (VNB) (Shell International B.V., PCT/EP2006/063220). All these compounds identify a heat-reactive olefinic liquid hydrocarbon obtained by partial polymerization of olefins. The proportions of the chemical stabilizer may be varied, depending on the end use of the cement. The chemical stabilizer used in the composition of aspects of this modification is any of the olefinic hydrocarbon polymers derived from petroleum having a non-volatile content greater than 50% by weight and a minimum Wijs iodine number of 100 cg/g, capable of reacting with sulfur, to form a sulfur-containing polymer. Typically, the chemical stabilizer is used in amounts up to 14% by weight of the total sulfur, and more especially in the proportion of 1-5% of the total sulfur by weight. The amount of such chemical stabilizer required depends on the end use of the cement and the properties desired.

These additives were generally utilized by various researchers: Leutner and Diehl (1977), Currell (1976), and Beaudoin and Feldmant (1984) for using DCPD to improve porous durability of porous system; McBee et al. (1977) for using a combination of DCPD and styrene; Blight et al. (1978) for using styrene and DCPD; Bordoloi and Pearce (1978) and Gillott et al. (1980) for using crude oil and glycol; Woo (1983) for using phosphoric acid to improve freeze-thaw resistance; Nimer and Campbell (1983) for using organosilane to improve water stability; Beaudoin and Feldmant (1984), McBee and Sullivan (1982a, b), and Lin et al., (1995) for using DCPD, or a combination of DCPD, cyclopentadiene and dipentene; Mohamed and El Gamal (2006, 2008) for using bitumen; and Shell International B.V., PCT/EP2006/063220 (2006) for using 5-ethylidene-2-norbornene (ENB) and/or 5-vinyl-2-norbornenen (VNB).

5.4.3 Modification Conditions

Modification in sulfur-based cement formulations have been achieved by adding a modifier component to the cement formulations. To optimize modification of sulfur, three conditions were tested. In the first condition, sulfur was melted in an oil bath, then an olefinic additive was added 130 to 140°C; the temperature

Table 5.1. Compressive strength of sulfur mortar prepared at different heating times and 140°C (adapted from Bahrami et al., 2008)

Sample	Reaction time (hr)	Compressive strength (MPa)
1	0.5	7.16
2	1	17.65
3	2	13.53
4	3	20.30

Table 5.2. Compressive strength of sulfur mortar prepared at different temperatures for 3 hr (adapted from Bahrami et al., 2008)

Sample	Reaction temp (°C)	Compressive strength (MPa)
1	160	10.19
2	160	10.98
3	160	8.04
4	160	12.36
5	140	19.91
6	140	20.40
7	140	14.32
8	140	36.48
9	140	31.38

maintained at 130 to 140°C during the mixing process that lasted approximately 3 hr (Blight 1978). In the second condition, an additive was added at 150 to 160°C and the mixture was heated in the same temperature for approximately 3 hr. In the third condition, after the addition of an olefinic additive to sulfur at 130 to 140°C, the mixture was heated for 0.5, 1, 2, and 3 hr to produce modified sulfur.

Reaction temperature and heating time are effective factors used to modify sulfur. Results in Tables 5.1 and 5.2 indicate that compressive strengths of sulfur mortar prepared at 140°C are greater than the sample prepared at 160°C. It should be noted that at 119 to 157°C molten sulfur exists essentially as cyclooctasulfane. Sulfur reacts with olefins at 90 to 160°C in the liquid phase to form several types of polysulfide products (Pryor, 1962). By increasing the temperature from 90°C, the S_8 rings progressively break up into reactive bi-radicals. If there is no olefin, these reach a sufficient concentration at around 160°C to polymerize spontaneously into μ-sulfur chains (Sander et al., 1984). Polymeric chains have lower affinities than other compounds thus, reactivity of sulfur with additive reduces at 160°C as compared with sulfur at 140°C; there is a linear relation between time and compressive strength.

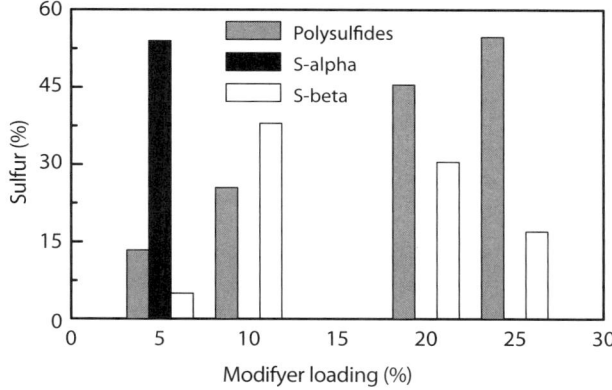

Figure 5.4. Effect of DCPD percentage on the composition of sulfur materials at a reaction temperature of 140°C and a heating time of 3 hr (data from Blight et al., 1978).

5.5 INDUSTRIAL MODIFIED SULFUR CEMENT

5.5.1 Sulfur Modified with Dicyclopentadiene

Sulfur Cement Composition

The beneficial use of DCPD to modify sulfur has been reported by a number of researchers (Currell et al., 1974; Sullivan el al., 1974; Diehl, 1976). Currell et al. (1974) showed that the interaction of DCPD and elemental sulfur at 140°C gives a mixture of polysulfides and free elemental sulfur that is held as a mixture of presumably monoclinic, orthorhombic sulfur crystal, and noncrystalline sulfur. The various fractions have been separated and identified.

The efficiency of DCPD in inhibiting the formation of orthorhombic sulfur is demonstrated in Figure 5.4. The fact that orthorhombic sulfur is formed with a loading of 5% and not with a 10 and 25% loading is in agreement with the minimum necessary levels reported by Sullivan et al. (1974) to stop the brittle nature of elemental sulfur permanently, which is 13% if the reaction temperature is less than 140°C and only 6% if the reaction temperature is greater than 140°C. Presumably, the polysulfide reaction products form a solid solution with the unreacted sulfur form that orthorhombic sulfur cannot crystallize.

The amount of polysulfide formed increases with the amount of DCPD added at 140°C. Figure 5.5 shows that as the reaction time proceeds, the amount of polysulfide formed increases. With a 25% loading after an hour, the reaction is incomplete with unreacted olefin still present; after 3 hr, 57.7% of the final

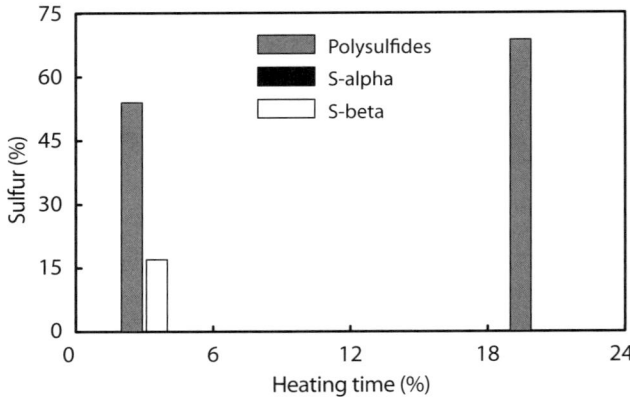

Figure 5.5. Effect of heating time on the composition of sulfur materials modified by DCPD; modifier percentage of 25% and a reaction temperature of 140°C (data from Blight et al., 1978).

product is polysulfide; and after 20 hr, 68.7% is polysulfide. Also, as reaction time proceeds, the molecular weight of the polysulfide increases.

The reaction products were examined by extraction with carbon disulfide to give, in many cases, a fraction soluble in carbon disulfide. All the unreacted free sulfur was soluble in carbon disulfide, indicating that it is nonpolymeric and presumably rings material. The amount of insoluble material formed increases with reaction time. The insoluble material fraction is a high molecular weight, cross-linked material that swells in carbon disulfide and has an elemental analysis that corresponds to $C_{10}H_{12}S_{11}$.

Chemical Structure of the Formed Product

As a result of the interaction between sulfur and DCPD, the formed product is a mixture of cyclic tri- and penta-sulfide, each formed by adding sulfur across the bicycloheptenyl double bond. The structure of penta-sulfide was confirmed by x-ray crystallography (Figure 5.6).

The position of the double bond, as C_6–C_7, may be caused by de-localization effects, pseudo-symmetry, or the fact that the product is actually a mixture of the two isomers. The polymer is formed by ring opening polymerization of the cyclic penta-sulfide (Figure 5.7). The nature of the soluble polysulfide was evaluated and it was determined that the relative proportions of the higher molecular weight fraction increase with respect to both reaction time and the amount of DCPD added. It was concluded that this polymer is formed by ring-opening polymerization of the cyclic penta-sulfide or other cyclic polysulfide.

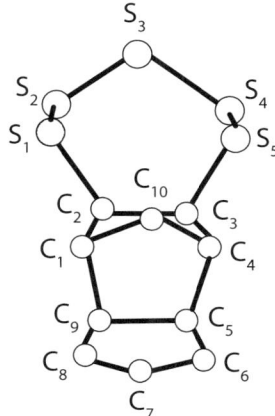

Figure 5.6. Structure of cyclic penta-sulfide formed by the interaction between sulfur and DCPD (adapted from Blight et al., 1978).

Viscosity

Blight et al. (1978) confirm the reports of Sullivan et al. (1974) that the reaction between DCPD and sulfur is exothermic. If the temperature rises above 150°C, the extreme viscosity increase causes the mixture to become nearly solid and the reaction is difficult to control. Diehl (1976) and Bordoloi and Pearce (1978) have reported quantitative studies of these viscosity changes. They show that there are large viscosity increases as the amount of DCPD, reaction temperature, and reaction time are increased. Blight et al. (1978) results show that these increases in viscosity are caused by the formation of high molecular weight polysulfide. Viscosity measurements have been made to follow the co-polymerization reaction and to analyze the viscosity behavior. In sulfur-olefin reactions, only the C-S bond is formed and no new C-C bond is formed. Therefore, in such a co-polymerization reaction, one expects to get a strictly alternating structure. The maximum rate for the logarithmic increase in viscosity for an equi-molar feed of S_8 and DCP (Figure 5.8) indicates that the co-polymer is probably alternating in structure. Also, a step growth co-polymerization mechanism supports such behavior of viscosity.

Surface Tension

The behavior of viscosity and surface tension as a function of time for sulfur-DCPD solution at 140°C with two feed compositions (5 and 20 wt% DCPD) was studied by Allan and Neogi (1970). Both surface tension and viscosity were measured as a function of time. The results indicated that the materials made from

156 Sulfur Concrete for the Construction Industry

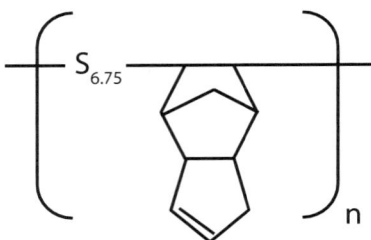

Figure 5.7. Ring-opening polymerization of cyclic penta-sulfide (adapted from Blight et al., 1978).

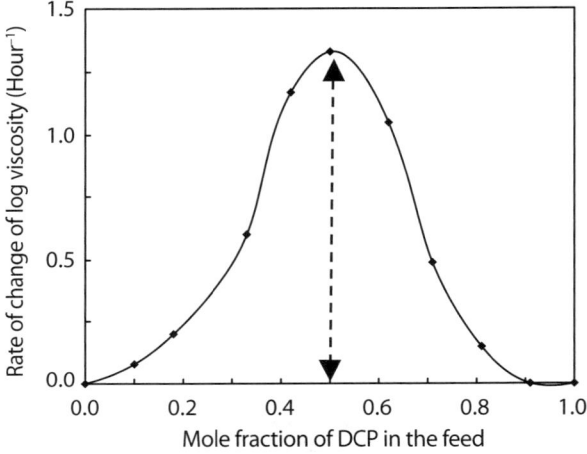

Figure 5.8. A plot showing the rate of increase of logarithmic viscosity with the feed composition in mole fraction of DCPD at 140°C (adapted from Bordoloi and Pearce, 1978).

sulfur with 20%–DCPD at 140°C for 3 hr contained 45.5% polysulfide and 54.5% unreacted sulfur; 30.5% crystallize as monoclinic and 24% remained as amorphous S_8. The results shown in Figure 5.9 clearly indicate that as the viscosity increases with time, surface tension increases, and the higher the rate of increase of viscosity, the higher the rate of increase of surface tension. A step growth copolymerization mechanism for the sulfur-DCPD solutions, as mentioned, will have an increase of molecular weight over time, and the surface tension behavior appears to support this mechanism.

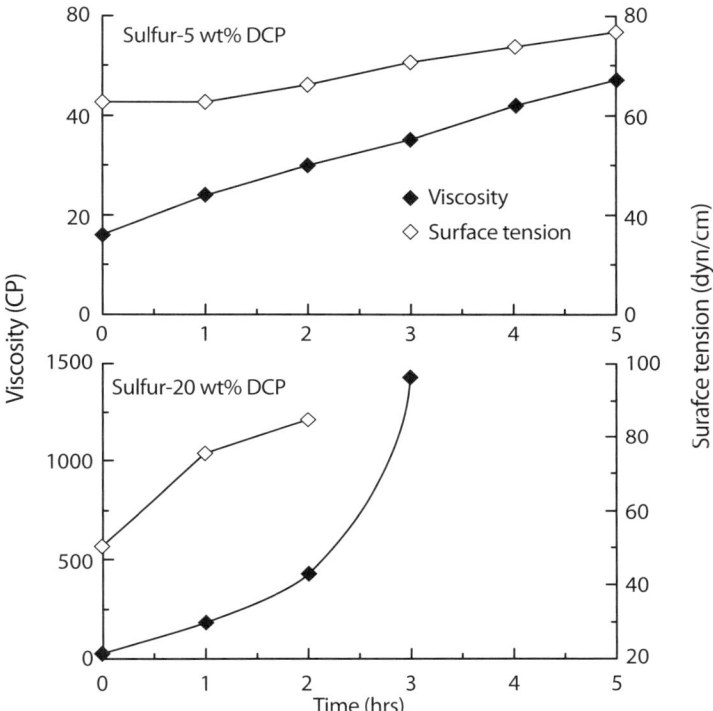

Figure 5.9. Measured viscosity and surface tension of sulfur-DCPD as a function of time for varying composition at 140°C (data from Bordoloi and Pearce, 1978).

Storage Effect

The DSC thermo-gram of sulfur-DCPD products made with 6 hr of reaction time at 140°C and after 6 months storage under ambient conditions (Figure 5.10) indicate that orthorhombic/monoclinic sulfur crystallized out in decreasing amounts with an increasing amount of DCPD, and no orthorhombic or monoclinic sulfur crystallized out from feeds of approximately equi-molar composition (34 wt% DCPD).

A correlation between the feed composition, melt viscosity at the end of the reaction at 140°C, and the crystallization of sulfur is shown in Table 5.3. The high viscosity for the equi-molar feed composition shows high conversion of sulfur to sulfur-DCPD co-polymer from which no crystallization of sulfur takes place. However, the sulfur-DCPD co-polymer materials thus obtained are brittle. It

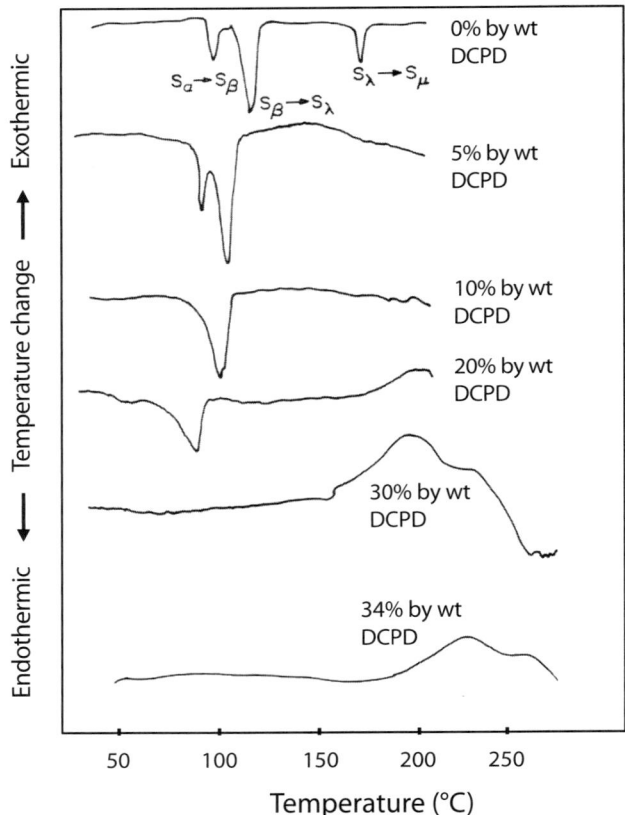

Figure 5.10. DSC thermo-grams of the sulfur-DCPD reaction product with varying feed composition after 6 hr of reaction at 140°C and after 6 months of storage under ambient condition (adapted from Bordoloi and Pearce, 1978).

appears that if a proper plasticizer can be developed, DCPD can be used successfully to stabilize and plasticize sulfur by co-polymerization with DCPD.

Limitations

Modification of sulfur by reaction with DCPD has been investigated (Leutner and Diehl, 1977; McBee et al., 1981a, b), but its practical use in commercial application has been limited because the reaction between sulfur and DCPD is exothermic and requires close control, and it is unstable when exposed to high temperature (> 140°C) especially when mixing with hot aggregates. It may react further to form an unstable sulfur product.

Table 5.3. Relationship of the sulfur-DCPD composition, melt viscosity at 140°C, and DSC thermal transition of the orthorhombic/monoclinic sulfur (Bordoloi and Pearce, 1987)

DCPD wt%	Melt viscosity at 140°C at 6 hr (cP)	DSC-endothermic of S_α / S_β transition of the product after 6 months storage
0	20	present
5	58	present
10	612	present
20	68000	present
30	>100000	absent
34	>>100000	absent

McBee and Sullivan (1982a) solved this problem through the development of a process for preparing modified sulfur that is stable and not temperature sensitive in mixing temperature range for production of sulfur concrete. This process utilizes the control reaction of cyclopentadiene.

5.5.2 Sulfur Modified with DCPD and an Oligomer of Cyclopentadiene

Chemical Reaction

Normally, the reaction is conducted without the presence of a solvent. However, such hydrocarbon materials as vinyl-toluene or styrene can be used as solvents. The reaction between sulfur and cyclopentadiene dimer to form the modified sulfur component of the cement must be carefully controlled because of the exothermic reaction between sulfur and DCPD. Cyclopentadiene is commercially available in the form of the dimer thereof. Liquid cyclopentadiene dimer will spontaneously de-polymerize to the monomer at room temperature. This de-polymerization reaction will accelerate greatly in the presence of sulfur at an elevated temperature of 120 to 140°C as shown by the chemical scheme represented in reaction numbers I, II and III in Figures 5.11 to 5.13, respectively (Sliva, 1996). Because of the exothermic nature of reaction I (Figure 5.11) it is difficult to control. However, when the dimer is present in the reaction mixture, the dimer reacts with the polysulfide product formed in reaction I as shown in reaction II (Figure 5.12).

Reaction II between cyclopentadiene dimer and the polysulfide-cyclopentadiene product is significantly less exothermic than reaction I. However, it is still difficult to control. It is evident that the combined exo-thermicity of reactions I and II poses significant control problems. If control of the reactions is not main-

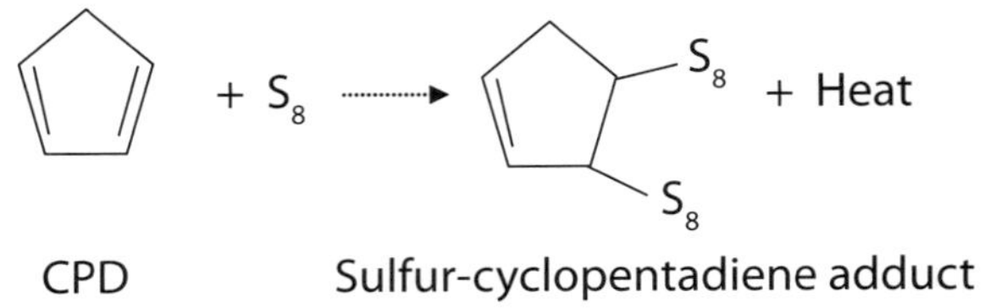

Figure 5.11. Reaction I between sulfur and cyclopentadiene dimmer; it is highly exothermic (McBee and Sullivan, 1979, 1982a).

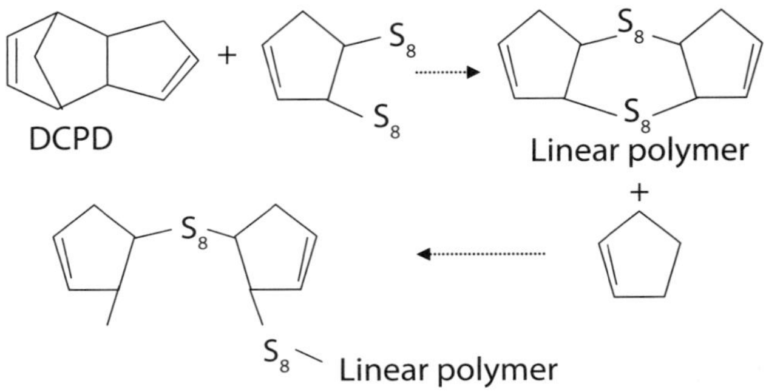

Figure 5.12. Reaction II between cyclopentadiene dimer and the polysulfide-cyclopentadiene product is significantly less exothermic than Reaction I (McBee and Sullivan, 1979, 1982a).

tained, extensive apparatus damage will occur and an undesirable, highly viscous rubber-like polymer is formed. On the other hand, when control of the reactions is maintained, the reactions result in the formation of linear polymeric polysulfides that are the essential components of the durable sulfur polymer cement (McBee and Sullivan, 1979, 1982a).

The desired control of the above exothermic reactions is achieved by conducting the reaction between sulfur and DCPD in the presence of a quantity of cyclopentadiene oligomer sufficient to achieve the desired linear polysulfide polymeric products.

Generally, the amount of oligomer in the reaction mixture, based upon the total amount of cyclopentadiene present, ranges from 10 to 100 wt%. If low levels of the reaction mixture, i.e., < 10%, are used in the reaction with sulfur, it may be necessary to add small amounts, i.e., 10 to 30%, of DCPD to the mixture in order to form the initial polysulfide-cyclopentadiene product that, in turn, reacts with the oligomer, although for some cements as little as 5% of the reaction mix-

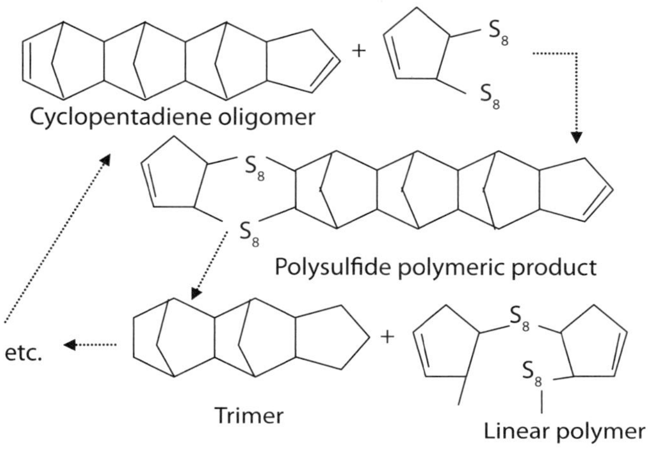

Sulfur Cement   161

Figure 5.13. Reaction III between oligomer and sulfur-cyclopentadiene product. It is non-exothermic reaction (McBee and Sullivan, 1979, 1982a).

ture can be reacted with sulfur. This is necessary because the oligomer contains minimal cyclopentadiene monomer. The reaction mixture generally contains 25 to 50% oligomer and 50 to 75% DCPD. With quantities of the reaction mixture above 10%, the reaction between sulfur and oligomer proceeds spontaneously because small but sufficient quantities of cyclopentadiene monomer are present to promote the reaction. The reaction between the oligomer and sulfur-cyclopentadiene product is illustrated in reaction III (Figure 5.13).

The reaction between the sulfur-cyclopentadiene product and the oligomer exhibits low exothermicity because the oligomer breaks down slowly to the final state of DCPD. The oligomer is used in the present reaction to moderate the polymerization of sulfur with cyclopentadiene.

In the reaction between sulfur and cyclopentadiene, virtually any source of cyclopentadiene-oligomer can be used. These sources range from virtually pure cyclopentadiene oligomer mixtures to oligomer sources contaminated with other olefinic materials. Normally, cyclopentadiene oligomer is obtained from the production of DCPD resin as steam oils. These oils are the generally undesirable low molecular weight components of the system that are commonly disposed of as a fuel.

In the manufacture of DCPD resins, generally a crude form of DCPD liquid is used as a feedstock for the reaction and is blended with crude vinyl aromatic streams rich in styrene, indene and alphamethylstyrene, as well as vinyl-toluene with a 30 to 40% pure liquid DCPD before polymerization. Thus, the actual sulfur-containing polymer material consists of DCPD co-polymers of vinyl aromatic compounds and some mixed vinyl aromatic polymers. A typical oligomer starting material is one that contains the following constituents: 5% cyclopentadiene, 10% each of dimer and trimer, 20% tetramer, 45% pentamer, and 10% traces of higher polymers such as alkyl naphthalenes, vinyl DCPD aromatic copolymers.

The reaction between sulfur and the combination is conducted at a temperature and for a time sufficient to promote and complete the reaction, which is normally from 120 to 160°C, and a reaction time within the range of 1 to 15 hr. When the proper combination of oligomer and DCPD is employed, the exothermicity of the reaction between sulfur and cyclopentadiene material can be controlled. In other words, the violent mass reaction and exo-therm exhibited on reacting sulfur with DCPD alone can be smoothly modulated into a controlled reaction by using cyclopentadiene oligomer with DCPD.

The Bureau of Mines initially studied this material to determine its value in repairing construction materials in bridges and roads. They studied and overcame shrinkage problems associated with cooling by finding admixtures that prevented this phenomenon. They added 5% admixtures (i.e., DCPD and oligomers of cyclopentadiene added in equal amounts) to the sulfur and created modified sulfur cement. This material possesses several interesting properties, the most important of which are its rapidly achieved high mechanical strength and its high resistance to corrosive environments. The product of the reaction is liquid above 115°C and, therefore, can be handled in liquid form at or above this temperature. The product cement is thermoplastic and solidifies below 115°C. The cement product can be stored in any type of storage container such as plastic or paper bags and any type of metal, glass, or plastic container.

Materials and Properties

The following results are based on the study of McBee and Sullivan (1983). The sulfur used is commercial-grade flake sulfur (99.9% minimum purity) from a secondary source. Technical grade DCPD was used to modify the sulfur together with an oligomer mixture of cyclopentadiene. The oligomer mixture used is the oligomer product known as steam sparge oil obtained from the production of DCPD resin. A typical oligomer starting material has the following composition: 5% cyclopentadiene, 10% each dimer and trimer, 20% tetramer, 45% pentamer and 10% of higher polymer such as alkyl naphthalenes and vinyl DCPD aromatic co-polymer.

The results in Table 5.4 show the amounts of modifier combined with sulfur and the percentages of DCPD and oligomer in the modifier in a series of test batches. The reaction in each between the sulfur and the modifier was conducted at 135 to 140°C. In each test batch sulfur was combined with 25 lb. of sulfur

Table 5.4. Test conditions with physical observation (McBee and Sullivan, 1983)

Modifier mixture (%)		Modifier concentration	Type	
DCPD	Oligomer	%	Cement	Observation
0	100	2–10	Rigid	Negligible reaction; no exothermic reaction
34	66	2–8	Rigid	Complete reaction; no exothermic reaction
50	50	2–8	Rigid	Complete reaction; no exothermic reaction
50	50	10	Rigid	Complete reaction; slight exothermic reaction
65	35	2–5	Rigid	Complete reaction; slight exothermic reaction
75	25	2–5	Rigid	Complete reaction; significant exothermic reaction
40	60	10–20	Flexible	Complete reaction; no exothermic reaction

Note that the percent of DCPD and oligomer expressed are of crude olefin mixtures.

cement product (present as an initiator for the reaction). Oligomer was then added to the reaction in three increments. Each addition of oligomer caused the temperature of the mixture to drop to approximately 125°C so that the temperature of the mixture was allowed to increase to 135°C before more oligomer was added. Therefore, the DCPD component was added to the reaction mixture in two increments the result of which was to increase the temperature of the mixture to 135°C after each addition. After completion of the reaction, the material was drained from the reactor into storage drums.

The total time for preparing the 500 lb test batches was 30 hr from the first addition of oligomer in order to assure a complete reaction and stability of the binder, as well as to obtain uniformity between batches for comparison purposes. The viscosity of each batch was monitored after the additions of all the materials and the reaction temperatures were held within ±5°C until completion.

The data shown in Table 5.4 suggest the following:

a. Oligomer alone does not react with sulfur at temperatures up to 180°C. Above 10% DCPD content, the reaction is complete.
b. When the modifier is present in amounts < 10%, at least one half of the modifier must be DCPD in order to obtain a complete reaction.

Table 5.5. Effect of modifier loadings on physical properties (McBee and Sullivan, 1983)

Modifier concentration %	Modifier mixture (%)		Softening point (°C)	Specific gravity	Viscosity @ 135°C
	DCPD	Oligomer			
5	100	0	>82	1.905	>450
5	50	50	>82	1.899	28
10	40	60	>82	1.818	40
20	25	75	>82	1.765	92
30	25	75	35	1.667	108
40	25	75	38	1.498	155

Note that the percent of DCPD and oligomer expressed are of crude olefin mixtures.

The modifier must contain at least 37% oligomer to control the exothermic reaction.

c. If the modifier concentration ranges between 10 and 20%, at least 60% of the modifier must be oligomer to adequately control the reaction temperature.

Furthermore, the results shown in Table 5.5 suggest that for the reaction between 5% DCPD (no oligomer present) and 95% sulfur, the viscosity (at 140°C) of the sulfur cement continues to increase during the reaction. This behavior results in a cement product that is virtually useless. For other samples that contain from 5 to 20% modifies, the softening point for the rigid cement product obtained in each case is > 82°C.

Thermal expansion properties of the tested sulfur and modified sulfur cements are shown in Table 5.6. The results indicated that the 50% oligomer cement exhibits the lowest thermal expansion coefficients of the cements tested.

5.5.3 Sulfur Modified with Styrene

Effect of Concentration and Reaction Time

The reaction of sulfur with styrene at 140°C gives a mixture of polysulfide and free elemental sulfur that held as a mixture of monoclinic, orthorhombic, and noncrystalline sulfur (Figures 5.14 and 5.15), depending on the styrene percentage and heating time (Blight et al., 1978).

Table 5.6. Thermal expansion coefficients ((McBee and Sullivan, 1983)

Sulfur cement	Thermal expansion coefficient (in/in °C)	Temperature range (°C)
Sulfur	46×10^{-6}	25–95
Sulfur	1000×10^{-6}	98–108
50% DCPD: 50% oligomer	59×10^{-6}	25–100
65% DCPD: 350% oligomer	97×10^{-6}	25–83
100% DCPD	98×10^{-6}	25–85

Materials formulated using 5% commercial grade chemicals.

Figure 5.14. shows the composition of sulfur materials modified by 25% of styrene as a function of heating time after storage for 18 months at ambient temperature. Figure 5.15 shows the composition of sulfur materials modified by styrene with different loadings after a heating time of 3 hr and storage for 18 months at ambient temperature.

The reaction of sulfur with styrene is exothermic; a temperature rise was observed approximately 30 minutes after addition (Blight et al., 1978). This rise is associated with a viscosity increase. However, in contrast to DCPD reaction, the viscosity decreased within 2 hr and the reaction mixture was stirred easily thereafter. Each reaction product is completely soluble in carbon disulfide and chloroform. Thus, after 45 min at 140°C, the 25% styrene reaction product results in a product of which 39% is insoluble in chloroform. This insoluble fraction comprises nearly the whole of the polysulfide fraction. But all longer heating times produce a completely soluble product. Thus the reaction produces initial formation of high molecular weight material (Currell et al., 1974), which at 140°C progressively de-polymerizes. This effect is shown also in the viscosity changes and in the molecular weight distribution of the chloroform soluble component.

Viscosity

As a consequence of the chemical modification, the formation of new compounds at certain amounts by the polymerization of sulfur, and especially the combination of it with the polymeric substances added as modifiers, the sulfur properties are affected, presenting a higher viscosity than unmodified one. This fact has an important effect in the crystallization of sulfur. In a more viscous liquid and one in which the molecules are more polymerized, the growth of the crystals will be more difficult and encounter more obstacles, thus, it will be slower. For example, if the styrene is added to sulfur melts, the viscosity rises to a maximum within a few minutes irrespective of the styrene concentration (Figure 5.16). After reach-

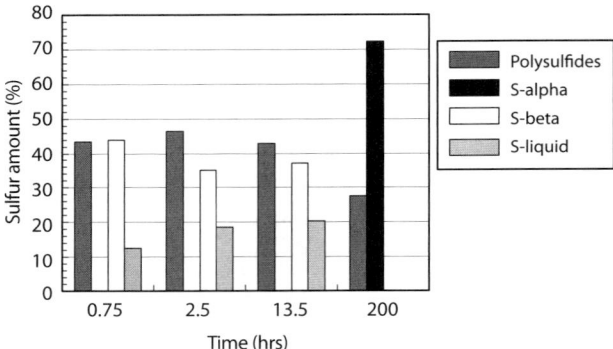

Figure 5.14. Composition of sulfur materials modified by 25% of styrene as a function of heating time after storage for 18 months at ambient temperature (data from Blight et al., 1978).

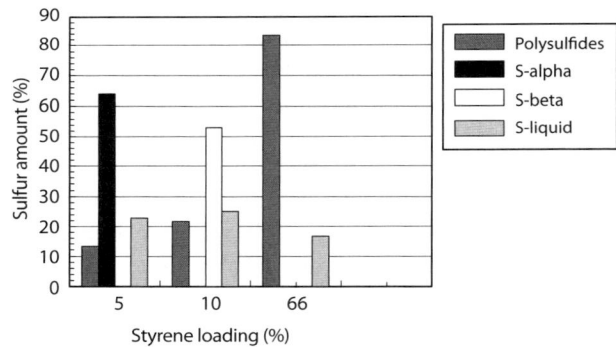

Figure 5.15. Composition of sulfur materials modified by styrene with different loadings after heating time of 3 hr and storage for 18 months at ambient temperature (data from Blight et al., 1978).

ing the maximum value, the viscosity decreases steadily to a constant final value that depends on the amount of styrene added. This characteristic behavior may be caused by thermal polymerization of the styrene followed by degradation of the polymer by the reaction with sulfur. When producing sulfur binder modifying with styrene, reaction times of at least 1.5 hr at 140°C were used. By then the viscosity maximum had been passed and a constant viscosity value had been reached. Sulfur melts plasticized by styrene have a comparatively high melt viscosity, making homogeneous mixing and processing especially difficult.

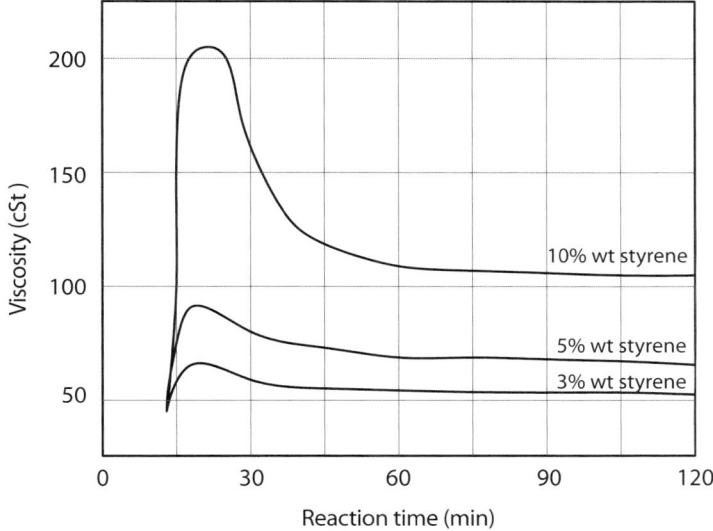

Figure 5.16. Variations of viscosity of sulfur styrene products with reaction time and styrene content (adapted from Gregor and Hackl, 1978).

Reaction Products

As the reaction time increased, with styrene loading of 25%, the proportion of the lower molecular weight component increases. Some components were identified based on elemental analysis, melting point, infrared, mass, and nuclear magnetic resonance spectroscopy.

The reaction products (Figure 5.17) appear to be chain styrene polysulfide of varying molecular weights. The two oligomers I and II have been obtained (Blight et al., 1978). On standing for one month under ambient conditions, oligomer II partially decomposed to 2,4-diphenylthiopene. Determination of molecular weight gives a degree of polymerization of approximately 5 and determination of the sulfur rank gives an average sulfur chain length of 4.4 (Blight et al., 1978). $LiAlH_4$ hydrogenation gave phenyl ethane-1,2-dithiol and 1-phenylethane-thiol, suggesting an average structure for oligomer III (Blight et al., 1978).

5.5.4 Sulfur Modified with a Combination of DCPD and Styrene

It has been found that the use of styrene as a plasticizer for the sulfur, in addition to the DCPD, generally results in further increased strength of the composite. Thus, when a higher strength material is desired, a combination of DCPD and up

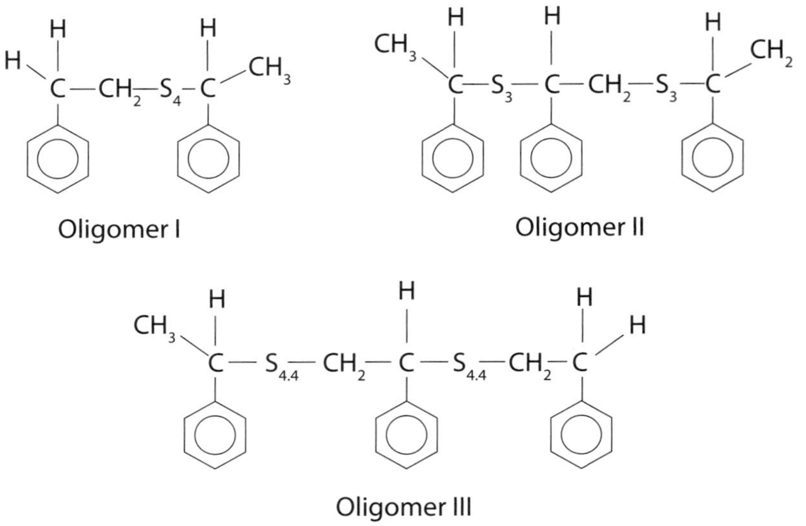

168 Sulfur Concrete for the Construction Industry

Oligomer I

Oligomer II

Oligomer III

Figure 5.17. Possible reaction products due to sulfur modification with styrene. Products appear to be chain styrene polysulfide of different molecular weights (Blight et al., 1978).

to 5 % styrene, based on the composite mixture, is used in plasticization reaction (McBee et al., 1977).

5.5.5 Sulfur Modified with Olefinic Hydrocarbon Polymers

Olefinic hydrocarbon polymer material is derived from petroleum and has a nonvolatile content > 50% by weight and a minimum Wijs iodine number of approximately 100 cg/g capable of reacting with sulfur to form a sulfur-containing polymer. Typically, the chemical stabilizer is used in amounts up to 10% by weight of the total sulfur, and especially in the proportion of about 1 to 5% of the total sulfur by weight. The amount of such a chemical stabilizer required depends on the end use of the cement and the properties desired.

Typical chemical stabilizers are those known by the trade names of RP220 and RP020, both products of Exxon Chemical.; CTLA, a product of Enjay Chemical; and Escopol, a product of Esso Chemical AB (Sweden). All of these approximately identify a heat reactive olefinic liquid hydrocarbon obtained by partial polymerization of olefins. The chemical stabilizer can be incorporated into the final sulfur cement mix by several reaction routes.

The hydrocarbon stabilizer (at approximately 25°C) was added to molten sulfur (at approximately 140°C) with vigorous stirring. Heat was applied only to maintain a reaction temperature of 140 to 150°C. At this temperature, reaction

Table 5.7. Characteristics of stabilizers (Vroom, 1981)

Properties	RP220	RP020	CTLA	Escopal
Flash point	150	138	150	125
API gravity	4	4	9.6	3
Iodine number (100cg/g) minimum	200	160	225	135
Non-volatile matter (wt.%) minimum	80	70	83	75
Density (gm/cc)	1.05	1.04	1.00	1.03
Viscosity (cSt/100°C)maximum	25	26	28	25

times were 15 to 30 min. The progress of reaction could be monitored by the degree of homogeneity of the mix, by careful observation of the temperature of the reaction mixture, or by observation of the increasing viscosity of the mixture. At sulfur stabilizer ratios of less than 4:1 by weight, control of the addition rate is required to prevent the exothermic reactions raising temperatures above 155°C, at which point hydrogen sulfide (H_2S) was evolved with subsequent foaming and degradation of the product. When reactions were completed, the product was a sulfur-containing polymer that, on cooling, possessed glass-like properties that were retained indefinitely. The properties of the olefinic hydrocarbon polymers used are given in Table 5.7 (Vroom, 1981). Reaction conditions for the preparation of seven containing polymers are shown in Table 5.8 (Vroom, 1981).

5.5.6 Sulfur Modified with Bitumen

Mohamed and El Gamal (2006, 2007, 2008, and 2009) reported that bitumen could be a viable modifying agent to molten sulfur. In their study, bitumen having the physico-chemical properties of: softening point of 48.8°C, specific gravity at 20°C of 1.0289 g/cm^3 and kinematics viscosity at 135°C of 431 cSt; a chemical analysis of C: 79, H: 10, S: 3.3, and N: 0.7% was used. The preparation method as described by Mohamed and El Gamal (2006) is detailed. In an oil bath 2.5 wt% of hot bitumen and 97.5 wt% molten sulfur containing emulsifying agent (0.025 gm/100 gm) were mixed and mechanically vigorously stirred at 140°C. At this temperature, reaction periods were in the order of 45 to 60 min. The progress of the reaction was monitored by recording the changes of temperature, viscosity, and degree of homogeneity of the mixture. After heating, samples were cooled at a rate of 8 to 10°C/min. The product is a sulfur containing polymer that on cooling possessed glass-like properties. The composition and properties of modified sulfur cement has been obtained in terms of chemical analysis, FT-IR spectra, x-ray diffraction, scanning electron microscopy, and thermal processing, including TGA and DSC.

Table 5.8. Reaction conditions for the preparation of sulfur-containing polymers stabilizer (Vroom, 1981)

Type of chemical stabilizer	Chemical stabilizer (wt%)	Sulfur (wt%)	Temperature (°C)	Time (min)	Color
CTLA	12.5	17.5	140	30	Dark brown
CTLA	8.3	41.7	140	40	Dark brown
Escopol	10.0	40.0	150	15	Light brown
Escopol	14.5	35.5	140	20	Light brown
RP220	8.3	41.7	150	15	Dark brown
RP220	12.5	37.5	140	20	Dark brown
RP020	12.5	37.5	140	15	Dark brown

Table 5.9. Chemical composition of the sulfur bitumen mixture (Mohamed and El Gamal, 2006)

Composition		Properties	
Material	Proportion, %		
Sulfur	97.4 ± 0.48	Specific gravity	1.89 gm/cm^3
Carbon	1.981 ± 0.074	Viscosity at 135°C	25 cP
Hydrogen	0.1 ± 0.067		

Physico-Chemical Composition

The chemical composition of sulfur bitumen mixture was determined using Finnigan Flash EA1112 CHN/S elemental analyzer for the determination of C, H, and S. The results (Table 5.9) indicated that the mixture contains various organic compounds (related to the presence of C, H). The manufacture's certification of sulfur bitumen cement analysis, listed in Table 5.9, agrees with the ASTM C1159 requirements for sulfur polymer cement chemical and physical properties.

Chemical Reactions

The addition of bitumen to sulfur in amounts of 2.5% initiates chemical reactions whose type depends on the bitumen content, heating temperature, and the time of the reaction. Some competing reactions could occur, including those with bitumen incorporation into sulfur molecules or dehydrogenation with the liberation of hydrogen sulfide. As discussed in Chapter 3, at T < 95°C sulfur exists as a cyclooctasulfane crown with an S-S bond length of 0.206 nm and an S-S-S bond angle of 108 degree; and at T < 119°C sulfur crystallizes. At 119°C (the melting point of sulfur), liquid sulfur is dispersed thoroughly in bitumen, forming an emulsion (role of emulsifying agent) and cyclooctasulfane turns partly into polymeric zigzag chains (bond length 0.204 nm) (Voronkov et al., 1979).

The crystallization features are affected by such factors as the chemical reaction of sulfur with bitumen components and its dissolution or dispersion. At the heating temperature of T < 140°C, elementary sulfur forms polysulfide that initiates formation of a network. Such structures differ considerably in the chemical and thermal stability from unmodified sulfur. At 119 to 159°C, molten sulfur essentially exists as cyclooctasulfane (λ-S). Above 159°C, an eight-member ring rapidly breaks down into bi-radicals. In their turn, bi-radicals recombine to form polymeric chains with the maximal length of approximately 10^6 sulfur atoms:

$$S_8 \Leftrightarrow S^0 - S_6 - S^0$$

$$S^0 - S_6 - S^0 + S_8 \Leftrightarrow S^0 - S_6 - S^0 - S_8 \quad \text{etc.}$$

However, above 140°C, de hydrogenation of saturated bitumen components can occur; also linear polysulfide can transform into stable cyclic thiophene structures (Syroezhko et al., 2003). It is noted that modified sulfur should be produced within its recommended mixing temperature range of 135 to 141°C.

Composition Analysis via FT-IR

To investigate the nature of chemical interactions between bitumen and sulfur, the completion of the bitumen sulfur reaction was strongly supported by FT-IR spectra measurement of the characteristic double bonds in bitumen that is the main modifying bitumen constituent. Figure 5.18 shows the band at 2975 cm^{-1} which is consistent with CH stretching. Since C-H is associated with C=C double bonds, the disappearance of C-H bonds in the modified sulfur spectra suggested the consumption of aliphatic C=C bonds that generally leads to the polymerization of sulfur with bitumen. Negative peak in 2400 cm^{-1} indicates removal of NC=O group indicating the presence of chemical reaction sites for polysulfide formation. An increase in relative intensity in spectra at 584 cm^{-1} is corresponding to disulfide S-S bond between adjacent thiol groups.

Composition Analysis via XRD

Powder XRD is an effective and accurate method for identifying the various crystalline phases in modified sulfur cement samples. The XRD diagrams for pure elemental sulfur and sulfur modified with a small amount of bitumen (2.5 wt%) is shown in Figure 5.19. XRD analysis confirms that the modified sulfur crystallizes in a structure different than the unmodified one. The XRD pattern of the elemental sulfur and sulfur modified with bitumen crystals are particularly noticeable.

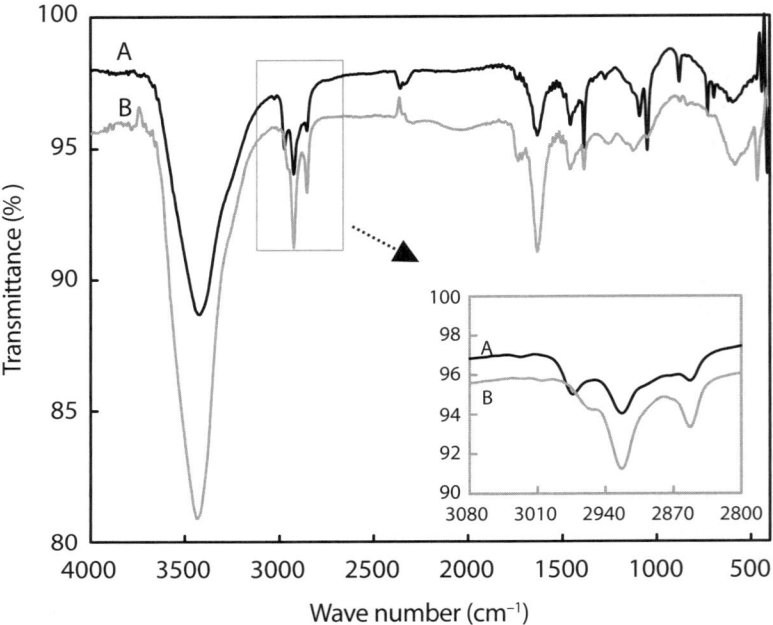

Figure 5.18. FTIR spectrum showing (a) bitumen spectra and (b) sulfur modified with bitumen at a loading of 2.5 wt.% and viewing the disappearance of CH characteristic double bond of the bitumen after the complete reaction with sulfur (Mohamed and El Gamal, 2006).

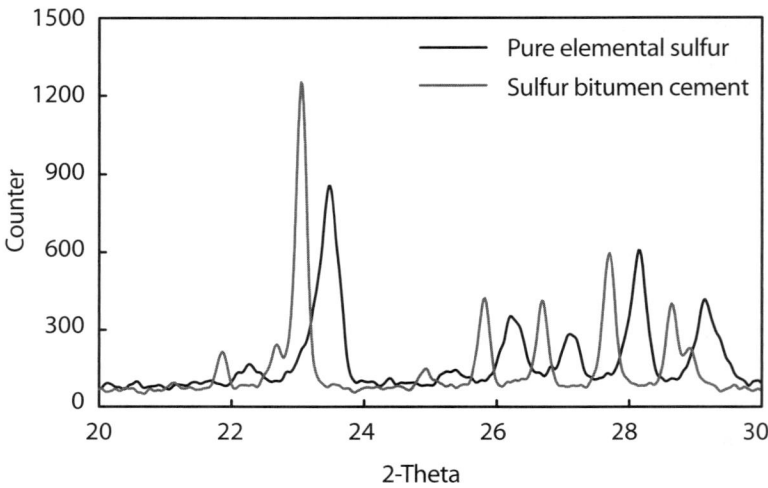

Figure 5.19. X-ray diffraction patterns for pure sulfur and sulfur bitumen cement (Mohamed and El Gamal, 2006).

Modified sulfur cement shifted to lower 2-theta that, in turn, indicates that the newly formed structure is finer in its microstructure.

Microstructure

Distribution of bitumen in sulfur and a free sulfur crystal type (orthorhombic or monoclinic) was evaluated using scanning electron microscopy, JSM-5600 Joel microscope, equipped with an energy dispersive x-ray detector (EDX). The specimen was sputter-coated with 12 nm gold to render them conductive. Scanning with an electron microscopic has revealed how the bitumen controls the crystallization of sulfur. Pure sulfur crystallizes and forms dense and large alpha sulfur crystals (S_α) with orthorhombic sulfur morphology as shown in Figure 5.20.

Orthorhombic sulfur is the stable form of sulfur at room temperature and atmospheric pressure. Orthorhombic sulfur is also called rhombic sulfur. With the addition of bitumen, the crystal growth is limited and controlled by the bitumen in such a way that all crystals are plate-like and of micron dimension as monoclinic sulfur crystal of beta form (S_β) as shown in Figures 5.21 to 5.23.

The formed microstructure confirmed the efficiency of bitumen in inhibiting the formation of orthorhombic sulfur, and the interaction between bitumen and sulfur gave a mixture of monoclinic sulfur and polysulfide. The presence of uniformly dispersed polymeric material in sulfur is evident, as seen by the formation of uniformly distributed monoclinic sulfur crystals and the bonds between these crystal plates. The smallest microstructure helps to resist cracking and to tolerate any thermal expansion. The distribution of organic components of bitumen, due to sulfur modification, can be detected by means of SEM and x-ray mapping of the elemental distribution in the hardened mortar as shown in Figures 5.24a and 5.24b. The formed microstructure is homogenous and much like sea shells.

The role of bitumen modification in sulfur modified cement is not one of stabilizing the crystalline phases, as claimed by McBee et al. (1982 and 1983), but controlling of the microstructure. The interlocked plate-like micro-sized microstructure helps to resist cracking and tolerate any thermal expansion mismatch in sulfur modified cement.

Degree of Sulfur Polymerization

In the process of producing modified sulfur, an olefin hydrocarbon polymer (bitumen), in combination with an emulsifying agent, modifies elemental sulfur by inducing sulfur polymerization. Thus, the resulting modified sulfur cement comprises polymerized sulfur. When it is present, the sulfur phase transformation (β to α) still occurs during cooling, but the polymerized sulfur acts as a

Figure 5.20. SEM of typical crystalline morphology of pure elemental sulfur (Mohamed and El Gamal, 2006).

Sulfur Cement 175

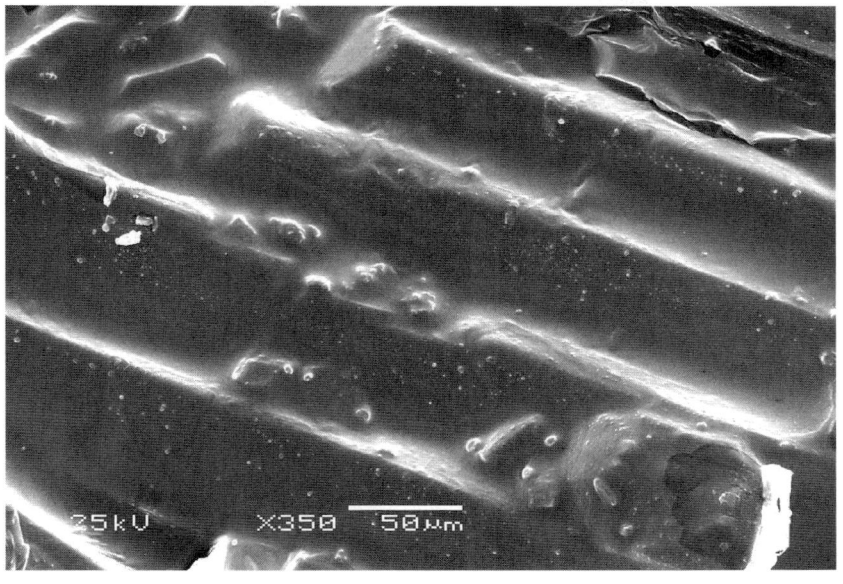

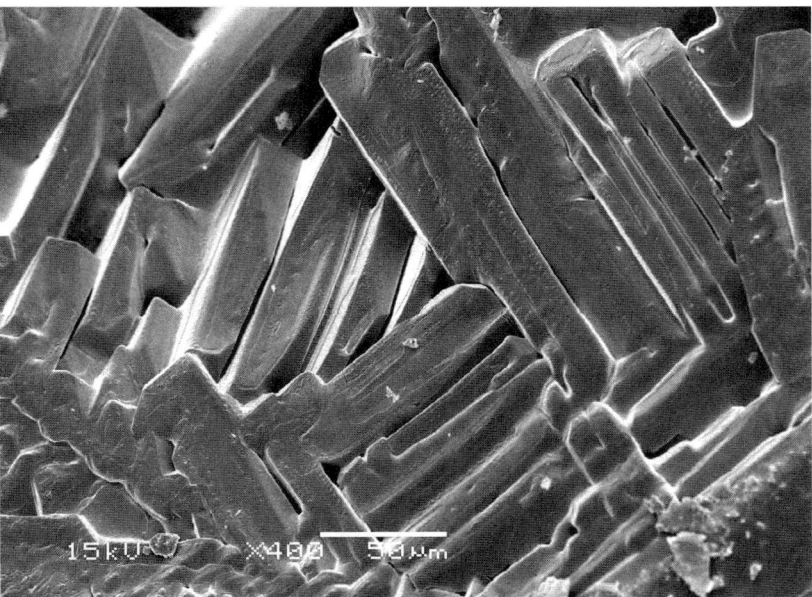

Figure 5.21. SEM micrographs for sulfur modified with 2.5 wt% bitumen at different magnifications (Mohamed and El Gamal, 2006).

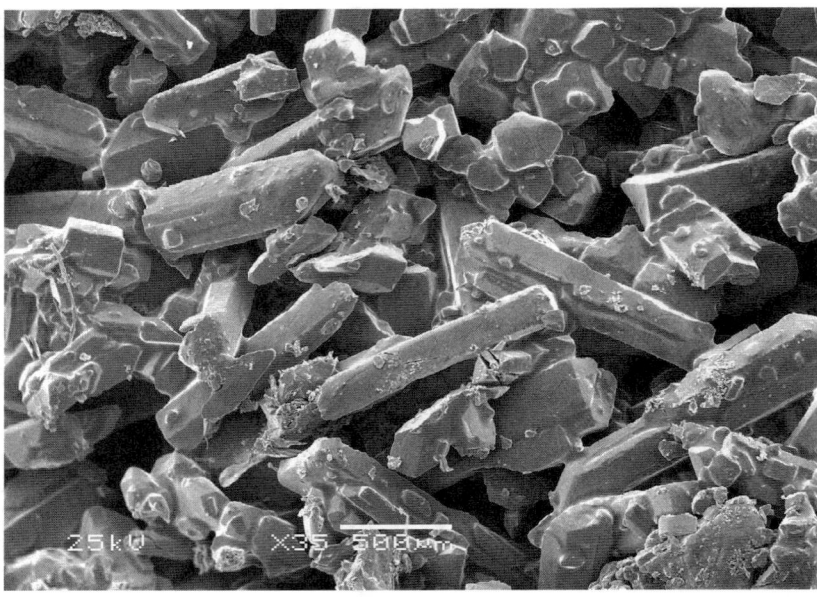

Figure 5.22. SEM micrographs for sulfur modified with 2.5 wt% bitumen, indicating that plate-like crystals of monoclinic beta sulfur crystal (S_β) were revealed and the polymer was likely present between plates and plate joints (interior images). (Mohamed and El Gamal, 2006).

Sulfur Cement 177

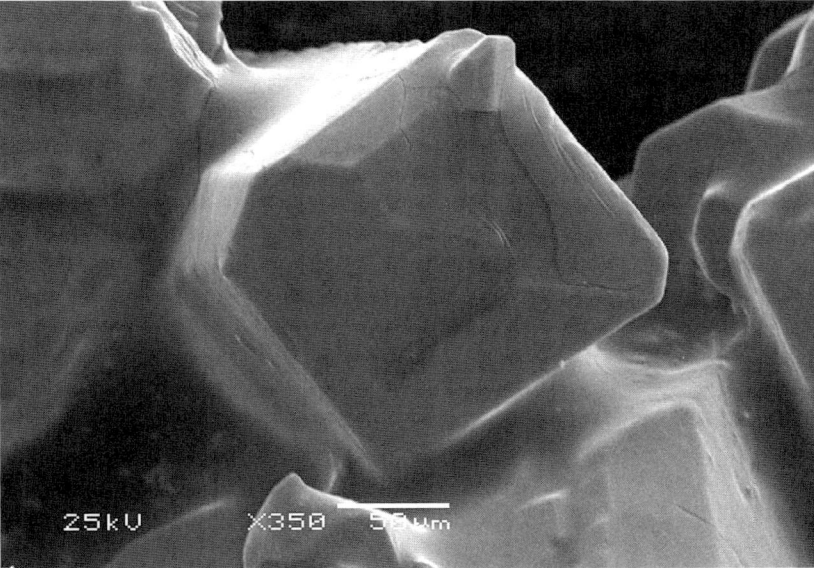

Figure 5.23. SEM micrographs of sulfur modified with 2.5 wt% bitumen showing interior images of plate-like crystals of monoclinic beta sulfur crystals (S_β) (Mohamed and El Gamal, 2006).

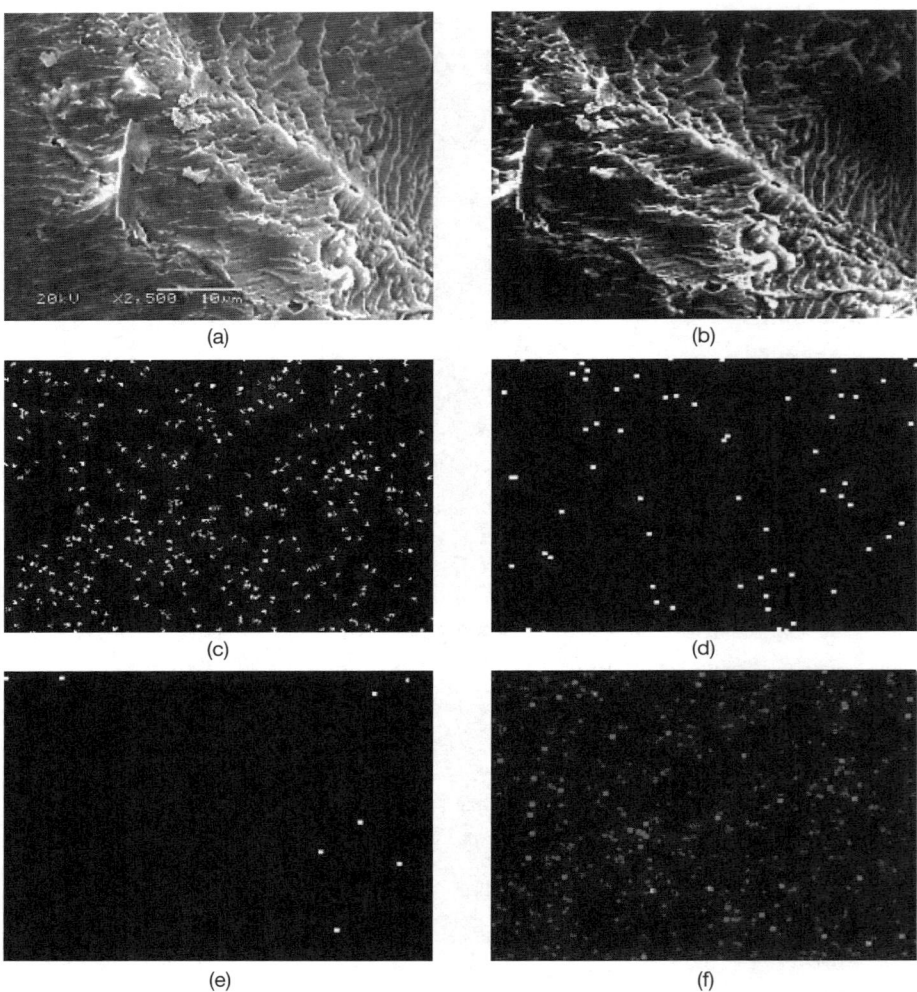

Figure 5.24a. SEM and EDX of modified sulfur: (a) secondary electron image at magnification of X2, 500; (b) back scattered electron image; (c) EDX sulfur distribution map; (d) EDX carbon distribution map; (e) EDX oxygen distribution map; (f) EDX sulfur, carbon, and oxygen distribution for the dominant microstructure of modified sulfur, indicating the homogeneity of the formed microstructure. These are similar to sea shells. (Mohamed and El Gamal, 2006).

Sulfur Cement 179

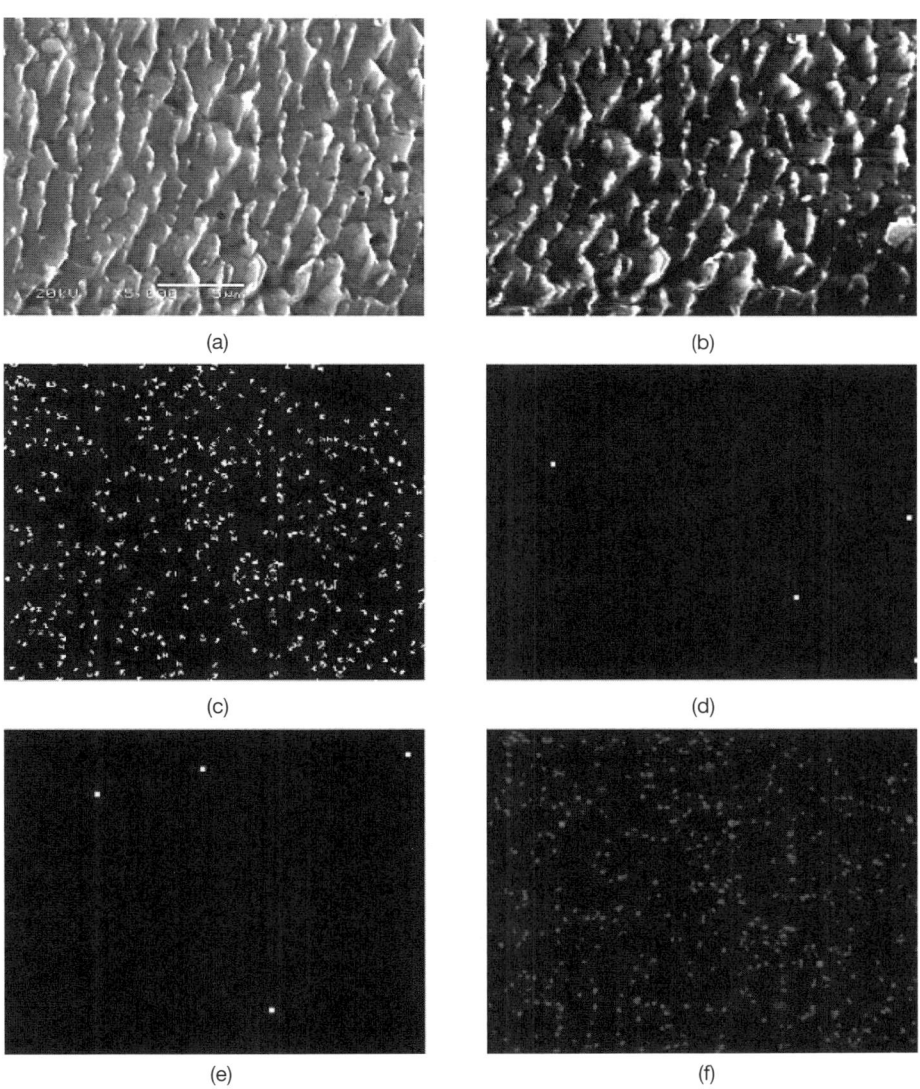

Figure 5.24b. SEM and EDX of modified sulfur: (a) secondary electron image at magnification of X5, 000; (b) back scattered electron image; (c) EDX sulfur distribution map; (d) EDX carbon distribution map; (e) EDX oxygen distribution map; (f) EDX sulfur, carbon, and oxygen distribution for the dominant microstructure of modified sulfur, indicating the homogeneity of the formed microstructure. These are similar to sea shells. (Mohamed and El Gamal, 2006).

compliant layer between the sulfur crystal, and serves to mitigate the effect of the phase transformation.

Modification of sulfur with bitumen depends on the degree to which they can disperse in each other and on how they interact. Types of interaction are: *pi-pi* bonding, polar or hydrogen bonding (polar interactions of hetero atoms), and Van Der Waals forces. A self-compatible mixture consisting of a variety of molecular species that are mutually dissolved or dispersed, typically this combination contains a continuum of polar and nonpolar material. This leads to areas of order or structure of polysulfides in the modified sulfur, depending on the amount of the polymer present, the reaction time, the reaction temperature, and the cooling rate.

Sulfur modified with the bitumen analysis test indicates that modified sulfur is comprised of 55 wt% monoclinic sulfur and 45% wt polysulfide, based on the total weight of the sulfur component. The structure of polysulfide was confirmed by analyzing the fraction that insoluble in CS_2 by HPLC column chromatograph. Data analysis indicated the presence of low- and high-molecular weight fractions of polysulfides with a weight average molecular weight of 17,417 and an average number molecular weight of 344. The poly-disperseability index that is a reflection of the broadness of the product molecular weight was determined to be 5.

As a consequence, sulfur's rheological properties are affected, hence, present a higher viscosity than an unmodified one. This fact has an important effect in the crystallization of sulfur. In a more viscous liquid, and one in which the molecules are more polymerized, the growth of the crystals will be more difficult and, thus, it will be slower.

Thermal Analysis

Thermal analysis techniques such as TGA and DSC have been used effectively for determining the temperature history of sulfur and sulfur bitumen mixtures (Mohamed and El Gamal, 2006). The TGA measures the weight loss of a material from a simple process such as drying or from more complex chemical reactions that liberate gases and structural decomposition. A thermal gravimetric analyzer (Perkin Elmer TGA7) was used to measure weight changes in sample materials as a function of the temperature. A furnace heats the sample while a sensitive balance monitors loss or gain of sample weight due to chemical changes and decomposition. TGA coupled with mass spectrometry and FT-infrared spectroscopy provides elemental analysis of decomposition products. Weight, temperature, and furnace calibrations have been carried out for the usable range of the TGA (100 to 600°C) at a scan rate of 20°C/min .

Figure 5.25 provides a comparison of the TGA curves of pure sulfur and bitumen modified sulfur. The maximal thermal effect was observed at 180 to 450°C

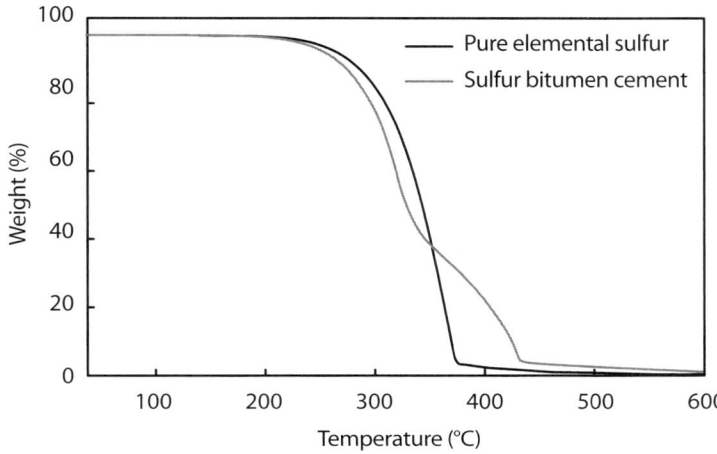

Figure 5.25. Thermal gravimetric analysis for pure sulfur and a sulfur bitumen mixture at a heating rate of 20 deg/min (Mohamed and El Gamal, 2006).

for pure sulfur that was accompanied by active liberation of hydrogen sulfide, whereas the maximal thermal effect was observed at 200 to 540°C for modified sulfur. This effect reflects the increase of thermal stability of modified sulfur.

Differential scanning calorimeter (Perkin Elmer DSC7) was used for the measurements of heat capacity through phase transitions on heating. The tested 10 mg sample was heated up to 150°C with a heating rate of 5°C/min. The sample was self cooled to room temperature for 24 hr and then reheated up to 150°C; the expansion coefficient varied between the first and second runs. DSC results of both sulfur and modified sulfur cement are shown in Figure 5.26.

As discussed in Chapter 3, for pure sulfur, the melting of S-alpha and S-beta was detected in the first run, whereas only S-alpha melting was detected in the second run. For modified sulfur, the melting of S-alpha and S-beta was detected in the first run, whereas only S-beta melting was detected in the second run. DSC thermographs, therefore, show the modified sulfur remains in the monoclinic modification of S-beta form and does not undergo a phase transformation to the orthorhombic form (S-alpha form) upon solidification. The bitumen does, however, prevent the growth of macrosulfur crystals. This is important because sulfur has a relatively high linear coefficient of thermal expansion, 7.4×10^{-5}/C and a low thermal conductivity, 6.1 gcal/cm^2/sec ($\Delta t = 1$ C), both at 40°C. When a material containing adjacent macrocrystals of sulfur is subjected to changing temperatures, there will be a constant movement between these macrocrystals, as one example, or contracts relative to its neighbor. This movement will gradu-

182 Sulfur Concrete for the Construction Industry

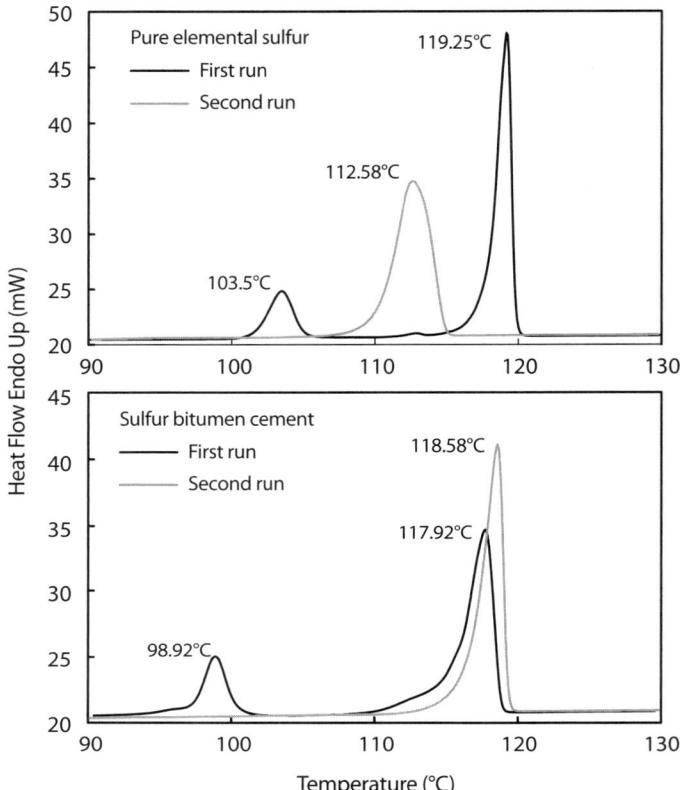

Figure 5.26. Differential scanning calorimetric results for sulfur and modified sulfur (Mohamed and El Gamal, 2006).

ally break the bonds with other cross-laid crystals, causing microfractures and the eventual formation of cleavage planes (Vroom, 1998). Because the modified sulfur does not go through the allotropic transformation upon solidification, it has less shrinkage and, hence, develops less residual stress on cooling (Lin et al., 1995).

Durability of Sulfur Bitumen Cement

It is worth noting that when molten sulfur is heated above 159°C, preferably 200 to 250°C, and then rapidly quenched to −20°C, a translucent, elastic dark brown material (plastic sulfur) is obtained (Tobolsky and MacKnight, 1965). It undergoes embrittlement, especially above −10°C, because of the rapid crystallization of octameric sulfur to orthorhombic sulfur. Also polymeric sulfur de-polymerizes

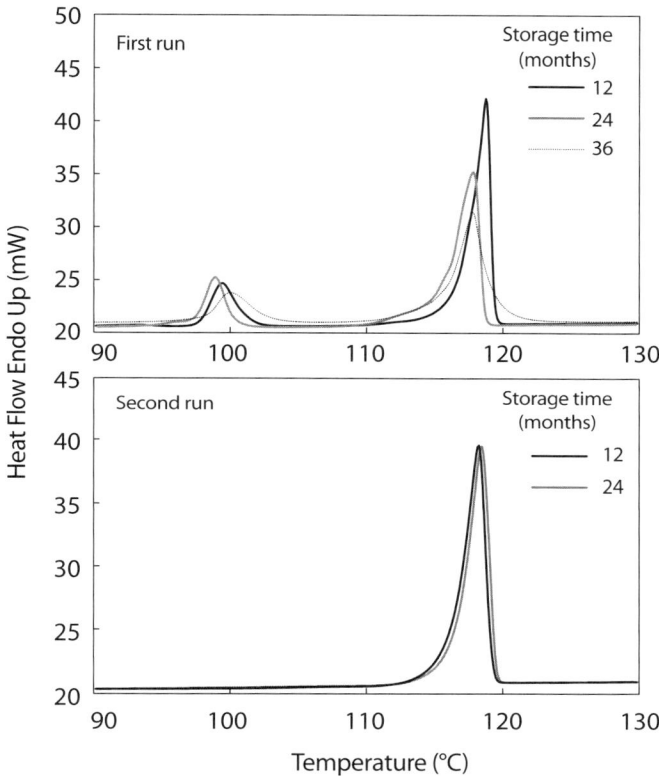

Figure 5.27. Differential scanning calorimetric results for sulfur bitumen cement, at different storage time (Mohamed and El Gamal, 2009).

and crystallizes to orthorhombic sulfur, slowly at ambient temperature and rapidly above 90°C (Tobolsky and MacKnight 1965).

The molecular weight analysis test of sulfur modified with bitumen indicates that sulfur bitumen cement contained 45% polysulfides and 55% unreacted sulfur crystallized out as monoclinic. Stability of the sulfur bitumen cement was studied by following the thermal stability at different times. The thermal stability of stored modified sulfur cement samples for 12, 24, and 36 months at room temperature was evaluated using the DSC thermograms under ambient conditions. The DSC analyses (Figure 5.27) indicated that sulfur bitumen cement is stable in the monoclinic form (S_β) and does not transform to orthorhombic sulfur (S_α) after 36 months of storage. The alpha to beta transition is one of the prime sources of failure associated with elemental sulfur cement.

5.5.7 Sulfur Modified with 5-ethylidene-2-norbornene

5-ethylidene-2-norbornene (ENB) is the most widely employed termonomer even though it is the most expensive. This is because ENB is the most readily incorporated during co-polymerization, and the double bond introduced has the greatest activity for sulfur vulcanization. Another unique characteristic of termonomer is that it makes it possible to prepare linear as well as branched polymers by varying the conditions under which the polymers are synthesized (Easterbrook et al., 1971). Branching has an important role in establishing the rheological properties of a polymer that is beneficial in certain application (Morton, 1987).

ENB is a bicyclic diene compound used as a co-polymer in the production of ethylene-propylene-diene monomer (EPDM) elastomers. ENB is manufactured by reacting 1,3-butadiene and dicyclopentadiene in a four-step process. In the first step, dicyclopentadiene is decoupled to cyclopentadiene and reacted with 1,3-butadiene via Diels-Alder condensation to vinyl-norbornene (VNB). This is followed by distillation to obtain refined VNB that is catalytically isomerizes to ENB. Finally, the crude ENB is purified.

The use of ethylidene norbornene or 5-vinyl norbornene as a sulfur modifier is well known. In a research disclosure by Reynhout et al. (USPTO Application: 20070068422–Class: 106287320, 1983), it was reported that ethylidene norbornene (ENB) may be used as a sulfur plasticizer. Plasticized sulfur was prepared by reacting elemental sulfur with 40 to 43 wt% olefinic plasticizers (as a blend, including ethylidene norbornene and 5-vinyl norbornene) based on the weight of sulfur. The resulting plasticized sulfur was a black, glassy solid and, thus, not suitable to be further processed into sulfur-bound products such as cement, mortar, or concrete.

Recently, Nippon Oil Corporation, Tokyo (JP 105-8412: EP 1961713A1 2008), reported a method for producing a binding material containing modified sulfur with ENB. The method for producing the binding material includes the steps of: (a) providing a starting material for modified sulfur composed of 100 parts by mass of sulfur and 0.1 to 25 parts by mass of ethylidene norbornene (ENB), (b) mixing the starting material in a molten state at 120 to 160°C, and (c) cooling the molten mixture down to a temperature not higher than 120°C when the viscosity at 140°C of the resulting molten mixture falls in the range of 0.050 to 3.0 Pa•s.

It has been reported that if the olefinic modifiers comprise at least 50 wt% 5-ethylidene-2-norbornene or 5-vinyl-2-norbornene and the total concentration of the olefinic modifier does not exceed 20 wt% of the weight of sulfur at any stage of the modified sulfur preparation process, a modified sulfur is obtained that, if used in sulfur-bound products, results in sulfur-bound products that are both acid and alkaline resistant. According to Shell International research

(WO/2006/134130), modified sulfur was prepared by admixing molten elemental sulfur with one or more olefinic sulfur modifiers that satisfy the following two conditions: (1) at least 50 wt% of the olefinic sulfur modifiers is 5-ethylidene-2-norbornene and/or 5-vinyl-2-norbornene and (2) the total amount of olefinic sulfur modifiers is in the range of 0.1 to 20 wt% based on the weight of sulfur. The use of this method of preparation in the production of sulfur concrete is discussed in Chapter 6.

The advantages of using ENB as a modifier are:

a. The use of 5-ethylidene-2-norbornene and/or 5-vinyl-2-norbornene as a modifier instead of the most common olefinic modifier, i.e., dicyclopentadiene, is that it is easier to process. The dicyclopentadiene dimer reverts to its volatile monomer during processing and, therefore, has to be reacted with sulfur under refluxing conditions. The reaction of either 5-ethylidene-2-norbornene or 5-vinyl-2-norbornene with sulfur can take place at a temperature below its boiling temperature and, thus, the modified sulfur preparation can be carried out without refluxing of the modifier.

b. The toxicity of either 5-ethylidene-2-norbornene or 5-vinyl-2-norbornene is much lower than that of dicyclopentadiene. Modified sulfur prepared with 5-ethylidene-2-norbornene and/or 5-vinyl-2-norbornene has high-alkaline resistance.

c. The obtained modified sulfur is light in color and can be pigmented in the desired color.

d. The melting temperature of the processed mixture is below the melting temperature of elemental sulfur.

The stability of modified sulfur by ENB was studied by x-ray spectroscopy (WAXS) as reported by Shell International B.V. (PCT/EP2006/063220). The modified sulfur was prepared by heating elemental sulfur and ENB in an amount of 5 wt% of the weight of sulfur at 140°C for one hour. The mixture was then poured into an aluminum mould and allowed to solidify at room temperature. From 30 min after pouring the mixture into the mould, the crystal structure of the resulting sample (15 × 10 mm) was analyzed by WAXS for 650 hr. The x-ray diffraction measurements showed that the modified sulfur has stable monoclinic crystals. Even after 650 hr, the monoclinic crystals (beta crystallinity) were not reverted to the orthorhombic form (alpha crystallinity).

5.6 SULFUR CEMENT

Sulfur cement is generally produced after the addition to the modified sulfur of a physical stabilizer such as fly ash or other fine substances. Therefore, sulfur cement production employs two complementary stabilizers:

a. The chemical stabilizer (sulfur modifying chemical reagent) such as DCPD, DCPD plus oligomer of cyclopentadiene, styrene, DCPD plus styrene, bitumen, RP220, RP020, CTLA, bitumen, and ENB.
b. The viscosity-increasing, surface-active stabilizer such as fly ash, furnace slag, cement kiln dust, talc, mica, silica, graphite, carbon black, pumice, insoluble salts (e.g., barium carbonate, barium sulfate, calcium carbonate, calcium sulfate, magnesium carbonate, etc.), magnesium oxide, and mixtures. Such fillers typically have a particle size less than 100 mesh (U.S. Standard Testing Sieves) and preferably less than 200 mesh. Such fillers generally act as thickening agents and improve the hardness or strength of the sulfur cement product.

Typically, the chemical stabilizer is used in amounts up to 14% by weight of the total sulfur, and in the proportion of 1 to 5% of the total sulfur by weight. The amount of such chemical stabilizer required depends on the end use of the cement and the properties desired.

The chemical stabilizer can be incorporated into the final cement mix by several reaction routes. Preferably, the chemical is pre-reacted at approximately 140°C for approximately 30 min with a smaller proportion of sulfur than is required in the final mix. The resulting concentrate can be either stored for future use or dissolved in the residual sulfur required for the final mix at the mixing temperature.

Fly ash from the burning of hydrocarbon fossil fuels and generally in the form of tiny hollow spheres called cenospheres consists of major amounts of silicon oxide and aluminum oxide, e.g., 55.61% silicon oxide, 26.30% aluminum oxide with smaller quantities of ferric oxide, calcium oxide, magnesium oxide, sodium oxide, potassium oxide, and carbon. It is particularly effective due to its small particle size, shape, and surface texture. It has been found to impart an extra measure of durability to the final sulfur cement, independent of its source, and serves the dual function of viscosity increaser and sulfur cement stabilizer. Depending upon the degree of fineness of the fly ash and the consistency desired, an amount up to one and one-half times the total weight of the sulfur may be added beneficially.

A notable feature of the sulfur cement thus modified is that it does not require high-purity sulfur and can be made with off-grade sulfur containing hydrocarbon impurities, blow dirt, and other contaminants.

The viscosity-increasing, surface-active stabilizers or fillers generally act as thickening agents and improve the hardness or strength of the sulfur cement product. When fillers are used, the sulfur cement typically contains 1 to 15% by weight and more generally, 5 to 10% of the filler based on the total weight of the sulfur. Various other additives can be added as desired to alter various properties of the sulfur cement. See for example, U.S. Pat. No. 4,188,230 (durability altered by the addition of certain petroleum products (Gillot et al., 1980) and U.S. Pat. No. 4,210,458 (viscosity altered by the addition of polyhydric alcohols (Simic, 1980).

5.7 FACTORS CONTROLLING FORMATION OF SULFUR CEMENT

The amorphous content plus the S-alpha and S-beta crystalline phase content in the sulfur modified mixtures are affected by many factors such as the type and percentage of modifiers as well as the thermal history, including the reaction and curing temperature, the curing time, and the cooling rate and storage.

Factors such as type, loading, heating time, and storage are discussed in the preceding sections and in the following section we discuss the effect of the cooling rate on the crystal formation and particle size.

Cooling Rate

Previous studies conducted at the Pacific Northwest National Laboratory (PNNL) revealed an effect of the thermal history and cooling rate on the crystalline structure independent of the chemical modifiers that are added. They claim that the role of the polymer modification is to assist in the control of the microstructure on cooling by facilitating the formation of plate-like microcrystals found in beta sulfur rather than the larger crystals found in alpha sulfur. They found that slow cooling (< 1.5°C/min) resulted in the formation of alpha sulfur regardless of the polymer modification, whereas more rapid cooling (> 1.5°C/min) yielded the more desirable beta form (Sliva et al., 1996; Spence, 2004).

Figure 5.28 shows the effect of cooling rate on the formation of the beta sulfur phase (Sliva et al., 1996). The cooling rate from the sulfur molten stage to solidification had a profound effect on the final phases present in sulfur cement. The normal equilibrium crystallization path from a higher temperature is from the melt to beta phase then to the alpha phase. Slower cooling rates (< 1.5°C/min)

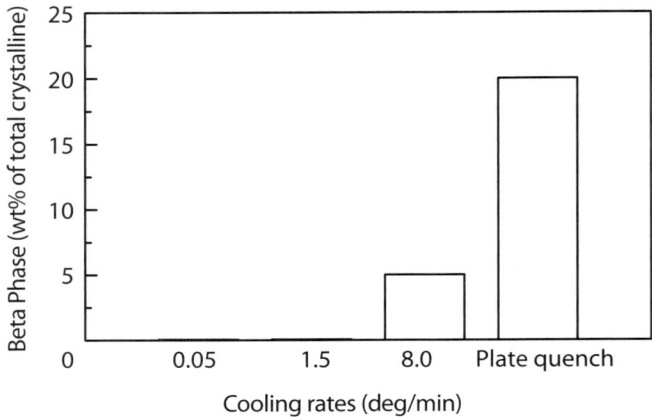

Figure 5.28. Beta formation as a function of cooling rates for sulfur polymer cement lot 030695 (adapted from Sliva et al., 1996).

resulted in nearly all alpha phases being at room temperature. Faster cooling rates increased the beta to alpha ratio; however, the amorphous content also increased due to the super-cooling nature of sulfur.

Studies related to the combined effect of the cooling rate and the polymer modification on the final phase in sulfur cement formation suggest that to preserve sulfur in the beta form, polymer modification is necessary; for pure elemental sulfur, no beta phase was detected using the cooling rates that ranged from 25°C/min to 0.05°C/min. When the polymers were fully reacted with sulfur in sulfur cement, a cooling rate approximately 8°C/min stabilized the beta sulfur phase so no alpha phase was present at room temperature. The polymer modification of sulfur has controlled the crystalline morphology and microstructure. Without polymer modification, alpha sulfur crystal was at least several millimeters. With polymer modification, the dominant microstructure was plate-like crystals of micron size (Sliva et al., 1996).

The effect of the cooling rate and the polymer modification on peak crystallization temperature was also significant. For a cooling rate near 25°C/min, pure sulfur showed a peak crystallization temperature at approximately 102°C, whereas polymer modified cement (SPC lot 030695) showed a peak crystallization temperature of about 56°C (a 46°C difference). The difference was even greater when the same sulfur cement was cooled at 1.5°C/min (Figure 5.29). Because chemical stability of various sulfur phases may differ significantly, knowing the degree of the polymer modification and controlling the cooling rates accordingly may result in optimized long-term stability through sulfur phase formation and control.

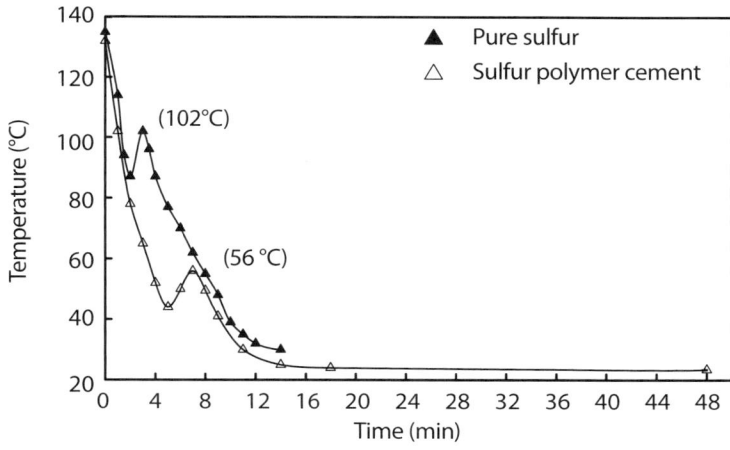

Figure 5.29. The effect of cooling rate on peak crystallization temperature (adapted from Sliva et al., 1996).

5.8 STANDARD TESTING OF SULFUR CEMENT

Testing procedures for sulfur modified mixtures as determined by ACI 548.2R are summarized:

a. Sulfur and carbon are determined by combustion of the sulfur modified mixtures using a carbon/sulfur analyzer
b. Specific gravity measurements are made in accordance with ASTM D70. Measurements are made at 25°C
c. Viscosity is determined at 135°C using a rotating spindle-type viscometer equipped with an electrically heated cell and temperature controller
d. The test material (0.040 g) is melted on a microscope slide at 130°C and is then covered with a square microscope slide cover slip. The slide is transferred to a hot plate and is kept at a temperature of 70 ± 2°C, as measured on the glass slide using a surface pyrometer. One corner of the melt is seeded with a crystal of test material. The time required for complete crystallization is measured. Thus, the rate of crystallization is determined.

5.9 ADVANTAGES AND DISADVANTAGES OF SULFUR CEMENT

5.9.1 Advantages

a. Sulfur cement has been used as a construction material because of its excellent resistance to acid and salt environments and its superior water tightness as compared with Portland cement concrete (McBee et al., 1985; Mattus and Mattus, 1994; Mohamed and El Gamal, 2006, 2009).
b. It has emerged as a possible alternative as a binding and stabilizing agent for the solidification and stabilization of hazardous and mixed wastes (Lin et al., 1995).
c. The thermal expansion coefficient for the modified sulfur is approximately one third that of the unmodified sulfur. Because the modified sulfur does not go through the allotropic transformation on solidification and has a lower thermal expansion coefficient, it has less shrinkage and, hence, has less residual stress upon cooling. (Lin et al., 1995).
d. Sulfur cement operational characteristics have greater flexibility that results in easier operations compared with Portland cement. Once water is added to Portland cement, only a few hours are available before the mixture becomes hard and cannot be used. With sulfur cement, even if the material is heated, it could remain hot in the mixer for extended periods. Cleaning of the mixer after one job should be easier than when using Portland cement. Even if some material remains in the mixer, it will be re-melted during the next job. Also, no contaminated rinse water is generated from cleaning operations (Darnell, 1991).

5.9.2 Disadvantages

a. Sulfur cement material will melt if exposed to temperatures above the melting point of 119°C and will lose its integrity (McBee and Weber, 1990).
b. The optimum processing temperature range of sulfur cement is between 127 and 139°C (McBee and Weber, 1990). Above the temperature range, a sharp rise in viscosity occurs because of additional polymerization within sulfur cement that makes material gummy and unpourable (Darnell el al., 1992). Hydrogen sulfide gas, which is poisonous and flammable, also forms. Additionally, if the pro-

cessing temperature is too low, the sulfur will be incompletely or partially melted.
c. Sulfur cement is not recommended for use with strong bases and oxidizing agents, aromatic or chlorinated hydrocarbons, or oxygenated solvents because of the risk of chemical corrosion due to H_2S formation (Van Dalen and Rijpkema, 1989). Sulfur cement deteriorates in hot, concentrated chromic acid solutions, hot organic solvent solutions, sodium chlorate-hypochlorite copper slimes, and strong alkalis (> 10%) (Darnell, 1991).
d. Operational disadvantages of sulfur cement include the following: an off-gas treatment system is needed to scrub vapors and gas released during mixing operations; handling, mixing, and the use of sulfur cement are more sensitive than for Portland cement; the mixing equipment is more complex than for standard Portland cement mixers; the mold that receives the molten mixture and its contents must be heated prior to and during the processing to control the cooling rate.

5.10 SUMMARY AND CONCLUDING REMARKS

In this chapter we have defined terminologies used in manufacturing sulfur cement. Clear distinctions have been between sulfur modifiers and sulfur cement. Sulfur modifiers were defined as materials or mixtures of materials that, when added to elemental sulfur, lower its melting point and increase its crystallization time. Sulfur modifying chemical reagents such as DCPD, DCPD plus oligomer of cyclopentadiene, styrene, DCPD plus styrene, bitumen, RP220, RP020, CTLA, etc., are generally used and reported in the literature.

Sulfur cement is produced via the addition of viscosity-increasing, surface-active stabilizer such as fly ash, furnace slag, cement kiln dust, talc, mica, silica, graphite, carbon black, pumice, insoluble salts (e.g., barium carbonate, barium sulfate, calcium carbonate, calcium sulfate, magnesium carbonate, etc.), magnesium oxide, and mixtures. Such fillers typically have a particle size less than 100 mesh and preferably, less than 200 mesh. Such fillers generally act as thickening agents and improve the hardness or strength of the sulfur cement product.

The proportions of the chemical stabilizers/additives may vary depending on the end use of the sulfur cement. Typically, the chemical stabilizer is used in amounts up to 14% by weight of the total sulfur, and more especially in the proportion of 1 to 5% of the total sulfur by weight. The amount of the chemical stabilizer required depends on the intended use of the cement as well as the properties desired. It is generally incorporated into the final cement mix by several reaction routes. Preferably, the chemical is pre-reacted at approximately 140°C for 30 min with a smaller proportion of sulfur than is required in the final mix. The resulting concentrate can be stored for future use or dissolved in the residual sulfur required for the final mix at the mixing temperature.

Controlling parameters such as the type and percentage of modifiers and the thermal history, including reaction and curing temperatures, curing time, cooling rate, and storage were discussed and evaluated. It is clear that these parameters have a strong impact on product crystallization and grain size development.

In the next chapter, we deal with the production of sulfur concrete and the evaluation of its strength and durability using chemical modifiers discussed in this chapter.

6

SULFUR CONCRETE

6.1 INTRODUCTION

Elemental sulfur concrete discussed in Chapter 4 is known to be practical, but failures caused by the internal stresses that result from changes in crystalline structure on the cooling of the material makes it unsuitable for many purposes. Modified or plasticized sulfur, discussed in Chapter 5, has been used in building construction, including flooring surfaces, as well as for roadway paving and roadway marking because of its excellent adhesion to substrates and its resistance to abrasion and corrosion. Sulfur concrete construction materials, based on *sulfur polymer cement* (SPC) and mineral aggregate, have been developed by the U.S. Bureau of Mines to utilize domestic mineral resources. *Modified sulfur concrete* (MSC) or *sulfur concrete* (SC), is a thermoplastic construction material composed of mineral aggregates with modified sulfur as a binder. The modified sulfur binder consists of elemental sulfur that has been reacted with a small percentage of a polymer modifier to improve physical properties and durability.

MSC is prepared by hot-mixing the sulfur cement and aggregate at a temperature of 125 to 150°C. The concrete is cast into the desired shape and, on cooling to ambient temperature, forms the final MSC structure. The quality of the aggregates has a substantial effect on the properties and durability of the finished sulfur concrete.

SPCs are high performance composite materials whose main properties are quick setting, high strength, low water absorption, and high resistance to acid and alkaline environments. For example, sulfur concrete has been used to fabricate corrosion-resistant tanks and as flooring surfaces in highly corrosive areas

(a) Sulfate-resistant hydraulic cement concrete specimen immersed in 50 wt% sulfuric acid

(b) Sulfur concrete specimen immersed in 50 wt% sulfuric acid

3 weeks   3 years

Figure 6.1. Acid effect on sulfate-resistant hydraulic cement concrete and sulfur concrete (adapted from Vroom, 1998).

as shown in Figure 6.1 (Vroom, 1998). SC offers many benefits as an alternative construction material, particularly in situations that require fast setting, placement in excessively cold or hot climates, corrosion resistance, and impermeability. Field experience with SC is making it increasingly more attractive for use in major projects.

This chapter describes the research related to the development of SPCs or MSCs, and discusses the basic ingredients of SC and its thermo-hydro-mechanical behavior.

6.2 TERMINOLOGY

In Chapter 5, we defined the term *modified sulfur* to indicate that chemical additives are added to the elemental sulfur and *sulfur cement* to indicate the product from the addition of mineral fillers to modified sulfur. In this chapter, *sulfur concrete* is used when aggregates are added to sulfur cement (Figure 6.2). Furthermore, to identify the type of chemicals used in modifying the sulfur properties, we could attach the name of the used chemicals to the term sulfur concrete, i.e., *dicyclopentadiene*-based sulfur concrete, *styrene*-based sulfur concrete, *bitumen*-based sulfur concrete, and *ENB*-based sulfur concrete and so on.

6.3 DEVELOPMENT OF SULFUR CONCRETE

Major advances in the development of sulfur concrete have taken place in the last decade. Research was based on the premise that for sulfur concrete to be a viable construction material, its durability would have to be improved, better sulfur cements would have to be developed, and better mixture designs would have to be formulated for the production of uniform products on a routine basis. The use of

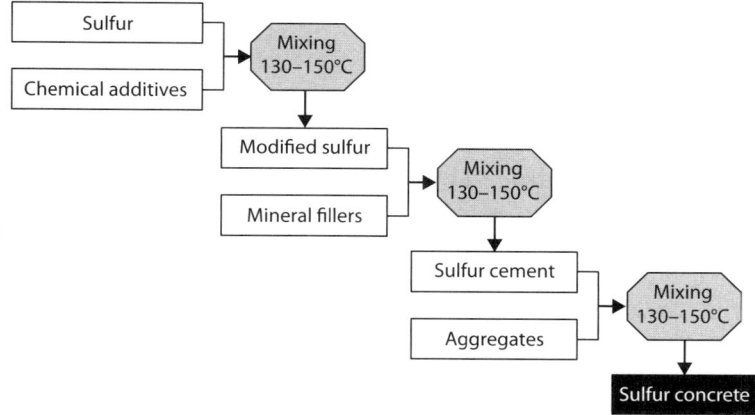

Figure 6.2. Sulfur terminologies used for sulfur concrete production.

the polymer additive provides a simple and effective method of modifying sulfur. Duecker (1934) was the first to modify a sulfur chemically with an olefin polysulfur, known commercially as Thiokol, that delayed the volume change tendency and the strength loss. The use of sulfur concrete for chemical plants increases to a certain extent, especially for construction in acidic environments in which the durability of the material was excellent. During the following 30 years, much effort was devoted to the study of modifiers and plasticizers for sulfur concrete as discussed in Chapter 5.

During the 1960s, there was a remarkable investment in environmental protection against the discharge of sulfur into the atmosphere, thus making sulfur a surplus commodity on the market, particularly in the United States and Canada. This was a crucial point that made interest in the use of sulfur as a structural binder to grow and initiated extensive research programs that became active in the 1970s and focused on various properties of the material, including durability.

In the late 1960s, Dale and Ludwig (1968) pioneered the work on sulfur-aggregate systems, pointing out the need for well-graded aggregates to obtain optimum strength. This work was followed by the investigations of Crow and Bates (1970) on the development of high-strength sulfur-basalt concretes.

At the beginning of the 1970s, successful projects in which sulfur concrete was used as a construction material were carried out on different levels. The U.S. Department of the Interior's Bureau of Mines and The Sulphur Institute (Washington, D.C.) launched a cooperative program in 1971 to investigate and develop new uses of sulfur. At the same time, the Canada Center for Mineral and Energy Technology (CANMET) and the National Research Council (NRC) of Canada initiated a research program for the development of sulfur concrete (Malhotra, 1973, 1974; Beaudoin and Sereda, 1973).

In 1973, the Sulfur Development Institute of Canada (SUDIC), jointly founded by the Canadian federal government, Alberta provincial government, and the Canadian sulfur producers, was established to develop new markets for the increasing Canadian sulfur stockpile. In 1978, CANMET and SUDIC sponsored an international conference on sulfur in construction (Malhotra et al., 1978). Also during this period, a number of investigators (McBee and Sullivan, 1979; Vroom, 1977, 1981; Sullivan et al., 1975; McBee and Sullivan, 1982a, b; Funke et al., 1982; McBee et al., 1981a, b, 1983a, b, 1986; Sullivan et al., 1986) published a number of papers and reports that investigated various aspects of sulfur concrete. Most of the studied substances were organic polymers of several types that, during a chemical reaction with sulfur, induced the formation of polysulfide and altered sulfur crystallization as discussed in Chapter 4 (Bright et al., 1978; Bordoloi and Pearce, 1978; Jordan et al., 1978).

McBee et al. (1981a, b; 1982; 1983a, b) published a number of papers and reports dealing with various aspects of sulfur and sulfur concrete. All of these

Table 6.1. Sample of early disclosed chemically-modified sulfur concrete

Patent number	Title	Issue date
3954480	Concrete compositions and pre-formed articles made therefrom	04 May 1976
4025352	Manufacture of sulfur concrete	24 May 1977
4058500	Sulfur cements, process for making same and sulfur concretes made therefrom	15 November 1977
4188230	Sulfur concretes, mortars, and the like	12 February 1980
4256499	Sulfur concrete composition and method of making shaped sulfur concrete articles	17 March 1981
4293463	Sulfur cements, process for making same, and sulfur concretes made therefrom	06 October 1981
4332911	Sulfur cement-aggregate compositions and methods for preparing	01 June 1982
4332912	Sulfur-aggregate compositions and methods for preparing	01 June 1982
4376831	Phosphoric acid-treated sulfur cement-aggregate compositions	15 March 1983
4981740	Acid-resistant concrete articles, especially sulfur concrete pipes, and a method of manufacturing said articles	01 January 1991
5004799	Pelletized sulfur concrete and method of preparing same	02 April 1991
5395442	Lightweight concrete roof tiles	07 March 1995

Table 6.2. Compositions of typical Portland cement concrete and modified sulfur cement concrete; concentrations are cubic feet/yard

Ingredient	Portland cement concrete	Modified sulfur concrete
Cement	2.5	None
Modified sulfur	None	3.7
Water	4.5	None
Air	1.5	1.6
Fly ash	None	2.2
Aggregate	18.5	19.8

Source: Pickard (1985) and Hammons et al. (1993).

activities led to an increased awareness of the potential use of sulfur as a construction material. As an example, Czarnecki and Gillott (1989) reported sulfur modifiers that improve the ductility and durability of sulfur concrete by strengthening the bonds between the sulfur matrix and the aggregate and between the sulfur crystals in the matrix. Makenya (2001) discussed a proprietary modifier called SRX (developed by Vroom) that cuts the orthorhombic crystalline chains of sulfur into smaller pieces, reducing shrinkage of the cement. Recently, Mohamed and El Gamal (2006, 2007, 2008, 2009) reported a new technology that improved the performance of sulfur concrete products by using bitumen as an additive for sulfur modification. Table 6.1 lists some of the disclosed chemically-modified sulfur concrete mixtures in various patents.

6.4 COMPOSITION

Sulfur concrete is designed to replace Portland cement concrete (ready-mix concrete) for many applications. The compositions of typical Portland cement concrete and modified sulfur concrete are compared in Table 6.2. The optimum mixture design for sulfur concrete must take into consideration the properties desired for its specific use. The usual objective of a mixed design is to prepare sulfur concrete with the following characteristics: (a) resistance to attack by most acids and/or salt solutions, (b) minimum moisture absorption, (c) mechanical strength properties equivalent to or better than those of Portland cement concrete, and (d) sufficient fluidity for good workability.

Because the composition of sulfur concrete is significantly different from Portland cement concrete, the material's selection, mix proportioning, production, and placement are unique. Sulfur concrete row materials are sulfur, chemical additives, mineral fillers, and aggregates. These materials are discussed later. Figure 6.3 shows alternatives for sulfur concrete composition.

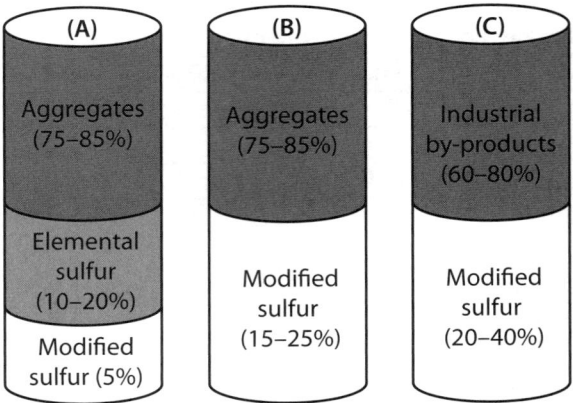

Figure 6.3. Alternatives for sulfur concrete composition.

The sulfur concrete can be made by adding 5% by weight of a chemical modifying agent to molten sulfur at 130°C. The modified sulfur is then added to aggregate that has been heated to 155°C and mixed thoroughly and then cast. Usually, the composition of the sulfur concrete is 15 to 25% modified sulfur and 75 to 85% aggregate by weight (Figure 6.3). Industrial byproducts such as fly ash, furnace slag, cement kiln dust, talc, mica, silica, graphite, carbon black, pumice, insoluble salts (e.g., barium carbonate, barium sulfate, calcium carbonate, calcium sulfate, magnesium carbonate, etc.), magnesium oxide, and mixtures have been reported as potential aggregate materials. Such materials, typically, have a particle size < 100 mesh (0.149 mm in diameter) per U.S. Standard Testing Sieves and, preferably, < 200 mesh (0.074 mm in diameter). In such a case, the amount of modified sulfur would increase to 20 to 40% (Figure 6.3).

6.4.1 Sulfur

Under standard conditions, elemental sulfur occurs in the environment as soft, yellow orthorhombic crystals with a density of 2.07 Mg/m³. The crystalline sulfur has a ring-type structure with the formula S_8. As the temperature increases, the molten sulfur becomes more viscous. The increase in viscosity is caused by the breakage of the ring structure to form chains that may polymerize as the temperature increases to form long molecules or cyclic compounds (Beall and Neff, 2005). The form and purity of the sulfur in sulfur concrete is quite unimportant, providing that the impurities do not represent more than approximately 4% clay or other water-expansive material. The sulfur may be in either solid or liquid (molten) form. For additional information about elemental sulfur, the reader should refer to Chapters 2 and 3.

6.4.2 Chemical Additives

Sulfur modifiers are required to keep the sulfur in a preferential crystalline form that resists concrete fracturing, through the allotropic transformation during cooling, resulting in sulfur concrete that is not highly stressed and is much more durable as discussed in Chapter 5. For example, Czarnecki and Gillott (1989) described sulfur modifiers that improve the ductility and durability of sulfur concrete by strengthening the bonds between the sulfur matrix and the aggregate and between the sulfur crystals in the matrix. The various types and concentrations of sulfur modifier in sulfur concrete have different degrees of resistance to environmental conditions. Most additives that are currently being used in the production of sulfur cement and sulfur concrete are discussed in Chapter 5.

6.4.3 Mineral Fillers

Sulfur concrete is a thermoplastic material prepared by hot-mixing sulfur cements and mineral aggregates. Mineral filler with aggregates and binder are necessary for the assurance of maximum density and limited absorption (Weber, 1993). The main functions of mineral fillers are to:

a. Control the viscosity of the fluid sulfur-filler paste, workability, and bleeding of the hot plastic concrete
b. Act as a thickening agent that reduces the separation tendency and helps to produce homogeneous products
c. Provide nucleation sites for crystal formation and growth in the paste and to minimize the growth of large needle-like crystals
d. Fill the voids in the mineral aggregate that would otherwise be filled with sulfur and to reduce shrinkage during hardening
e. Act as a reinforcing agent in the matrix and, hence, increase the strength of the formation

Therefore, to meet these functions, the filler must be reasonably dense-graded and possibly finely divided, as discussed in Chapter 4, to provide a large number of particles per unit weight, especially to meet the second function, provision of nucleation sites.

Materials such as crushed dust, silicate flour, and other inorganic fillers, including fly ash, have been used as mineral fillers. The amount of filler required in the mix, although typically 5% percent by weight, will ultimately be determined by the presence of < 200 (75µm) mesh-sized fines in the aggregate source. As with the aggregate, the mineral filler should be tested for chemical resistance to confirm its suitability for the intended environment (Crick and Whitmore, 1998; Mohamed and El Gamal, 2009).

Mineral fillers, including silica powder, coke, carbon black, and some types of fly ashes can be used in acid environments. Calcium carbonate, fly ash, and all acid-resistant fillers are reported to be suitable for use in sulfur concrete exposed to a saline environment (Soderberg, 1983). It must be emphasized again that the suitability of a mineral filler is highly dependent on the potential leachability of its chemical constituents to the surrounding environment. Therefore, standard leachability tests should be performed on the solidified matrix prior to its use for construction purposes.

6.4.4 Aggregates

Aggregate is crushed mineral material, usually natural rock of various types. The aggregates used in Portland cement concrete are similar to those used to manufacture sulfur concrete. Three sizes of aggregate are generally used in preparing the dense-graded product: coarse aggregate, fine aggregate, and mineral filler. Waste solids such as fly ash, steel slag, and cement kiln dust can be used as aggregate or filler in sulfur concrete. The quality of the aggregates has a substantial effect on the properties and durability of the finished sulfur concrete.

The selection of quality aggregates that will be appropriate for each particular application is necessary for a sulfur concrete material. To meet the requirements of durability and cleanliness, and to limit harmful substances, the composite aggregates must meet the ASTM C33 specification as discussed in Chapter 4. The influence of the grain shape and size distribution on the strength development of elemental sulfur concrete is discussed in Chapter 4. As per the ACI-guide for mixing and placing sulfur concrete (1988), aggregates should conform to and be classified under the following requirements:

a. *Aggregate gradation:* Dense-graded aggregates are used in the production of sulfur concrete to minimize binder requirement.
b. *Corrosion resistant aggregates:* Should not show any effervescence when tested in acid of a given concentration and at a given temperature. Aggregates that are insoluble in acids such as quartz should be used for preparing acid-resistant sulfur concrete; salt-tolerant sulfur concrete can be prepared from crushed quartz or limestone. Crushed rock should be used because it bonds better with sulfur than weathered, rounded sand and gravel.
c. *Moisture absorption of aggregates:* Should be highly impervious and highly resistant to freeze-thaw. Clay contained within the solidified sulfur concrete is believed to have an absorptive capacity which will allow water to permeate through the material. When clay absorbs water, expansion occurs and will result in the deterioration of the

Table 6.3. Range of sulfur cement levels for maximum sizes of dense-graded aggregate as recommended by ACI Committee report 548.2R-5

Maximum aggregate size (mm)	Sulfur cement (percent by weight)
25.4	12 to 15
19.1	13 to 16
12.7	14 to 17
9.5	16 to 19

product. Therefore, clay containing aggregates should not be used in producing sulfur concrete. Hence, porous aggregates, in general, should not be used. Chemical additives used to control the moisture absorption capacity of clays are discussed in Chapter 7.

6.5 SULFUR CONCRETE REQUIREMENTS

6.5.1 Binder Requirements

The role of the modified sulfur cement binders in the corrosion-resistant sulfur concrete technology is threefold: (1) to bond the aggregate particles together; (2) to fill the voids in the mineral aggregate to minimize moisture absorption, and (3) to provide sufficient fluidity in the mix to give a workable sulfur cement mixture. When sulfur cement containing 5% modifier was used to prepare sulfur concretes, less binder was required to fill the voids in the mineral aggregate than that required for an unmodified sulfur binder.

Determination of the sulfur cement quantity, in combination with specific aggregates that provide the most desirable balance between mechanical strength, high specific gravity, low absorption, and good workability, was discussed in Chapter 4. For example, ACI 548.2R-5 has recommended the ranges of sulfur cement levels for the maximum sizes of dense-graded aggregate as shown in Table 6.3. As the maximum aggregate size increases, the required amount of sulfur cement decreases.

The next step of developing the mix proportioning is to determine the necessary amount of sulfur and sulfur cement that is necessary to control the expansion and contraction of the cement during thermal cycling. This is best accomplished through small, laboratory-sized batches to determine which minimum level of sulfur will provide a workable mix.

6.5.2 Mix Design Requirements

To obtain durable sulfur concrete with sufficient strength, the mix is designed for moisture absorption < 0.1% by mass. Criteria used to determine the best mix design are based on adjusting the binder level to provide the best balance between maximum compressive strength, maximum specific gravity, minimum absorption, and workable mixture. An optimal sulfur concrete mix proportioning will allow for sufficient workability of the mix while minimizing sulfur cement content and, as a result, limit excessive shrinkage as the sulfur concrete goes through its phase change from liquid to solid. Excessive shrinkage will lead to the formation of unnecessary cracks in finished products.

Figure 6.4 shows the variations of specific gravity, voids, compressive strength, and absorption as a function of the loading of sulfur cement for dense-graded quartz aggregate with diameters < 9 mm. It is worth noting that the range of sulfur cement levels should be adjusted according to the type, size, and gradation of the aggregate used; the information displayed in both Table 6.3 and Figure 6.4 should be used as guidelines.

6.6 MANUFACTURING EQUIPMENT AND METHODS

Sulfur concretes are manufactured in a modified asphalt batch plant (or continuous mix). The process involves first drying and heating the aggregates to a temperature higher than the melting point of sulfur. The hot aggregates, sulfur (either liquid or solid), modified sulfur, and mineral filler are then combined in the pug mill or other heating mixer.

Because of its low melting point and low-melt viscosity, sulfur concrete can be successfully processed by a variety of simple, heated mixing vessels. Bench- and pilot-scale development work was conducted using several types of heated batch mixers, including low- and high-shear blade mixers and double planetary orbital mixers (Kalb and Colombo, 1985). The latter provides a highly efficient mixing pattern at relatively low mixing speeds, thereby diminishing air and gas entrainment within the molten mixture. Heating can be provided by thermocouple-controlled electric-resistance band heaters, steam generators, or hot oil circulation heaters (Kalb et al., 1991; Kalb et al., 1999).

Proper temperature control in the mixer and the subsequent handling is important because the modified sulfur cement will suffer degradation if exposed to temperatures above 150°C with a consequent loss of quality in the end product. The normal working range of temperatures for mixing, transporting, and placing the mix is 125 to 145°C. The mix design is of critical importance and is determined by the aggregate and application requirements.

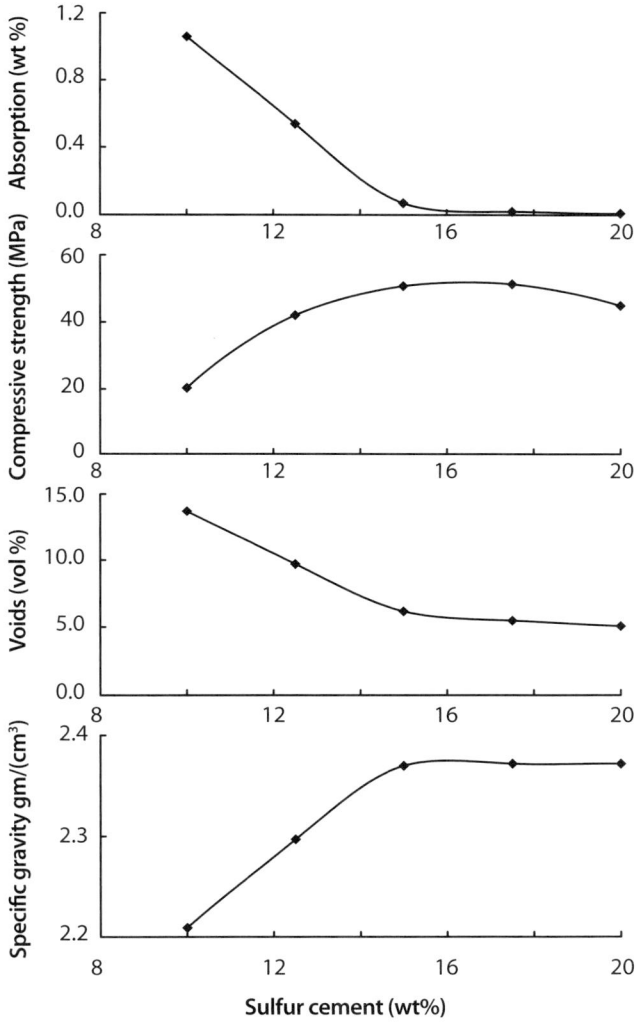

Figure 6.4. Effect of sulfur cement quantity on the absorption, mechanical strength, voids, and the specific gravity for sulfur concrete mix design; for dense-graded quartz aggregate with diameters less than 9 mm (data from ACI 548.2R-5).

The workability of the mix may be adjusted by adding slightly more or less sulfur and modified sulfur. The hot mix may be transported in insulated, ready-mix concrete trucks equipped with heaters or in oil-jacketed mixer transporters. The mix may be held in such equipment for many hours without deterioration (Gjørv et al., 1998).

Sulfur concrete specimens are cast from sulfur concrete materials in the mixing and working temperature range (132 to 141°C) in standard ASTM molds as specified in ASTM C31, using steel molds. The molds may be preheated to approximately 138°C before adding the sulfur concrete. The material is compacted as the sulfur concrete is added to the mold by tamping with a heated 16-mm rod having a hemispherical tip. Samples are cast in an upright position and allowed to cool to room temperature before being removed from the molds. Before testing, samples are allowed to cool a minimum of one day at room temperature.

6.7 RECOMMENDED TESTING

The ACI committee report 548.2R-5 recommended a guide for sulfur concrete testing as follows:

1. *Compressive strength*: Should be in accordance with ASTM C39 or ASTM C109. Specimens generally should be tested no less than 24 hr after they are made.
2. *Flexural strength*: Determination is made in accordance with ASTM C78.
3. *Splitting tensile strength*: Tensile strength measurements are made in accordance with ASTM C496.
4. *Voids*: Content may be determined by finding the specific gravity of sulfur concrete using ASTM C642. Voids are determined by calculation of the void content from the measured specific gravity and the theoretical specific gravity of the aggregate cement mixture. A distinction between discrete and continuous voids should be stated as provided in ASTM C642. It also may be found by inspecting a section cut from a sulfur concrete specimen microscopically, in accordance with ASTM C457, using the linear traverse method.
5. *Absorption*: Sulfur concrete specimens are immersed in water at 20°C for 24 hr and then the surface is dried and weighed. From the difference in weight, moisture absorption is calculated.
6. *Coefficient of thermal expansion*: Linear thermal expansion determinations are made on $13 \times 13 \times 25$ mm long specimens cut from a 76×152 mm cylinder of sulfur concrete. The expansion is measured over the 25 to 100°C temperature range at a constant heating rate of 3 ± 0.1°C per min.

7. *Freeze-thaw durability*: Measurements are determined in accordance with ASTM C666, using Procedure A *rapid freezing and thawing in water* on 76 × 76 × 356 mm bars of sulfur concrete.
8. *Modulus of elasticity*: Determination is made on 76 × 152 mm cylinders of sulfur concrete in accordance with ASTM C469. Dynamic modulus of elasticity is determined by ASTM C215.
9. *Aggregate gradation*: Determination is made on 76 × 152 mm cylinders of sulfur concrete after burning off the cement in a muffle furnace. Initial combustion is at 150°C. When the bulk of the cement has been removed by combustion, the temperature is raised to 440°C and held until a constant weight is achieved. After cooling to ambient temperature, the gradation of the aggregate is determined in accordance with ASTM C136.
10. *Swelling clays*: Sulfur concrete should not contain swelling clays because these clays can cause premature failures. An effective test for swelling clay is to immerse a 152 × 152 × 25 mm sample of sulfur concrete in water. The sample is first weighed dry then dipped in water, towel dried, and weighed again. The sample is then immersed in hot water at a 82°C minimum for 24 hr, removed, examined, towel dried, and weighed. Following this, the sample is re-immersed in the hot water. Removal, examination, towel drying, weighing, and re-immersion are repeated daily until failure is observed. The first sign of failure is usually a gain in mass (or weight) of 1% or more. The weight gain will soon be followed by spalling, hairline cracks, and an additional weight gain of 3 to 5%. If large amount of swelling clays are present, cracking may be so extensive that the sample cannot be recovered from the hot water.

It is worth noting that the testing of prepared sulfur concrete specimens should generally be done at least 24 hr after they are made. However, in applications in which high early strength is desirable, representative specimens may be tested within a shorter time period.

6.8 ADVANTAGES OF USING SULFUR CONCRETE

The main advantage of sulfur concrete is its use as a highly durable replacement for construction materials, especially Portland cement concrete, in locations within industrial plants or other locations in which acid and salt environments result in premature deterioration and failure of Portland cement concrete. There are several advantages to using sulfur concrete for construction in areas exposed to highly corrosive acids. Whereas ultimate life or durability of sulfur concrete has not been completely established in many end-user applications, enough evidence of its corrosion resistance and durability has been accumulated to show that it has several times the life of other construction materials now being used in corrosive environments (McBee, 1983). Other advantages of sulfur concrete are its fast setting time and rapid gain of high strength. Since it achieves most of its mechanical strength in less than a day, forms can be removed and the sulfur concrete placed in service without a prolonged curing period. Generally, sulfur concrete has the following useful characteristics:

a. Low permeability and porosity; properties that produce the impervious nature of this material to water.
b. Mechanical properties, including tensile, compressive, and flexural strengths, and fatigue life are greater than those obtained with normal Portland cement concrete (*Concrete Manual*, 7th ed, 1963). High mechanical strength is achieved rapidly on solidification; about 80% of the final strength is achieved in only a few hours and full strength is achieved within one day (McBee, 1986).
c. Comparable density with that of hydrated Portland cement; this material should provide similar radiation-shielding properties.
d. Excellent resistance to attack by most acids and salts, some at high concentrations; also resists corrosive electrolyte attack (McBee, 1983; McBee and Weber, 1990; Mohamed and El Gamal, 2009). Table 6.4 summarizes test results obtained with sulfur concrete materials. Additional information is provided later in this chapter and in Chapter 7.
e. Could be placed year-round in below-freezing temperatures.
f. Compositional materials could be recycled and used again (the greatest handicap in concrete).
g. Impurity of used materials does not have any effect on the final strength properties.
h. No need of water for production (best for dry lands).
i. Possibility of color coding without any problem.
j. Utilization as construction materials for floors, walls, and sump pits in the chemical, metallurgical, battery, fertilizer, food, pulp, and paper industries.
k. Utilization as road paving and bridge decking in which salt corrosion problems are encountered (Pickard, 1981).
l. Utilization as barrier systems for the containment of hazardous waste (Mohamed and El Gamal, 2009).

Table 6.4. Results of industrial testing of sulfur concrete material (McBee and Weber, 1990)

Testing Environment	Performance
Sulfuric acid	Nonreactive
Copper sulfate-sulfuric acid	Nonreactive
Magnesium chloride	Nonreactive
Hydrochloric acid	Nonreactive
Nitric acid	Nonreactive
Zinc sulfate-sulfuric acid	Nonreactive
Copper slimes	Attacked by organics in processing
Nickel sulfate	Nonreactive
Vanadium sulfate-sulfuric acid	Nonreactive
Uranium sulfate-sulfuric acid	Nonreactive
Potash brines	Nonreactive
Manganese oxide-sulfuric acid	Nonreactive
Hydrochloric acid and nitric acid	Nonreactive
Mixed nitric-citric acid	Nonreactive
Ferric chloride-sodium chloride	Nonreactive
Boric acid	Nonreactive
Sodium hydroxide	Attacked by > 10% NaOH
Citric acid	Nonreactive
Acidic and biochemical	Nonreactive
Sodium chlorate hypochlorite	Attacked by solution at 50 to 60°C
Ferric-chlorate ion	Nonreactive
Sewage	Nonreactive
Hydrofluoric acid	Nonreactive with graphite aggregate
Glyoxal-acetic-acid formaldehyde	Nonreactive
Chromic acid	Deteriorated at 80°C and 90% concentration; marginal at lower temperature and concentration

Nonreactive means that test results show no sign of corrosion or deterioration for a test period of 6 to 9 years.

6.9 DICYCLOPENTADIENE-MODIFIED SULFUR CONCRETE

A well-known chemical additive that has been involved in a number of investigations is the modification of sulfur with dicyclopentadiene (DCPD). The strength of DCPD-modified sulfur concretes depends on DCPD content, aggregate type, mix proportions, storage time, and thermal loadings.

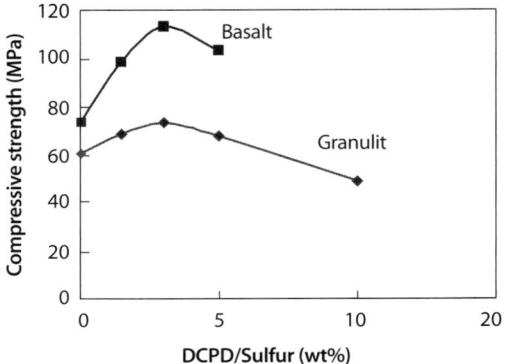

Figure 6.5. Compressive strength of sulfur concrete samples made with basalt and granulit aggregates with different DCPD content in sulfur cement (data from Gregor and Hackl, 1978).

6.9.1 DCPD Loadings and Aggregate Types

The influence of DCPD content on the compressive strength of sulfur concrete made with basalt and granulit aggregates is shown in Figure 6.5 (Gregor and Hackel, 1978). The DCDP content ranges from 1.5 to 10 wt%. The highest values for compressive strength were obtained for both granulit and basalt sulfur concrete with 3 wt% DCPD in sulfur cement. The highest compressive strength obtained with DCPD-based sulfur concrete amounted to 113 MPa. For comparison, normally constructed Portland cement concrete has a compressive strength of approximately 20 MPa.

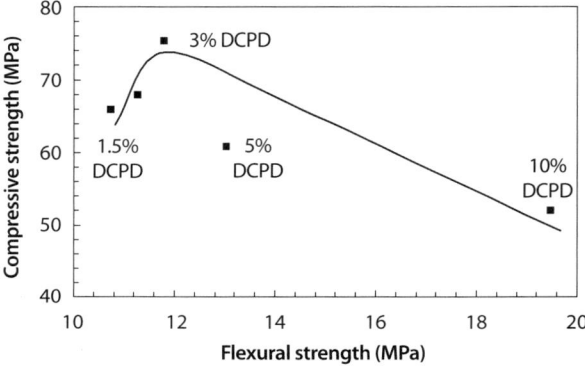

Figure 6.6. Compressive strength versus flexural strength of sulfur concrete samples with various DCPD content in sulfur cement (data from Gregor and Hackl, 1978).

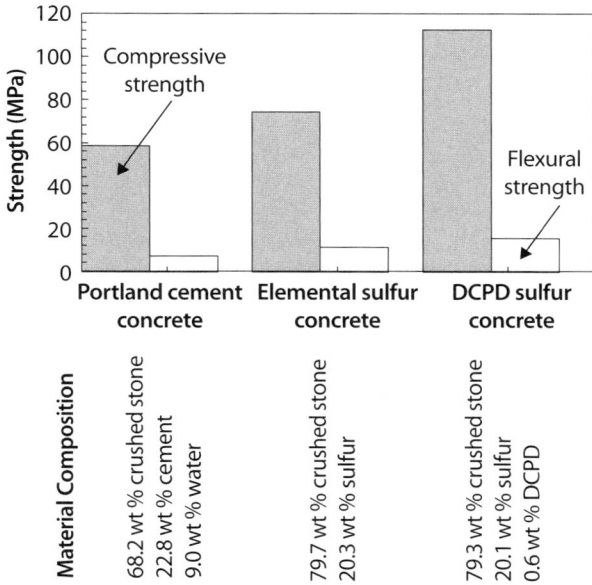

Figure 6.7. Comparison of strength properties of Portland cement, elemental sulfur concrete, and DCPD sulfur concrete (data from Gregor and Hackl, 1978).

The changes in flexural strength are shown in Figure 6.6. The flexural strength rises proportionally with an increased DCPD content. Whereas the elemental (i.e., unmodified) sulfur concrete has a ratio of flexural to compressive strength of approximately 1:6 (Chapter 4), this ratio is 1:2 for sulfur concrete modified with 10 wt% DCPD. According to the expected loading, certain strength ratios may be adapted for modified sulfur concrete.

Figure 6.7 shows the comparison of strength properties of Portland cement, elemental sulfur concrete and DCPD sulfur concrete. It is clear that DCPD sulfur concrete has the highest compressive and flexural strength among the three tested concretes. It is higher than Portland cement concrete by approximately 195% and elemental sulfur concrete by 152% with regard to compressive strength. For flexural strength, it is higher than Portland cement concrete by approximately 213% and elemental sulfur concrete by 134%. DCPD sulfur concrete should be preferred to elemental sulfur concrete because of its more favorable mechanical, chemical, and physical properties.

6.9.2 Storage Time

Figures 6.8(a) and (b) show the effect of storage time on the compressive and flexural strength of sulfur concrete modified with various DCPD loadings. These

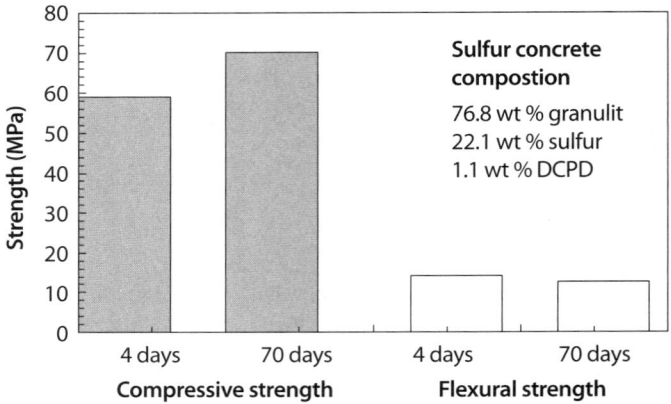

Figure 6.8(a). Influence of the storage time on the compressive and flexural strength of sulfur concrete modified with 5 wt% of DCPD (data from Gregor and Hackl, 1978).

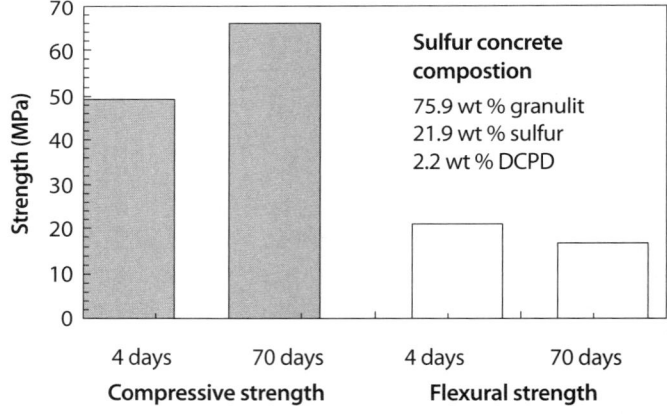

Figure 6.8(b). Influence of the storage time on the compressive and flexural strength of sulfur concrete modified with 10 wt% of DCPD (data from Gregor and Hackl, 1978).

changes in strength can be attributed to the unreacted sulfur, part of which, in the beginning, is in the amorphous form and recrystallizes proportionally with increased storage time and reinforces the polymer matrix. After 70 days storage at room temperature of the compressive strength increased by 35% and the flexural strength decreased by 18% for the sulfur concrete samples whose binder was modified with 10 wt% DCPD. In the cases where the binder was modified with 5 wt% DCPD, the average change in compressive and flexural strength was a 15.5% increase and a 5.6% decrease, respectively, in the same storage period.

Also, Syroezhko et al. (2003) studied the plasticizing effect in the course of prolonged storage. They found that dissolved sulfur gradually crystallizes thus increasing the degree of crystallinity of the material. Hence, with time, the initially plastic coagulation structure turns into a rigid coagulation crystalline structure that is accompanied by changing properties of the binders; they turn from one rheological status into another typical of each type of the disperse structure.

6.9.3 Thermal Stability

Figure 6.9 shows the comparison of thermal conductivity and stability against thermal loadings of Portland cement, elemental sulfur concrete, and DCPD sulfur concrete. From the thermal aspects (Figure 6.9), DCPD sulfur concrete has the lowest thermal conductivity and it is about 37% less than Portland cement concrete. However, in view of stability against thermal loading, DCPD sulfur concrete, elemental sulfur concrete and Portland cement concrete, are stables up to 80°C, 95°C, and 400–500°C, respectively. This means that DCPD sulfur con-

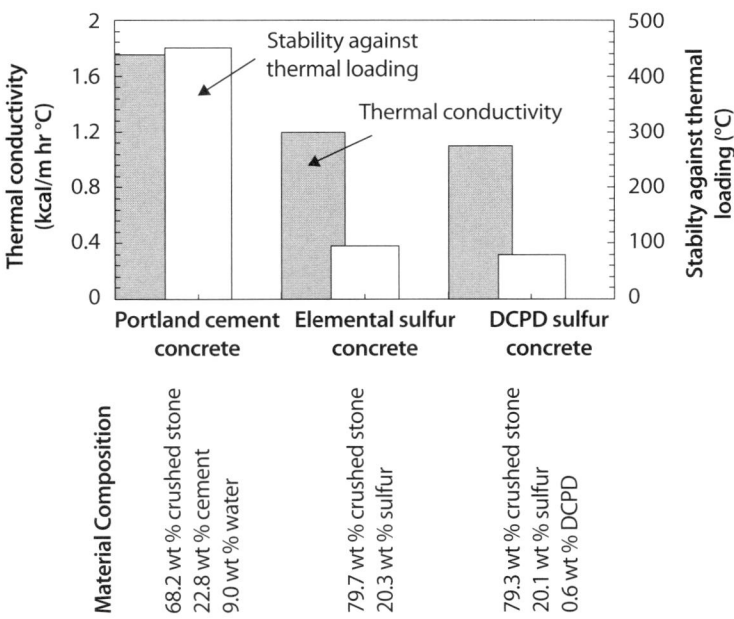

Figure 6.9. Comparison of thermal conductivity and stability against thermal loading of Portland cement, elemental sulfur concrete, and DCPD sulfur concrete (data from Gregor and Hackl, 1978).

crete and elemental sulfur concrete are combustible material whereas Portland cement concrete is not.

6.10 DICYCLOPENTADIENE-CYCLOPENTADIENE OLIGOMER-MODIFIED SULFUR CONCRETE

It is well known that the reaction between sulfur and cyclopentadiene dimer to form the modified sulfur component of the cement must be carefully controlled because of the exothermicity of the reaction between sulfur and DCPD. The desired control of the exothermic reactions is achieved by conducting the reaction between sulfur and DCPD in the presence of a sufficient quantity of cyclopentadiene oligomer (CPDO) to achieve the desired linear polysulfide polymeric products and to maintain a workable cement formulation. Manufactured sulfur concrete can be attained by a modified sulfur cement formulation comprising the polymeric reaction of sulfur with 2 to 20 wt% of a DCPD-CPDO containing modifier in which the CPDO content is at least 37 wt% (McBee and Sullivan, 1983). Such modified sulfur cement possesses excellent strength and freeze-thaw stability characteristics. Suitable sulfur concrete can be prepared by blending 7

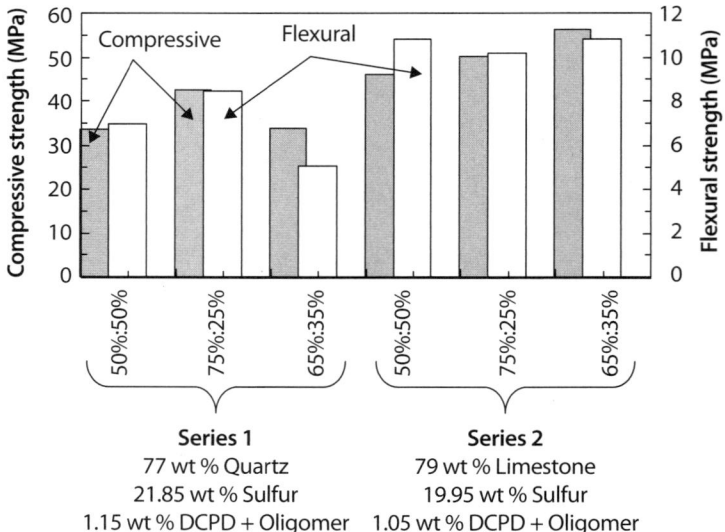

Figure 6.10. Changes in compressive and flexural strength with aggregate type, sulfur, and DCPD–CPDO loadings (data from McBee and Sullivan, 1983).

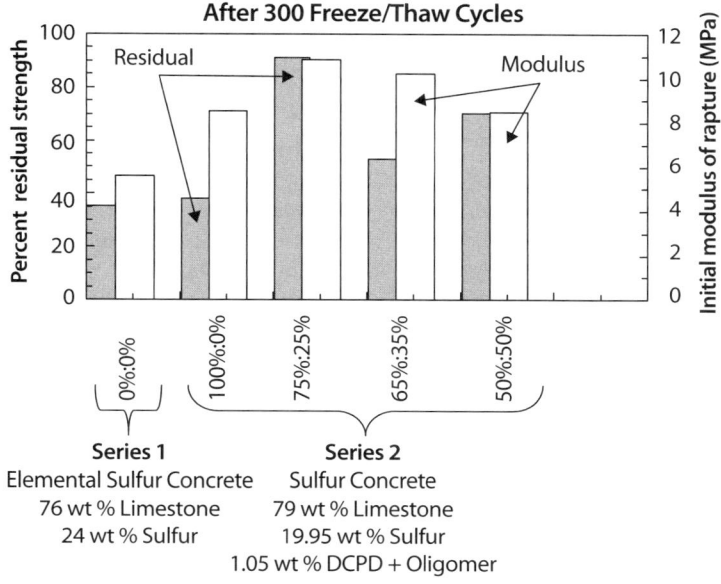

Figure 6.11. Changes in residual strength and modulus of rapture with aggregate, sulfur, and DCPD-CPDO loadings after 300 freeze-thaw cycles (data from McBee and Sullivan, 1983).

to 80% by wt modified sulfur cement with 20 to 93% by wt aggregate. Modified sulfur cements were prepared by the reaction of sulfur, oligomer and DCPD at 130°C for 24 hr. The modified sulfur cements were then blended with the aggregates. The physical properties of the resulting concrete are discussed.

6.10.1 Effect of Mix Composition on Strength

A series of modified sulfur concrete materials were prepared by employing the amounts of ingredients shown in Figure 6.10 (McBee and Sullivan, 1983). Modified sulfur cements were prepared by re-acting the amounts of sulfur, oligomer, and DCPD indicated in Figure 6.10 at 130°C for 24 hr. The modified sulfur cements were then blended with the aggregates at 140°C. The changes of compressive and flexural strength with aggregate type, sulfur, and DCPD–CPDO loadings are shown in figure 6.10.

6.10.2 Effect of Freeze-Thaw on Strength

A series of limestone-based sulfur concretes were prepared in the same manner as described previously employing the quantities of raw materials shown in Figure

6.11 (McBee and Sullivan, 1983). The sulfur concrete materials were employed in a series of freeze-thaw durability tests under prescribed test conditions (ASTM C666-73, Procedure A). The results shown in Figure 6.11 indicate that concrete samples with 25 and 50% oligomer withstood 300 freeze-thaw cycles while maintaining 70 and 90% of the original dynamic modulus values. Also, the results indicated that modified sulfur concretes prepared from the reaction between sulfur and oligomer-DCPD exhibit degrees of durability and residual strength superior to the elemental sulfur concrete—concrete prepared from sulfur cements unmodified with oligomer-DCPD-based materials or modified with DCPD.

6.11 OLEFINIC HYDROCARBON POLYMER-MODIFIED SULFUR CONCRETE

The olefinic hydrocarbon polymer material is derived from petroleum and has a nonvolatile content > 50% by weight and a minimum Wijs iodine number of approximately 100 cg/g. According to Vroom (1981), the prepared sulfur concrete consists of:

(I) *Sulfur cement* consisting of:
   a. *Sulfur.*
   b. *An olefinic hydrocarbon polymer (i.e., chemical stabilizer).* Typical chemical stabilizers are those known by the trade names of RP220 and RP020, products of Exxon Chemical; CTLA, a product of Enjay Chemical; and Escopol, a product of Esso Chemical AB (Sweden); all identifying a heat reactive olefinic liquid hydrocarbon obtained by partial polymerization of olefins. The chemical stabilizer can be incorporated into the final sulfur cement mix by several reaction routes. Preferably, the chemical is pre-reacted at approximately 140°C for about 30 min with a smaller proportion of sulfur than is required in the final mix. The resulting concentrate can then be either stored for future use or dissolved in the residual sulfur (liquefied) required for the final mix at the mixing temperature.
   c. *A viscosity-increasing surface-active finely-divided particulate stabilizer.* Examples are fly ash, gypsum, dolomite, pulverized limestone, a mixture of pyrites and pyrrhotites, or rock dust of a size up to minus 100 mesh (0.149 mm in diameter); preferably of a size of minus 200 mesh. Fly ash, from the burning of hydrocarbon fossil fuels and generally in the form of tiny hollow spheres called cenospheres and consisting of major amounts of silicon oxide and aluminum oxide (55.61% silicon oxide, 26.30% aluminum oxide) with small quanti-

ties of ferric oxide, calcium oxide, magnesium oxide, sodium oxide, potassium oxide, and carbon, is particularly effective in this regard due to its small particle size, shape, and surface texture. It has been found (Vroom, 1981) that in addition to supplying an extra measure of durability to the final sulfur cement, independent of its source, fly ash serves the dual function of a viscosity increaser and a sulfur cement stabilizer. Depending upon the degree of fineness of the fly ash and the consistency desired, an amount up to one and one-half times of total weight of the sulfur may be added beneficially.

(II) *Natural or manufactured aggregates.* A wide range of aggregates can be used. Among the conventional useful aggregates are sand, crushed cinders, brick dust, foundry sand, crushed quartzite gravel, crushed limestone, siliceous tailing sand, expanded shale, expanded clay, crushed barite, crushed brick, crushed Portland cement concrete, and crushed granite. Preferably, the aggregate particles are of an angular shape and of a rough surface texture as obtained by crushing.

The sulfur concrete prepared as described tend to be self-extinguishing with ash contents approaching two-thirds the weight of sulfur. It can also be made fire resistant and/or to inhibit the formation of SO_2 when heated by the addition of suitable additives, e.g., 1,5,9-cyclododecatriens or the reaction product of diphenoxy-dithiophosphinic acid with sulfur and alpha methyl styrene (Vroom, 1981).

The experimental results shown in both Table 6.5 and Figure 6.12 consist of two series. For the first series (1 to 9), samples were prepared with the addition of the required sulfur (less than contained in the pre-reacted material), the pre-reacted material, and fly ash to achieve the desired consistency of the sulfur cement. Then aggregate was added to provide the sulfur concrete.

The components were mixed at approximately 130°C in a heated 1/3 cubic foot concrete mixer for 15 minutes before pouring into the molds. Compaction was obtained through vibration or tapping of the molds. RP220 as a chemical stabilizer was used in these experiments.

For the second series (10 to 13), samples were prepared in a manner analogous to that of the first series with the exception that raw stabilizer, i.e., not pre-reacted with sulfur, was added directly to the mix at the previous point of pre-reacted material addition. To allow for a complete reaction, the mixing time was increased to 20 min.

Table 6.5. Compressive strength variations with aggregate type and material loadings of olefinic hydrocarbon polymer-modified sulfur concrete (data from Vroom, 1981)

No.	Aggregate type	Aggregate %	Sulfur %	Fly ash %	RP220	Density	Strength MPa
1	Crushed quartzite	64.4	19.6	15.5	0.49	2.38	60.10
2	Crushed brick	54.4	27.3	17.5	0.68	2.17	58.85
3	Expanded clay	38.6	32.0	29.2	0.38	1.77	57.62
4	Crushed granite	65.9	20.0	12.6	0.50	2.39	53.54
5	Crushed barite	79.1	13.2	7.4	0.33	3.28	52.79
6	Crushed barite	78.7	15.0	5.9	0.38	3.18	51.89
7	Crushed quartzite	71.7	20.4	7.4	0.50	2.38	49.40
8	Siliceous tailings	60.9	30.8	7.5	0.77	2.23	45.06
9	Expanded shale	50.0	26.6	22.7	0.66	1.70	32.71
10	Crushed Portland cement	64.1	24.1	11.2	0.60	2.23	39.26
11	Crushed limestone	73.5	26	—	0.50	2.41	36.50
12	Siliceous tailings	63.5	34.8	—	1.70	2.21	36.50
13	Expanded s hale	38.6	37.6	23.2	0.59	1.73	31.80

6.12 5-ETHYLIDENE-2-NORBORNENE-MODIFIED SULFUR CONCRETE

According to Shell International B.V. (PCT/EP2006/063220), modified sulfur was prepared as follows:

Sample 1: An amount of elemental sulfur was weighted in a glass tube. The sulfur was melted by placing the tube in an oil bath of 135°C. An amount of 5-ethylidene-2-norbornene (ENB) (5 wt% based on the weight of sulfur) was added and the fluid was stirred for 3 hr. The tube was then taken out of the oil bath and the fluid was poured in a cylindrical mold and allowed to solidify at room temperature.

Sample 2: An amount of elemental sulfur was weighted in a glass tube. The sulfur was melted by placing the tube in an oil bath of 150°C. An amount of ENB (10 wt% based on the weight of sulfur) was added and the fluid was stirred for 1 hr. The tube was then taken out of the oil bath and the fluid was poured in a cylindrical mold and allowed to solidify at room temperature.

Sulfur Concrete 217

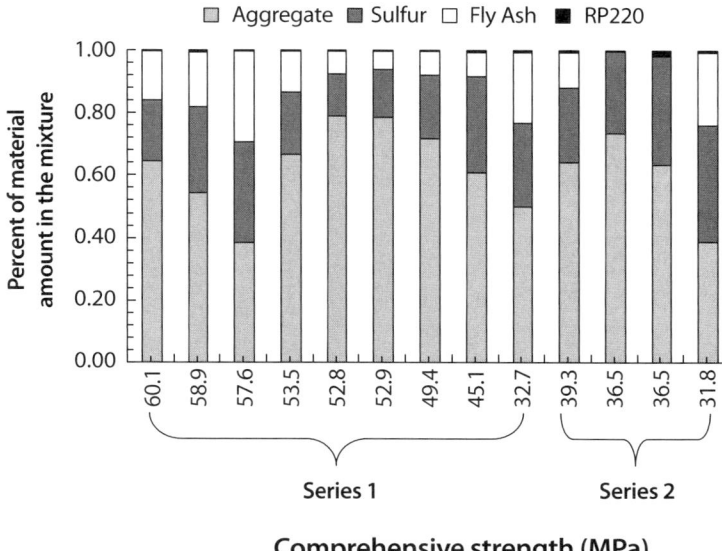

Figure 6.12. Compressive strength variations with aggregate type and material loadings of olefinic hydrocarbon polymer-modified sulfur concrete (data from Vroom, 1981).

Samp 3-6: Modified sulfur samples comprised of 1.0, 2.5, 5.0, and 7.5 wt% of ENB were prepared by mixing Sample 2 with further elemental sulfur at a temperature of 130°C. Each mixture was stirred at this temperature for 5 min and then poured in a cylindrical mold and allowed to solidify at room temperature.

Sample 7: Modified sulfur comprised of 10 wt% (based on the weight of sulfur) of a commercially available sulfur modifier STX™ (STARcrete Technologies) in a tube that was placed in an oil bath heated to 150°C. The mixture was stirred for 10 min. The tube was then taken out of the oil bath and the fluid was poured in a cylindrical mold and allowed to solidify at room temperature.

Sample 8: A sample of unmodified sulfur was prepared by melting elemental sulfur and placing a tube with elemental sulfur for 10 min in an oil bath that was heated, while stirring, to 150°C. The molten sulfur was then poured in a cylindrical mold and the sulfur was allowed to solidify at room temperature.

Table 6.6. Weight loss of modified sulfur upon immersion in 5M NaOH solution (adapted from Shell International B.V. PCT/EP2006/063220)

Sample	Modifier	Weight loss after 15 days (wt%)	Weight loss after 20 days (wt%)	Weight loss after 30 days (wt%)
1	5.0 wt% ENB[a]	2.2	2.7	
2	10.0 wt% ENB[a]			< 1
3	1.0 wt%[b] ENB[a]			< 1
4	2.5 wt%[b] ENB[a]			< 1
5	5.0 wt%[b] ENB[a]			< 1
6	7.0 wt%[b] ENB[a]			< 1
7	10.0 wt% STX™	34	57	
8	None	20	80	100

[a]ENB: 5-ethylidene-2-norbornene
[b]prepared from 10 wt% ENB sample (Sample 2)

The physico-chemical properties of the ENB-based sulfur concrete mixes are discussed below.

6.12.1 Weight Loss in Alkaline Environment

The alkaline resistance of the modified sulfur that was prepared as described was determined by placing the cylinders in a solution of 5M NaOH and water. The weight loss (wt% based on the initial weight of the sample) of the cylinders was measured after 15 and 20 days in the 5M NaOH solution. The results are shown in Table 6.6.

6.12.2 Compressive Strength in Alkaline Environment

According to Shell International B.V. (PCT/EP2006/063220), modified sulfur concrete comprising 50 wt% dried sand (Norm-sand), 30 wt% dried filler (quartz), and 20 wt% modified or unmodified sulfur were prepared by mixing the ingredients at 150°C until a homogeneous mixture was obtained. The mixture was then poured in a steel mold that was preheated to 150°C. Pressure was applied (0.25 to 0.5 tons) until droplets of sulfur were visible at the bottom of the mold. The thus-formed mortar cylinders were then demolded. Three different compositions were prepared as follows:

Composition 1: *Unmodified elemental sulfur:* 20 wt% of unmodified elemental sulfur, 50 wt% of dried sand, and 30 wt% of dried quartz were mixed.

Table 6.7. Compressive strength of sulfur compositions after immersion in 5M NaOH solution (adapted from Shell International B.V. PCT/EP2006/063220)

Composition	Sulfur	Compressive strength (MPa)		% loss in strength
		Initial	After 30 days in 5M NaOH solution	
1	Unmodified	60	7	88.33
2	Modified with 11 wt% STX™	58	34	41.38
3	Modified with 2.5 wt% ENB	69	51	26.09

Composition 2: *Sulfur modified with 11 wt% of STXTM*: 50 wt% of dried sand, 30 wt% of dried quartz, 18 wt% of elemental sulfur, and 2 wt% of STX™ modifier were mixed.

Composition 3: *Sulfur modified with 2.5 wt% of 5-ethylidene-2-norbornene (ENB):* 50 wt% of dried sand, 30 wt% of dried quartz, 15 wt% of elemental sulfur, and 5 wt% of modified sulfur prepared with 10 wt% of 5-ethylidene-2-norbornene were mixed.

These compositions were immersed in a 5M NaOH solution for a period of 3 days. After 30 days, Composition 1 was significantly degraded more than Composition 2, and Composition 2 was significantly degraded more than Composition 3. The compressive strength of the three compositions before and after immersion in 5M NaOH solution is shown in Table 6.7. The results indicated that sulfur concrete modified with 5-ethylidene-2-norbornene (ENB) lost approximately 26% of its strength after 30 days of immersion in 5M NaOH solution.

6.12.3 Ignition and Biological Oxidation

According to Morihiro et al. (2008) (European Patent Application, EP 1961713A1), elemental sulfur was modified with 5-ethylidene-2-norbornene (ENB) and prepared as detailed:

Sample 1: *Sulfur modified with ENB:* In this case, 970 g of elemental sulfur was placed in a stirring mixer, melted at 140°C, and held at 135°C. Then, 30 g of ENB was added slowly and stirred carefully for about 5 min. When no temperature rise was confirmed, the mixture was heated to 140°C. The mixture started reacting and the viscosity gradually rose. When the viscosity reached 0.06 Pa.s in 3 hr, the heating was immediately stopped, and the resulting material was poured into a suitable mold or a container and cooled at room temperature. Next, aggregate composed of 1120 g of No. 3 silica sand, 1127 g of No. 7 silica sand, and 413 g of coal ash was preheated to 140°C, and 480 g of modified sulfur was reheated to 140°C to melt into a molten state and introduced into a kneader held at 140°C. The molten mixture was kneaded for 20 minutes, while its viscosity at 140°C was maintained within the range of 0.05 to 0.07 Pa.s, poured into a cylindrical mold of 50 mm in diameter and 100 mm in height, and cooled to room temperature.

Sample 2: *Unmodified sulfur:* The sample was prepared in the same way as that of Sample 1 except the amount of elemental sulfur was 1000 g and ENB was not used. The viscosity at the termination of heating during the preparation of the binding material was 0.06 Pa.s.

Sample 3: *Sulfur modified with DCPD:* The sample was prepared in the same way as that of Sample 1 except the amount of sulfur was 950 g and 30 g ENB was replaced with 50 g of DCPD. The viscosity at the termination of heating during the preparation of the binding material was 0.06 Pa.s.

Sample 4: *Sulfur modified with ENB and styrene monomer:* The sample was prepared in the same way as that of Sample 1 except the amount of sulfur was 970 g and 30 g ENB was replaced with 20 g of ENB and 10 g of styrene monomer. The viscosity at the termination of heating during the preparation of the binding material was 0.06 Pa.s.

The experimental results of these samples are shown in Table 6.8. The results indicated that, for ignition resistance, Samples 1 and 3 were evaluated as not dangerous while, Samples 2 and 4 were evaluated as inflammable. For compressive strength, Sample 1 achieved the highest strength. In view of resistance to sulfur oxidizing bacteria, the results indicated that Samples 1 and 3 resisted the activities of the sulfur oxidizing bacteria whereas Samples 2 and 4 failed to resist the bacterial activities.

6.13 BITUMEN-MODIFIED SULFUR CONCRETE

6.13.1 Production of Bitumen-Modified Sulfur Concrete

Bitumen-modified sulfur concrete (BMSC), consisting of elemental sulfur, modified sulfur, fly ash, and sand was prepared according to the procedure described in Mohamed and El Gamal (2007). The physical additives or aggregates (sand and fly ash) were heated in an oven to 170 to 200°C for a period of 2 hr. The specified

Table 6.8. Test results of modified and unmodified sulfur concrete according to Morihiro et al. (2008) (European Patent Application, EP 1961713A1)

			Sample 1	Sample 2	Sample 3	Sample 4
Binding material	Starting material	Elemental sulfur	970 g	1000 g	950 g	970 g
		Modifier	30 g of ENB	—	50 g of DCPD	30 g of ENB & 10 g of SM
	Duration of reaction		3 hours	3 hours	3 hours	3 hours
Molded product	Aggregate		Silica sand & coal ash	Silica sand & coal ash	Silica sand & coal ash	Silica sand & coal ash
Ignition resistance			Not dangerous	Inflammable	Not dangerous	Inflammable
Compressive strength (MPa)			77	66	55	35
Resistance to sulfur oxidizing bacteria after 35 days			pH drop: slight	pH drop: considerable	pH drop: slight	pH drop: considerable

amount of sulfur was melted in a heated mixing bowl that was placed in an oil bath with controlled temperature in the range of 132 to 141°C. Fly ash was then transferred to the heated mixing bowl and properly mixed with the molten sulfur for approximately 20 min to insure a complete reaction between the sulfur and fly ash. Modified sulfur was added and mixing continued for an additional 5 min. Finally, desert sand was added and mixing continued for approximately 20 min more. In this process, the proportions of 0.25 wt% for modified sulfur, 0.9 wt% for sulfur to fly ash, 1 for sulfur to sand were used. At this stage of preparation, the mixture is viscous and can be easily bored into specified molds for casting. For the specimen preparation, cubical steel molds were used. The molds were preheated to approximately 120°C before adding the viscous BMSC. During the boring of BMSC into the specified molds, the mixture was compacted for 10 sec on a vibrating table. The surface of each specimen was then finished and the molds were placed in an oven at 100°C with a controlled cooling rate of 5° per min. After 24 hr, specimens were demolded.

6.13.2 Thermal Stability

A differential scanning calorimeter (Perkin Elmer DSC7) was used for the measurements of heat capacity through phase transitions on heating. Ten mg of the tested sample was heated to 150°C with a heating rate of 5°C/min. The sample was allowed to cool to room temperature for 24 hr, then the sample was reheated to 150°C. Sulfur crystal forms are obtained for elemental sulfur concrete (ESC) and BMSC using the differential scanning calorimetry (DSC) as shown in Figure 6.13. The addition of modified sulfur to the mix at small percentages not only directly

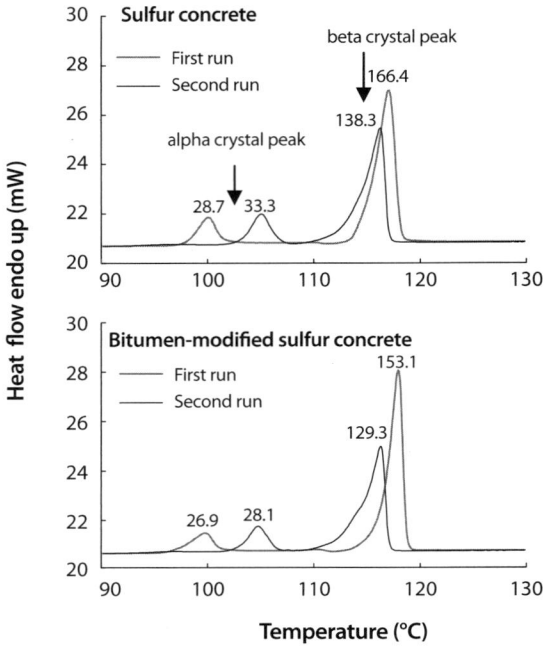

Figure 6.13. DSC curves for elemental sulfur- and bitumen-modified sulfur concrete showing the peaks for alpha and beta sulfur crystals at a heating rate that was 5°/min.

affects the crystal form of sulfur (making it crystallize in a structure different than the stable one), but also alters the behavior in the hot liquid mix period.

Crystallization of sulfur in this case is controlled by the relative percentages and distribution of space between the aggregates and the sulfur binder. The effect of modification seems to be related to the increase of sulfur polymerization. Sulfur itself possesses a tendency to polymerize; the modified sulfur would thus increase this tendency or maintain it for a longer time. In a more viscous liquid, and when the molecules are more polymerized, the growth of the crystal will be more difficult. This fact has an important effect in the crystallization of sulfur. The DSC results for BMSC show a significant reduction in the α and ß sulfur crystal forms, which are illustrated by the reduction of the areas under melted peaks, confirming that BMSC has a low order of crystalline structure compared to ESC.

A Thermo Gravimetric Analyzer (Perkin Elmer TGA7) was used to measure weight changes as a function of temperature in sample materials. Thermo gravimetric analyses (TGA) of ESC and BMSC are shown in Figure 6.14. The results indicated that the decomposition of ESC is faster than that of BMSC. This behavior can be considered as further evidence for the thermal stability of BMSC.

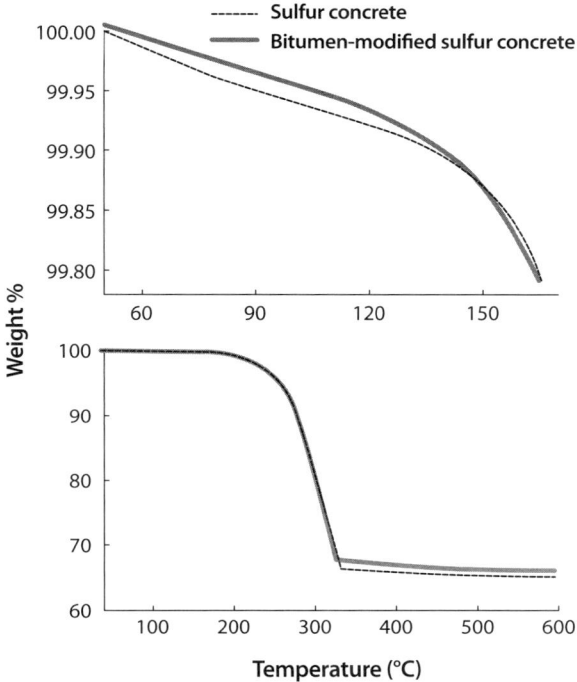

Figure 6.14. TGA curves for elemental sulfur- and bitumen-modified sulfur concrete for a heating rate of 20°/min.

6.13.3 Effect of Sulfur Ratio and Loading

Basic mechanical properties were measured for the SC samples to ensure that the developed materials met the properties found in literature for these materials. Figure 6.15 shows the desired optimum amount of sulfur to fly ash ratio according to the mechanical strength of the mortar. The obtained strength tended to increase as the sulfur/fly ash ratio increases up to 0.9, where all particles are coated by a thin layer of sulfur. However, with a large sulfur addition, the compressive strength decreased, because further increment of sulfur content increases the thickness of sulfur layers around the aggregate particles that leads to the increase of the brittleness of the formed composite material (Mohamed, 2002, 2003; Mohamed et al., 2002, 2003, 2005).

To have proper criteria for evaluating the difference between the performance of ESC and BMSC, samples were prepared using various sulfur loadings. The effect of the modified sulfur incorporated into the mortars is shown in Figure 6.16. The compressive strength decreased linearly with increasing amounts of

224 Sulfur Concrete for the Construction Industry

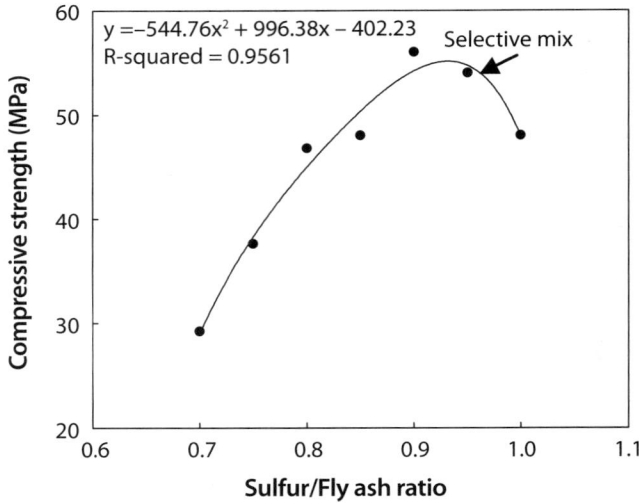

Figure 6.15. Effect of sulfur ratio on the compressive strength of sulfur concrete (Mohamed and El Gamal, 2006).

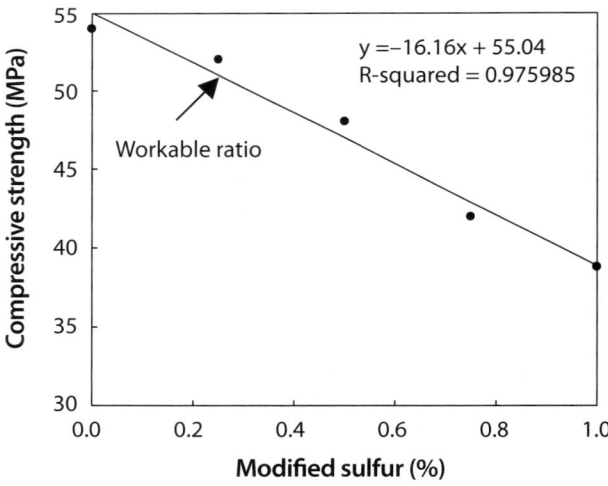

Figure 6.16. Effect of sulfur loadings on compressive strength.

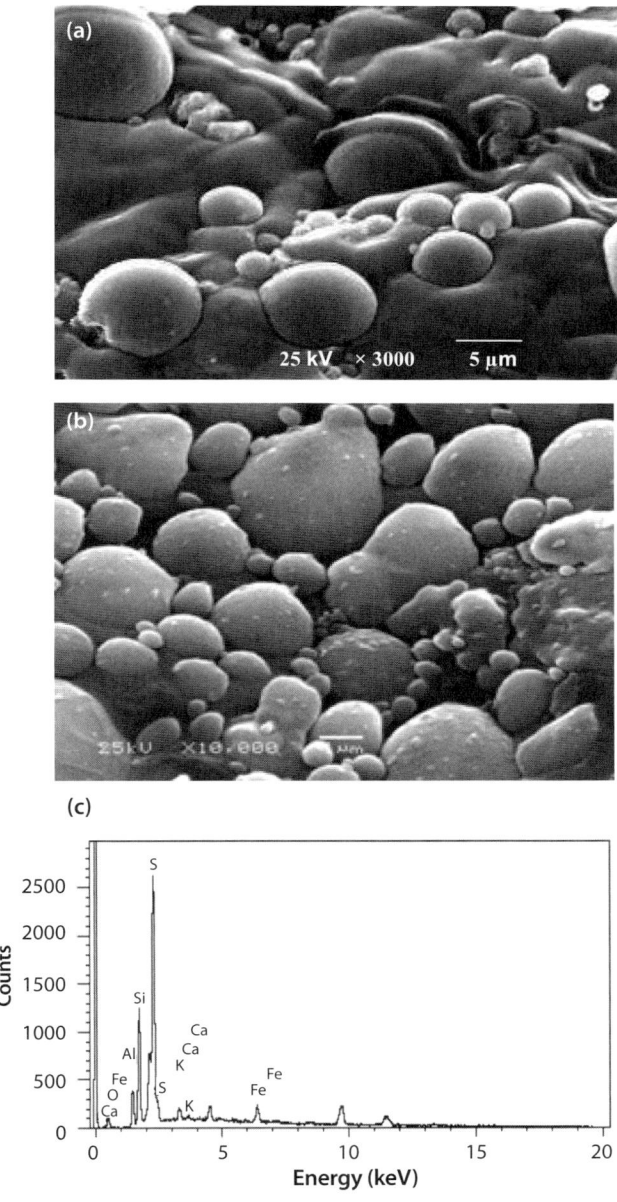

Figure 6.17. SEM micrograph showing: (a) how sulfur and modified sulfur bind, coat, and penetrate deep and between the aggregates; (b) a variety of particle shapes, sizes, and voids for bitumen-modified sulfur concrete; and (c) EDX spectrum showing the residue main chemical element for BMSC (adapted from Mohamed and El Gamal, 2009).

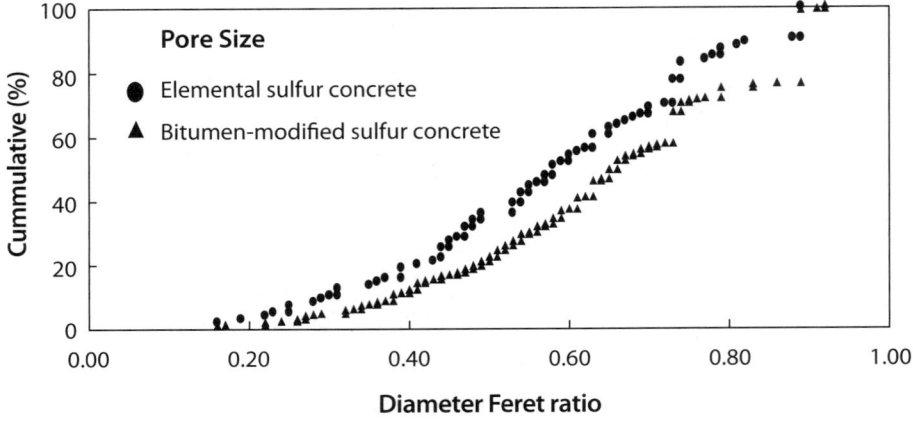

Figure 6.18. Cumulative percentage of pores sizes for elemental sulfur- and bitumen-modified sulfur concrete (BMSC).

modified sulfur due to the partial inhibition of the crystallization through the addition of modified sulfur. These results are in agreement with those reported by Vroom (1981). Mortars with modified sulfur show higher viscosity than unmodified ones. This fact has an important effect on crystallization of sulfur. In a more viscous liquid, the growth of the crystals will be more difficult and slow, causing partial reduction in compressive strength.

6.13.4 Microstructure Characterization

The microstructure characterization, performed by scanning electron microscopy (SEM), provides important data on sulfur concrete studies. Figures 6.17a and b show sulfur binding the aggregates by filling the inner spaces in such a way that the voids are discrete and homogeneous. This, in turn, increases the strength of the developed matrix. The chemical analysis performed by energy dispersive spectroscopy for bitumen sulfur concrete as shown in Figure 6.17c indicates that the samples are composed mainly of sulfur- and silicon-containing compounds. A low crystalline structure was obtained from the lower end of the signal intensity and the increase of atomic sulfur percentage proved to be the lower porosity.

It was thought that porous bodies were impregnated with modified sulfur that may have surface interactive forces different from those of ordinary sulfur (Beaudoin and Feldmant, 1984). The cumulative percentage of pore sizes studied with an image analyzer (Figure 6.18) proved lower porosity in BMSC, leading to high strength. The presence of modified sulfur has contributed to the increase of homogeneity, preventing the growth of large-size crystals and the decrease of porosity.

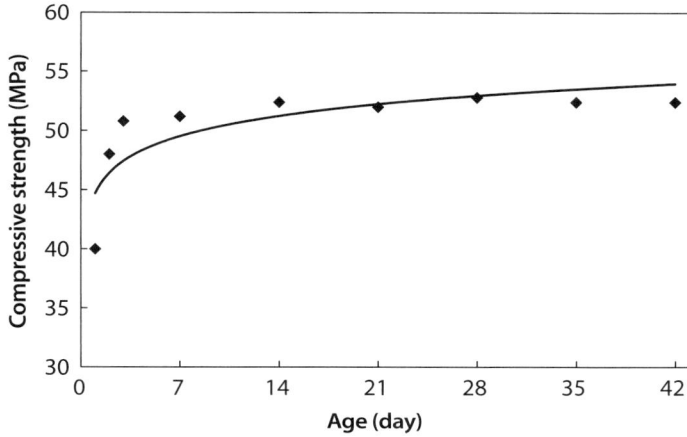

Figure 6.19. Variations of compressive strength with time for bitumen-modified sulfur concrete (BMSC) at 40°C.

6.13.5 Strength Development

Compressive strength was determined using 50 × 50 × 50 mm cubes. BMSC samples were prepared at 135°C, cured in the oven with a gradual cooling rate of 5 degrees/min, demolded after 24 hr, and then cured at the tested temperature. The raw materials used in these preparations possess some specific known characteristics. The main aspects of the concrete performance that will be improved by the use of fly ash are increased long-term strength and reduced hydraulic conductivity of the concrete, resulting in potentially better durability. The use of fly ash in concrete can also overcome some specific durability issues such as sulfate attack and alkali silica reaction. The desert sand was found to be of good quality, showing a low concentration of impurities and grains with an irregular geometry. This quality of sand grains may reduce the workability of the mortar but, on the other hand enables the molten sulfur to adhere more easily on the surface of the sand grains (Gemelli et al., 2004). The molten sulfur acts as a binder for these aggregates. Mortar with modified sulfur shows high viscosity that has an important effect on the crystallization of sulfur. The modified sulfur is efficient in binding and strengthening the aggregates and has been found to impart extra durability to the final BMSC.

Compressive strength development over time, for BMSC specimens cured in an oven with a gradual cooling rate of 2 degrees/min, demolded after 24 hr, and stored in incubators at 40°C, is shown in Figure 6.19. It is clear that BMSC develops about 76% of its ultimate compressive strength within one day and 97% after 3 days.

228 Sulfur Concrete for the Construction Industry

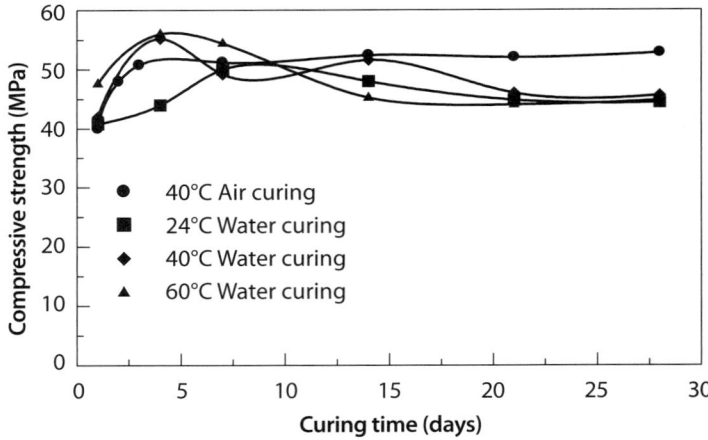

Figure 6.20. Influence of curing condition on the strength of BMSC samples (adapted from Mohamed and El Gamal, 2009).

However, up to 42 days, there was no clear trend on the compressive strength development, suggesting that the maximum strength was gained during the earlier days. Similar results were reported by many researchers. For example, Vroom (1981) reported that 80% of the ultimate concrete strength was developed in 1 day and virtually 100% of the ultimate strength was realized after 4 days. McBee et al. (1983) showed that the sulfur concrete developed approximately 70% of its ultimate strength within a few hours after cooling, 75 to 85% after 24 hr at 20°C, and the ultimate strength was commonly obtained after 180 days at 20°C.

To evaluate the effect of curing temperatures, specimens were prepared as indicated and cured for known temperatures and time periods in a specified environment (air and de-ionized water). Specimens were cured at 24, 40, and 60°C. Figure 6.20 shows the variations of compressive strength in the air and de-ionized water with various times and temperatures. The results, shown in Figure 6.20, indicate that for the first 3 days of immersion, the higher the water temperature and strength gain. It is also shown that after 3 days the strength loss occurred up to 15 days and then the strength remains nearly constant up to 28 days. A comparison between the losses in strength, after 28 days of BMSC cured in air and BMSC immersed in water, indicates that curing of BMSC in water resulted in some decrease in compressive strength.

It is clear that BMSC reaches hardening and gains its strength over a short time, resulting in a high strength material with an average compressive strength of 54 MPa and modules of elasticity of 1603 MPa. The strength increase could be attributed to the formation of cementing agents due to water availability. Considering the special chemical characteristics of fly ash and sand used in the

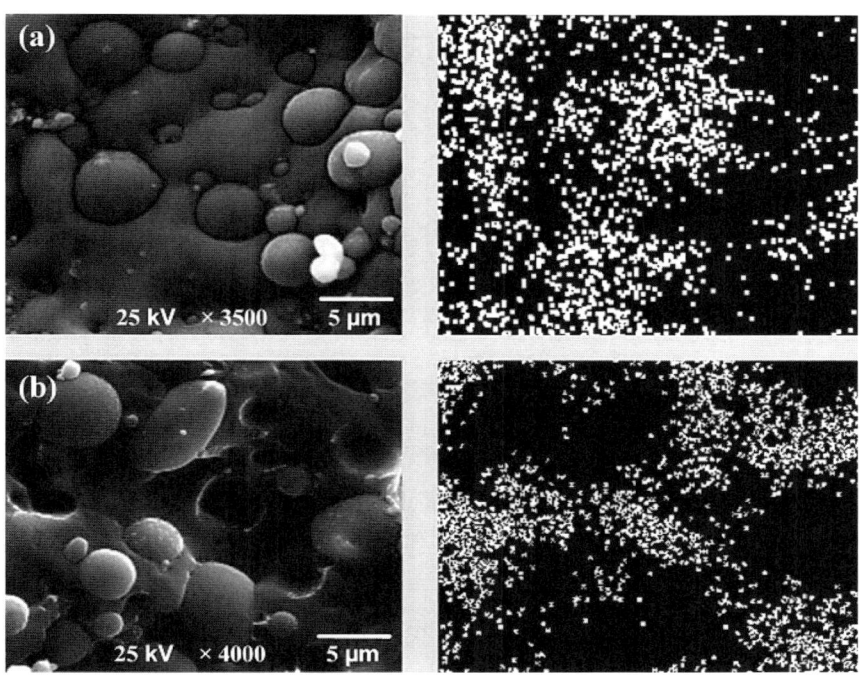

Figure 6.21. SEM micrograph of fractured surface of BMSC and its x-ray mapping for the sulfur elements of the same area; (a) air cured and (b) water cured after age time of 28 days at 40°C (adapted from Mohamed and El Gamal, 2009).

manufacturing process that indicates that sand has a high percent of SiO_2 (74.4%) and CaO (16.35%), and low amounts of Al_2O_3 (0.47%) and Fe_2O_3 (0.676%) whereas fly ash has high amounts of SiO_2 (60.9%) and Al_2O_3 (32.4%), and low amounts of Fe_2O_3 (4.34%) and CaO (0.46%). The availability of these types of oxides could lead to the formation of tricalcium silicate ($CaO.SiO2$), tricalcium aluminate ($CaO.Al_2O_3$), and tetracalcium aluminoferrite ($CaO.Al_2O_3.Fe_2O_3$).

With water availability, the following reactions will take place:

$$2(3CaO.SiO_2) + 6H_2O \rightarrow 3CaO.SiO_2.3H_2O + 3Ca(OH)_2$$
Tricalcium Silicate + Water → Tomberconte gel + Calcium Hydroxide
$$2C_3S + 6H \rightarrow C\text{-}S\text{-}H + 3CH \quad (6.1)$$

$$3CaO.Al_2O_3 + 6H_2O \rightarrow 3CaO.Al_2O_3.6H_2O$$
Tricalcium Aluminate + Water → Tricalcium Aluminate Hydrate
$$C_3A + 6H \rightarrow C_3AH_6 \quad (6.2)$$

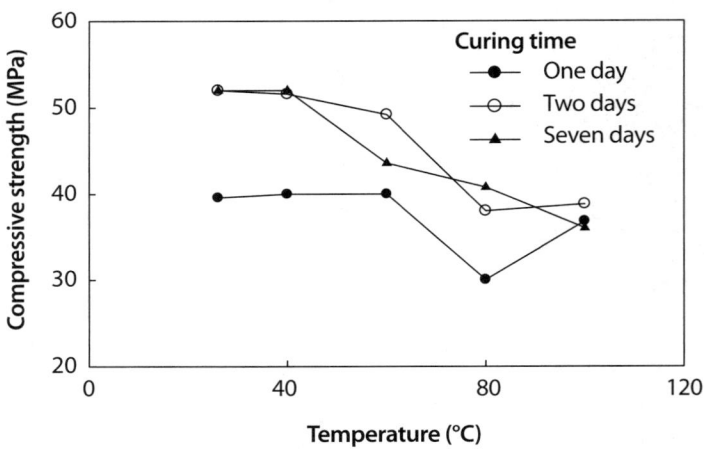

Figure 6.22. Variations of compressive strength with temperature and curing time for bitumen-modified sulfur concrete (BMSC) (adapted from Mohamed and El Gamal, 2009).

$$4CaO.Al_2O_3.Fe_2O_3 + 10H_2O + 2Ca(OH)_2.3H_2O \rightarrow 6CaO.Al_2O_3.Fe_2O_3.12H_2O$$
Tricalcium Aluminoferrite + Water + Lime → Calcium Aluminoferrite Hydrate
$$C_4AF + 10H + C_2OH \rightarrow C_6AFH_{12} \qquad (6.3)$$

and new minerals or cementing agents such as calcium silica hydrate (CSH), tricalcium aluminate hydrate (C_3AH_6) and calcium aluminoferrite hydrate (C_6AFH_{12}) are formed.

Furthermore, strength reduction was investigated via microstructure analysis of the tested specimens. Figures 6.21a and b show an SEM micrograph of the fractured surface of BMSC and its x-ray mapping for the sulfur elements of the same area for specimens cured in air and de-ionized water after the aging time of 28 days at 40°C. The fractured surfaces of BMSC reveal a different morphology. BMSC cured in the air-dry condition results in a homogenous sulfur distribution with good binding and coating of the aggregates as shown in Figure 6.21a. For specimens cured in de-ionized water, one observes that there is no homogenous distribution of sulfur where parts of the aggregate surfaces were uncoated with sulfur as shown in Figure 6.21b. This observation appears to explain the small strength reduction of BMSC cured in water, which is in complete agreement with reported studies by various investigators.

To evaluate the effect of both temperature and curing time, compressive strength was measured for BMSC samples cured at various temperatures of 26, 40, 60, 80, and 100°C and for time periods of 1, 2, and 7 days. The variations of compressive strength with temperature and curing time for BMSC are shown

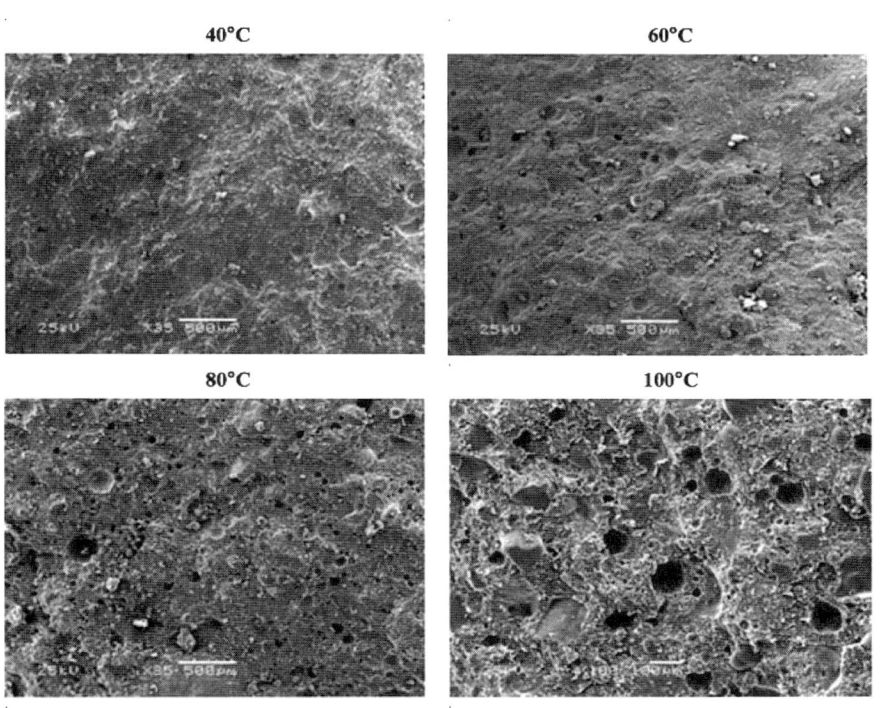

Figure 6.23. SEM images for bitumen-modified sulfur concrete at various temperatures.

in Figure 6.22. The results indicate that for the same curing time the temperature increases and compressive strength decreases. A systematic reduction in compressive strength with temperature was obtained especially at high temperatures (80 and 100°C). For the same temperature, as curing time increases the compressive strength increases. At temperatures below 60°C, the strength increase over time ranges from 20 to 30% whereas at higher temperatures, the rate of the strength increase with time is low. Similar results were reported by McBee and Sullivan (1979) who found that the strength gain is slower at elevated temperatures and faster at lower temperatures. The results also indicate that the compressive strength of BMSC for temperatures below 60°C is 20 to 30% more than for Portland cement concrete and for higher temperatures (60 to 100°C), the compressive strength remains within the same range.

Because mechanical strength is directly related to the defects in mortar microstructure, it is important to characterize such defects using SEM. Microscopic studies were determined from a section cut at a distance of 5 mm from the surface of BMSC samples. Figure 6.23 shows a comparison between the

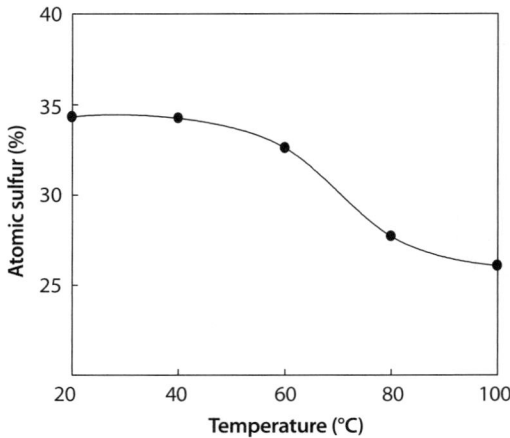

Figure 6.24. Changing of sulfur contents in BMSC with temperature as obtained from EDX.

mortar microstructure of BMSC cured after 7 days at different temperatures of 40, 60, 80, and 100°C.

It is obvious that for BMSC cured at 40 and 60°C, the voids entrained within the BMSC are small and discontinuous. However, for BMSC cured at 80 and 100°C, the voids are large without any noticeable cracking. The voids in BMSC serve as sites for stress relief and for improving the durability of the material (McBee et al., 1983). Also, the presence of voids in BMSC reduces the quantity of sulfur cement required to coat the mineral aggregate thereby minimizing the cement-related shrinkage.

The energy dispersive x-ray spectroscopy (EDX) spectrum of BMSC at different temperatures presents the intensity of the main chemical elements present in BMSC. The sulfur peak intensity decreased with increasing temperatures, which in turn confirms the SEM observations. This lower intensity means a lower degree of mineral crystalline formation, which in turn led to lower compressive strength. To visualize this set of results, the data were presented in terms of % atomic sulfur as a function of temperature as shown in Figure 6.24. The results indicated that the percentage of atomic sulfur in BMSC tended to decrease as the curing temperatures increased that are compatible with the presence of voids in BMSC at elevated temperatures as previously discussed.

To further evaluate whether the sulfur was lost due to gas formation or in combination with other existing heavy metals to form metal sulfide forms, loss of sulfur weight as a function of temperature was studied as shown in Figure 6.25. The gravimetric changes in elemental sulfur with temperature show no loss in sulfur weight at 40 and 60°C, whereas the loss of 0.25% at 80°C and 1% at 100°C

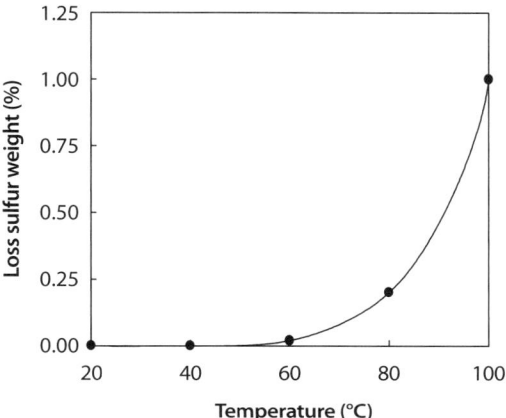

Figure 6.25. Percentage loss of sulfur weight varied with temperature; data obtained from gravimetric analysis.

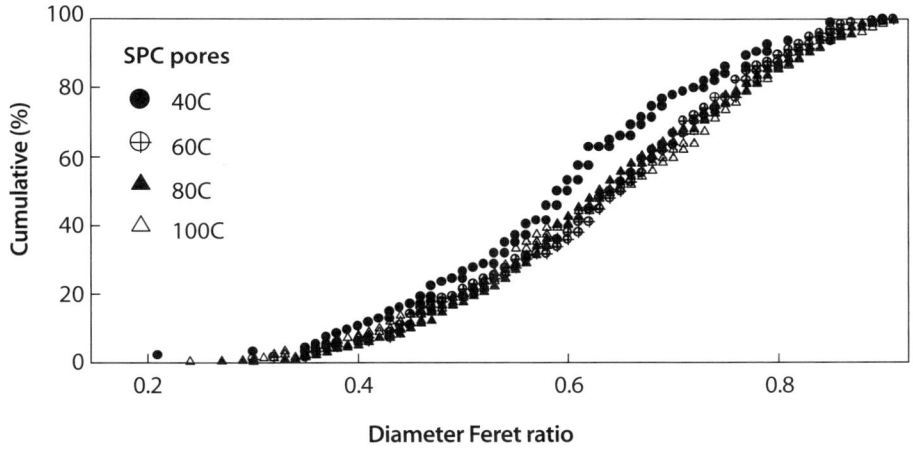

Figure 6.26. The cumulative percentage of voids for BMSC cured for seven days at various temperature using an image analysis program.

after 7 days indicates that the possible increase in voids at 80 and 100°C could be attributed to sulfur gas formation.

Furthermore, to evaluate the size of the voids from the results of SEM, image analysis software was utilized. The results are shown in Figure 6.26 in terms of the cumulative percentage of voids for BMSC at various temperatures. It can be seen that for the same cumulative percent, the diameter Feret ratio increased with

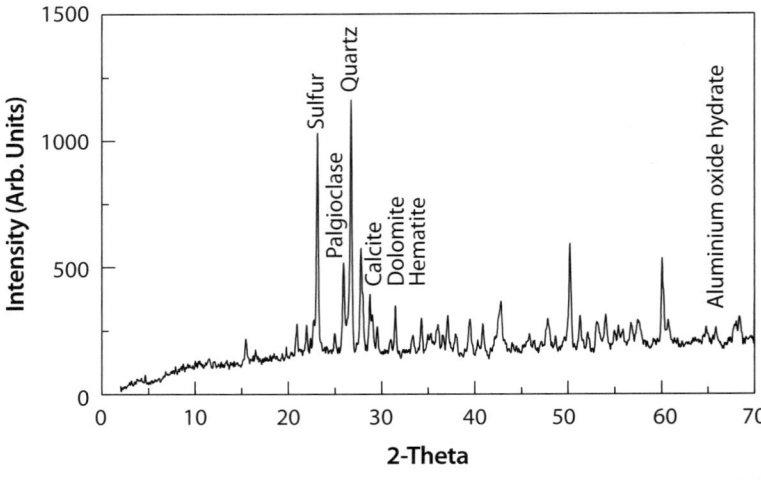

Figure 6.27. X-ray diffraction of BMSC cured for 28 days in de-ionized water (adapted from Mohamed and El Gamal, 2009).

increased temperatures, indicating that the size of the voids increased as temperature increased. It is worth noting that when diameter Feret ratio (r_{min}/r_{max}) of pore sizes reaches one, the pore sizes are spherical.

6.13.6 Reaction Products

To evaluate the potential formation of minerals during the manufacturing of BMSC, specimens were tested using x-ray diffraction analysis. Figure 6.27 shows the mineral composition of BMSC samples treated with de-ionized water for 28 days at room temperature and made up of major constituents in decreasing order; quartz (SiO_2) and sulphur. The minor constituents in decreasing order are aluminium oxide hydrate ($5Al_2O_3 \cdot nH_2O$), calcium silicate hydrate ($Ca_{1.5}SiO_{3.5} \cdot nH_2O$), plagioclase ($CaAlSi_3O_8$), calcite ($CaCO_3$), hematite ($Fe_2O_3$), dolomite ($CaMg(CO_3)_2$), and calcium aluminium oxide hydrate ($Ca_3Al_2O_6 \cdot nH_2O$).

The analyzed BMSC samples constitute variable percentages of sulfur, fly ash, and desert sand. Fly ash composition detected by inductively coupled plasma-atomic emission spectroscopy (ICP-AES) analysis in decreasing percentages are silica (60%), Al_2O_3 (32.4%), Fe_2O_3 (4.34%), and traces of Mg and Zn. On the other hand, the chemical composition of the sand dunes in decreasing percentages are made of SiO_2 (68%), CaO (10%), MgO (5%), Al_2O_3 (2.2%), and a trace of K_2O (0.8%).

Table 6.9. Percent formed mineral compositions with time

Mineral type	Curing time		
	1 day	2 days	7 days
Sulfur	27.9-35.4	28.0-35.2	28.4-31.5
Quartz (SiO_2)	44.0-56.2	42.7-50.0	44.0-50.5
Aluminium oxide hydrate ($5Al_2O_3 \cdot nH_2O$)	2.9-6.1	4.4-7.0	1.7-7.9
Calcite ($CaCO_3$)	2.2-2.9	2.5-5.0	2.0-3.8
Calcium silicate hydrate ($Ca_{1.5}SiO_{3.5} \cdot nH_2O$)	2.3-4.4	3.7-4.8	1.8-5.0
Plagioclase ($CaAlSi_3O_8$)	3.9-7.5	3.2-3.6	2.5-4.7
Hematite (Fe_2O_3)	1.1-3.5	1.0-2.5	2.2-2.5
Dolomite ($CaMg(CO_3)_2$)	1.0-1.4	1.0	1.5-3.9
Calcium aluminium oxide hydrate ($Ca_3Al_2O_6 \cdot nH_2O$)	0.0-1.9	1.0-1.3	0.0-1.2

It is likely that the source of detected minerals by x-ray diffraction analysis is the following: (a) the silica derived from two sources; quartz (SiO_2), plagioclase (calcium aluminum silicate), and newly formed calcium silicate hydrate mineral from the reaction of tricalcium silicate with water; (b) aluminum is detected in the three phases; plagioclase (calcium aluminum silicate) derived from desert sand, aluminum oxide hydrate derived from the reaction of aluminum oxide with water, and calcium aluminum oxide hydrate derived from the reaction between tricalcium aluminates with water; (c) iron appeared in the form of hematite (Fe_2O_3) and was derived from fly ash and desert sand; (d) calcium appeared in the phase of calcite ($CaCO_3$), dolomite ($CaMg(CO_3)_2$), calcium silicate, calcium aluminum oxide hydrate, and plagioclase (calcium aluminum silicate); and (e) magnesium found in minor proportions and concentrated mainly in dolomite ($CaMg(CO_3)_2$). Such minerals are stable and contribute to the observed BMSC compressive strength increase.

In Figure 6.28a, BMSC samples were subjected to various temperature (40, 60, 80, and 100°C) for 1, 2, and 7 days, respectively. As an example, Figure 6.28b shows that the mineralogical compositions of BMSC samples, when subjected to these temperatures for one day, are consisted of major components: sulfur (decreasing percentage of sulfur with increased temperature) and quartz (the opposite trend of sulfur).

The minor constituents include the following in decreasing percentages; plagioclase, aluminum oxide hydrate ($5Al_2O_3 \cdot nH_2O$), calcium silicate hydrate ($Ca_{1.5}SiO_{3.5} \cdot nH_2O$), calcite ($CaCO_3$), hematite ($Fe_2O_3$), dolomite ($CaMg(CO_3)_2$),

236 Sulfur Concrete for the Construction Industry

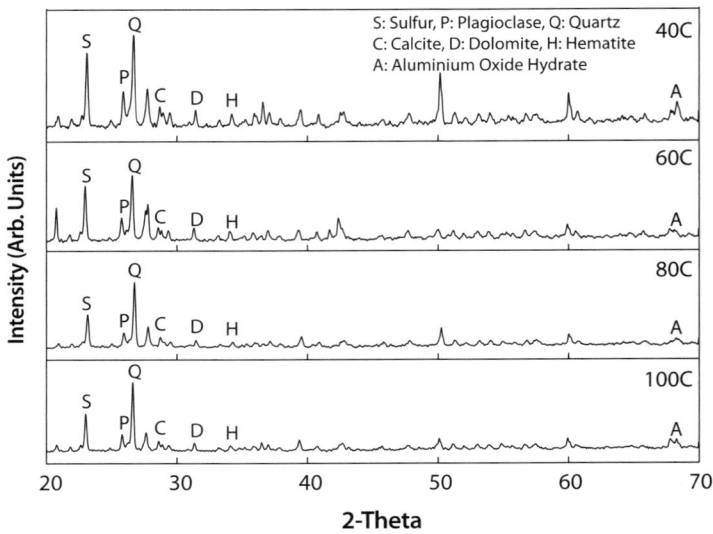

Figure 6.28(a). X-ray diffraction of BMSC cured for one day at various temperatures.

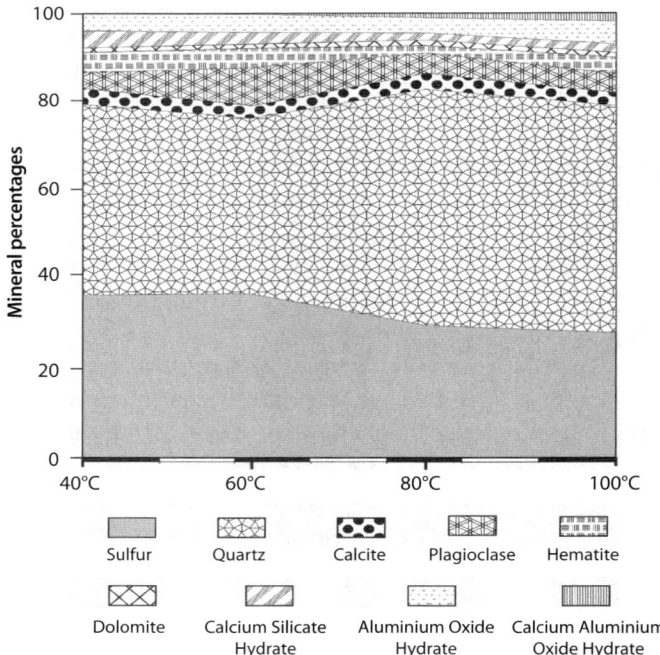

Figure 6.28(b). Lateral variation of mineral composition of BMSC sample cured for one day at various temperatures.

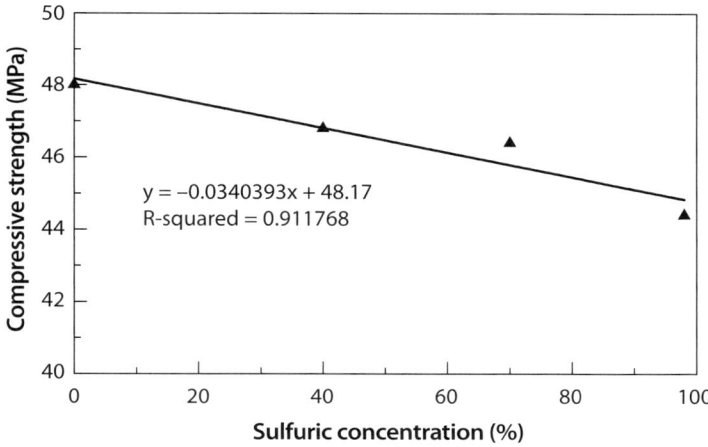

Figure 6.29. Compressive strength as a function of acid concentration at a constant temperature of 24°C and a curing time of 7 days.

and calcium aluminum oxide hydrate ($Ca_3Al_2O_6 \cdot nH_2O$). The percent composition of these minerals is shown in Table 6.9.

It is worth noting that the amount of water molecules associated with the formed hydrated compounds is unknown because the process does not utilize water at all. This leads one to pose the following question. Where is the water coming from? Is it from the humidity of the atmosphere or from the bonded water in the recycled aggregates? These issues need further investigation.

6.13.7 Durability

The durability of BMSC in different chemical environments and temperatures has been studied. Specimens were immersed in distilled water at different temperatures of 24°, 40°, and 60°C; 3% saline solution at 40° and 60°C; and 70% sulfuric acid solution at 40°C. All experiments were run for 28 days.

Moisture Absorption

The water absorption percent for PCC in water and saline solution reached 12.5 and 7.7%, respectively, which are considered high. However, for BMSC, the recorded water absorption values for water and saline solutions were 0.17 and 0.25%, respectively, indicating that BMSC has low water absorption characteristics in comparison with PCC. To explain such results, one has to realize that the existing sulfur and the added bitumen repel water because they are hydrophobic in nature. Also, the majority of the matrix is composed of sulfur coated materials (fly ash and sand) and sulfur in the voids between particles. This, in turn, will lead

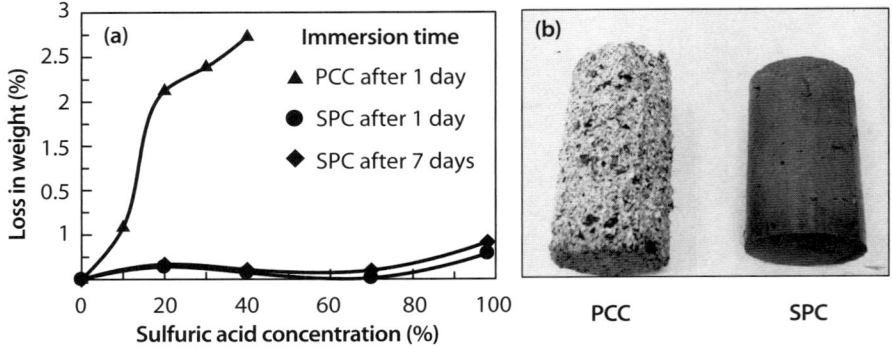

Figure 6.30. (a) Weight loss and (b) surface morphology of PCC and BMSC due to sulfuric acid attack.

to what is known as BMSC impermeability characteristics (Darnell, 1991). These results are in agreement with recommended specifications (ACI Committee 548, 1993) in which the maximum moisture absorption of BMSC should be less than 1% for coarse aggregates and less than 2% for fine aggregates.

Compressive Strength Development in Acidic Environment

The potential deterioration of BMSC specimens in acidic solutions was evaluated via compression test results after specimen's immersion in various concentrations of sulfuric acid solutions for one week. The results shown in Figure 6.29 indicate a slight decrease in compressive strength with the increase of acid concentration. This is ascribed to the consumption of small amounts of sulfur due to the slow reaction between sulfur and sulfuric acid (Eq. 6.4). Hence, the sulfur binding strength between aggregates is reduced leading to the decrease in BMSC strength. The fact that BMSC maintains its high strength, even in a concentrated acidic environment (98% H_2SO_4), is an indication that BMSC is acid resistant.

For an acidic solution of 40% H_2SO_4, the behavior is completely the reverse, whereby water was released from BMSC specimens. This attributed to the reaction of concentrated sulfuric acid with sulfur, forming sulfur dioxide and water according to the reaction shown by Eq. 6.4:

$$S(s) + 2H_2SO_4(l) \rightarrow 3SO_2(g) + 2H_2O(l) \quad (6.4)$$

To evaluate the behavior of both BMSC and PCC further in an acidic environment, Figure 6.30a shows the results of the loss in weight for concrete samples after immersion in sulfuric acid concentrations (20, 40, 70, and 98% H_2SO_4) for 1 and 7 days. There is no significant loss in weight for BMSC samples especially

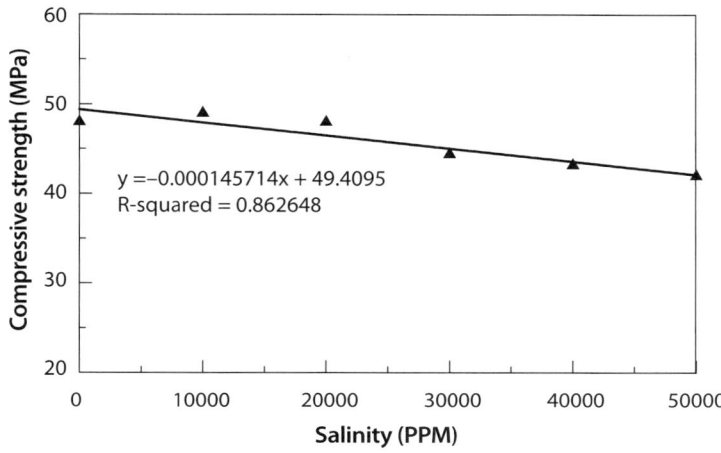

Figure 6.31. Compressive strength for BMSC as a function of sodium chloride concentration after immersion of 7 days at a constant temperature of 24°C (adapted from Mohamed and El Gamal, 2009).

for that immersed in dilute sulfuric acid, and a small loss in weight (0.19%) was detected after immersing the samples for 24 hr in 98% H_2SO_4 solution. No more loss in weight was observed for immersed BMSC samples for 7 days at different sulfuric acid concentration.

For PCC, the behavior with respect to acid attack is different. It is known that PCC is a highly alkaline material and is not resistant to attack by acids. Immersed PCC specimens in sulfuric acid solutions showed high effervescence and a loss in weight of 2.7% for samples immersed 24 hr in 40% H_2SO_4 solution was recorded. In addition to the observed weight changes, the color and surface morphology was observed. Figure 6.30b shows the corroded PCC surface after 24 hr of immersion in 40% H_2SO_4 solution that becomes soft and white due to the formation of calcium sulfate hydrate, according to the chemical reaction shown by Eq. 6.5. This indicates that BMSC offers better protection from the acid attack compared to PCC:

$$Ca(OH)_2 + 2H_2SO_4(aq) \rightarrow CaSO_4 \cdot 2H_2O(s) \tag{6.5}$$

Compressive Strength Development in Saline Environment

The BMSC durability in saline environments, after immersion of BMSC specimens in NaCl solution for 7 days at a constant temperature of 24°C, was evaluated. The results indicated that there is no loss in weight after immersion. Higher

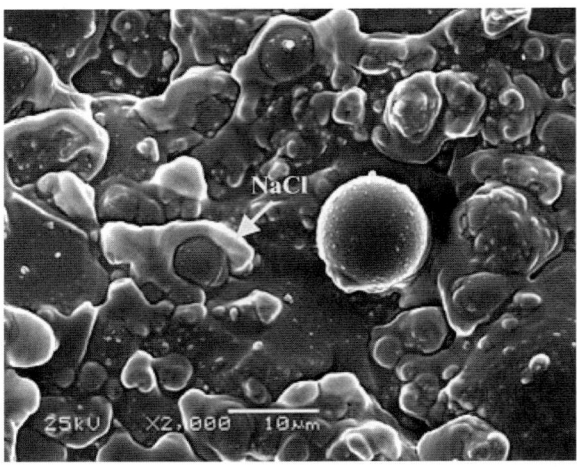

Figure 6.32. SEM micrograph of BMSC after immersed 7 days in 5% sodium chloride solution, at 24°C (adapted from Mohamed and El Gamal, 2009).

resistance to saline environment was achieved even with a small loss in compressive strength at high salinity as shown in Figure 6.31.

Concerning the microstructure of BMSC specimens, Figure 6.32 shows a high-magnification of surface image for the contact regions between sulfur and physical aggregates (fly ash and sand). The growth of sodium chloride crystals leads to a partial detaching between sulfur and physical aggregates that, in turn, results in compressive strength reduction. The chemical analysis performed by ICP, for BMSC specimens immersed for 7 days in 5% NaCl solution, indicates the presence of 0.049% Na atoms at the surface (i.e., 0.125% NaCl molecules). However, for the BMSC section at a depth of 10 mm from the surface, the analysis does not show any sodium atoms, indicating that NaCl penetration is only limited to the outer surface of BMSC specimens. The implication of such findings is that BMSC is a corrosion-resistant material and could be reinforced. In addition, hydraulic conductivity measurements, for BMSC specimens immersed in 3% NaCl solution, indicated that BMSC has low values in the range of 10^{-11} to 10^{-13} m/s that are less than that for the case of de-ionized water by about two orders of magnitude. Such low hydraulic conductivity makes BMSC a good candidate for stabilization of hazardous waste and barrier system design (Mohamed and El Gamal, 2008, 2009).

The variations of compressive strength with time for BMSC samples cured in 3% NaCl solution at 40 and 60°C are shown in Figure 6.33. It is clear that the higher the water temperature, the lower the strength gain is. Therefore, BMSC

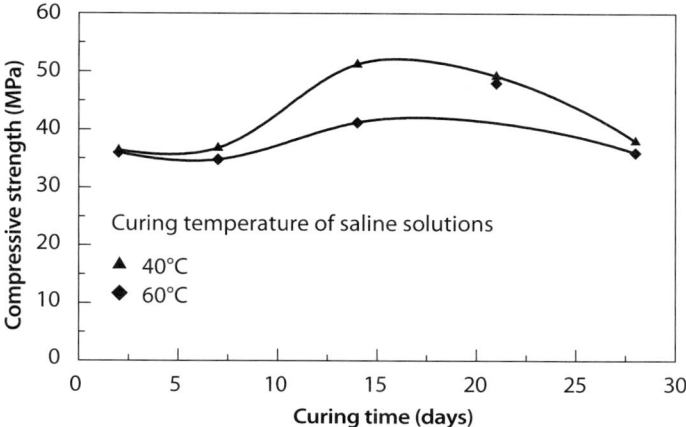

Figure 6.33. Change in compressive strength of BMSC cured in 3% sodium chloride solution at 40 and 60°C with curing time (adapted from Mohamed and El Gamal, 2009).

specimens have maintained their acceptable strength, indicating good resistance to saline environment.

6.13.8 Hydraulic Conductivity

Measured hydraulic conductivities of BMSC and PCC indicated that BMSC has low water permeation. Under a pressure of 2.2 MPa, BMSC has a hydraulic conductivity in the order of 1.456×10^{-13} m/s, whereas PCC has a hydraulic conductivity of approximately 8.39×10^{-8} m/s after being immersed in water. It is interesting to note that some researchers (Vroom, 1998) have evaluated the volume of pore spaces of BMSC and PCC and have indicated that they have approximately the same volume; however, the pores in BMSC concrete are not connected, providing low hydraulic conductivity characteristics, whereas the pores of PCC concrete are connected. It should be noted that the hydraulic conductivity of a material is highly dependent on the size of pore spaces, degree of connectivity between pores, grain shape, degree of packing, and cementation (Yong, Mohamed, and Warkentin, 1992).

To evaluate the effect of acid attack on the hydraulic conductivity, BMSC specimens were immersed in 98% sulfuric acid, 50% phosphoric acid, 30% boric acid, and 10% acetic acid at 24°C. Experimental results revealed that hydraulic conductivity values of BMSC specimens are in the range of 10^{-11}–10^{-13} m/s, indicating that BMSC is an impermeable material. Table 6.10 summarizes the

Table 6.10. Effect of acid type on hydraulic conductivity measurements, weight loss, and compressive strength loss of SPC after 7 days immersion in corrosive acids (adapted from Mohamed and El Gamal, 2009).

Acid type	Hydraulic conductivity (m/s)	Weight loss (%)	Strength loss %
Water	1.456×10^{-13}	0.00	0.0
98% sulfuric	7.660×10^{-11}	0.23	13.5
50% phosphoric	3.103×10^{-12}	0.08	7.9
30% boric	8.176×10^{-13}	0.07	4.0
10% acetic	2.196×10^{-12}	0.14	16.0

Table 6.11. Effect of acid concentration on the % element of BMSC composition; data obtained from the EDX (adapted from Mohamed and El Gamal, 2009)

Element %	Sulfuric acid concentration			
	20%	40%	70%	98%
O	34.9	26.6	20.44	25.06
Al	7.64	11.98	8.06	4.84
Si	28.21	29.76	21.93	11.51
S	24.6	26.85	44.49	56.01
K	0.93	0.89	1.25	0.54
Ca	1.87	1.02	0.39	0.42
Fe	1.84	2.89	3.43	1.63

experimental results of the hydraulic conductivity, loss in weights due to chemical reaction, and compressive strength loss. These data have revealed that BMSC exhibits high resistance to aggressive acidic environment; under the same conditions, Portland cement concrete specimens, in most cases, were destroyed.

Evaluation of the effect of the acid concentration on the durability of BMSC was studied through immersion of BMSC specimens in 20, 40, 70, and 98 wt% sulfuric acid solutions for 7 days at 24°C. Specimens were then washed, dried, and examined by SEM as well as by chemical analysis using EDX. The results shown in Table 6.11 indicated that BMSC samples were composed mainly of silicon- and sulfur-containing compounds and various metals such as calcium, aluminum, iron, and potassium.

The EDX observation revealed that with an increase of sulfuric acid concentration, the percentage of elemental sulfur increases because of the formation of metal sulfates due to the reaction of the basic oxides (in the fly ash and sand) with sulfuric acid. It is known that most bases dissolve in water, releasing hydroxide ions (OH^-) that react with acids to form salts. A calcium oxide base

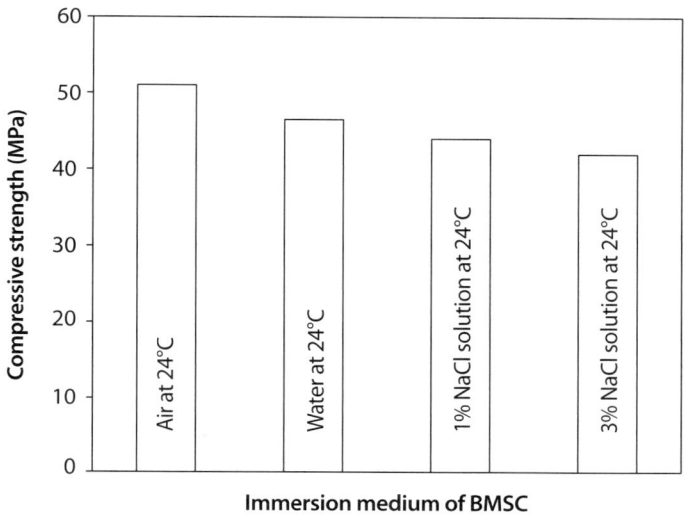

Figure 6.34. One year immersion test of BMSC in various saline solutions at 24°C.

accepts hydrogen ions, therefore, one could say that a base is a proton acceptor as indicated by Eqs. 6.6 to 6.8. With an increase of sulfuric acid concentration, a reduction in the capillary porosity of the system could be obtained due to salt formation that was precipitated on the BMSC surface and in the pore spaces:

$$\text{Acid} + \text{Basic Oxide} \rightarrow \text{Salt} + \text{Water} \tag{6.6}$$

$$H_2SO_4(aq) + CaO(s) \rightarrow CaSO_4(aq) + H_2O(l) \tag{6.7}$$

$$Ca^{2+} + Al^{3+} + 2SO_4^{2-} + 12H_2O \rightarrow CaAl(SO_4)_3 \cdot 12H_2O(s) \tag{6.8}$$

6.13.9 Long-Term Hydromechanical Behavior

BMSC specimens were further tested to determine their durability in hydrates and saline environments after one year at room temperature (24 ± 2°C). Figure 6.34 shows the compressive strength variations of specimens after one year of immersion in distilled water and in different saline solutions. The results indicated that there is no loss in weight and no adverse effects in compressive strength. BMSC is corrosion-resistant and could be used in hydrated and salt environments. It did not exhibit any deterioration; only limited compressive strength loss was observed.

6.13.10 Leachability

Sulfur-based concrete resembles Portland cement concrete in general composition except that a modified sulfur binder replaces the Portland cement. Portland cement concrete is considered toxicologically nearly inert and usually does not cause more than a physical alteration to habitats where it is used. The calcium salts in Portland cement can leach from the concrete and increase the pH of the surrounding soil. This is a localized effect and affects only acid-loving plants near a concrete installation such as a concrete block house foundation. However, significantly altering the pH of soil can impact the fate and transport of other pollutants. For example, mercury becomes soluble under the high pH conditions present in a leached solution from Portland cement concrete (Beall and Neff, 2005).

Resistance of BMSC to corrosive environments was evaluated by testing the specimens in various aqueous environments (Mohamed and El Gamal, 2007, 2008, 2009). Laboratory tests were performed by immersing specimens with a dimension of 50 × 50 × 5 mm in a transparent container filled with 1000 ml of tested aqueous environment. One ml from each aqueous solution was used for analysis by ICP for the determination of the total leaching sulfur as sulfate and metals such as calcium, magnesium, aluminum, and iron salts.

Influence of pH and Solution Composition

Leaching tests were executed at several constant pH values to evaluate the influence of a pH medium on the leaching of sulfur and metal oxides from the solidified matrix. Universal buffer solutions were used that were prepared by modifying the method reported by Britton (1952) by mixing equal volumes of acids (acetic acid, phosphoric acid, and boric acid) in bottles; the total morality of the acid mixture was maintained at 0.4 M for the three acids. The desired pH was reached by mixing the acid mixture with the required 1M sodium hydroxide solution. A constant ionic strength of the three buffer solutions, pH 4, 7, and 9, was maintained and adjusted using a pH meter. BMSC specimens were immersed in a transparent container filled with tested buffer solution. Aliquots were sampled and submitted for ICP analysis for total sulfur and metals leaching.

Chemical leaching of sulfur from BMSC specimens was measured as sulfates using ICP-AES. Because elemental sulfur exists in various allotropic forms with different densities, which are sensitive to cooling rates, it may cause microcracking and surface imperfections that provide excellent spots for oxidation. In the presence of oxygen and water, sulfur is slowly oxidized to sulfite and then to sulfate as shown by Eqs. 6.9 and 6.10 (Mattus and Mattus, 1994):

$$S^0 + O_2 + 2H_2O \rightarrow H_2SO_3 \qquad (6.9)$$

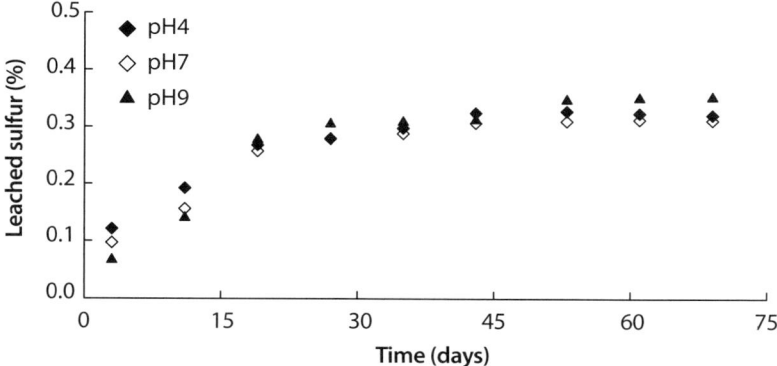

Figure 6.35. Variations of leached sulfur from BMSC with time and solution pH.

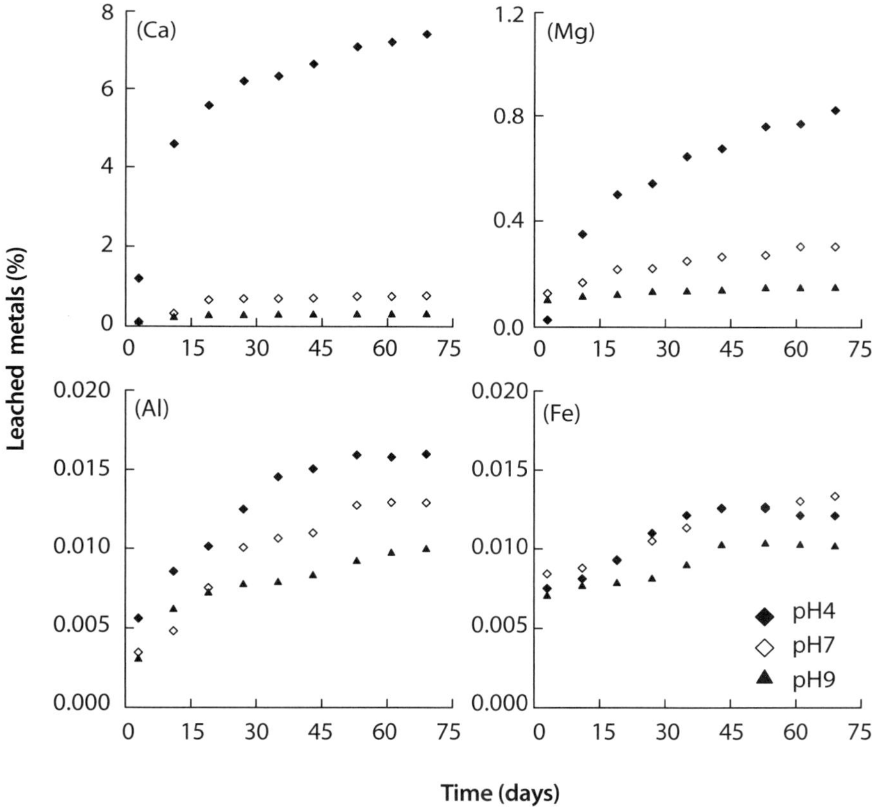

Figure 6.36. Variations of amount of Ca, Mg, Al, and Fe leached from BMSC.

$$H_2SO_3 + \frac{1}{2}O_2 \rightarrow H_2SO_4 \quad (6.10)$$

The leaching experiments were performed in accordance with the Accelerated Leach Test procedure that was developed for evaluating the potential leachability from solidified matrices. The test protocol specifies changes in pH medium, temperature, surface area to volume ratio, and testing time. The results obtained included the incremental and cumulative sulfur fraction leached. The incremental leaching data as a function of time are shown in Figures 6.35 and 6.36.

The results shown in Figures 6.35 and 6.36 indicate the following:

1. The leaching rates of sulfur are extremely low irrespective of pH variations of the aqueous solutions. The amount of leached sulfur from the solidified matrix in acidic (pH4), neutral (pH7), and alkaline (pH9) mediums is approximately the same as shown in Figure 6.35. This may suggest that the stability of the solidified matrix in an aqueous environment is independent of the pH of the solution; similar results were reported by Sliva (1996). A small increase of leached sulfur over time was observed, but the released sulfur could not overcome the buffer capacity because the solution pH was reported to be constant with time.

2. Materials used such as sands and fly ash contain leachable or extractable metallic pollutants, therefore, it is of prime importance to evaluate the potential leachability of these metals. The results shown in Figure 6.36 indicate that the main leached metals are Ca with fewer amounts of Mg, Al, and Fe. This could be explained in view of the electronegativity of atoms. It is known that metals have different tendencies to gain electrons. The greater the electronegativity of an atom, the greater is its affinity for electrons. The electronegativities of Ca, Mg, Al, Fe, and sulfur are -1.00, -1.55, -1.61, -1.83, and -2.58, respectively. This, in turn, highlights that the electronegativity increases, moving from left to right in the periodic table and decreases moving down a group (Masterton and Hurley, 1997).

3. It can be seen from Figures 6.35 and 6.36 that irrespective of pH values, all curves follow the same trend. Metal ions have lower solubility at alkaline than acidic pH values. This is not surprising because of the basic nature of the oxides of these metals that are different from one another which leads to different leaching:

    - Many metal oxides react with water to form alkaline hydroxides, e.g., calcium oxide (lime) reacts with water to form calcium hydroxide (Brown et al., 2000):

$$\text{Metal Oxide} + \text{Water} \rightarrow \text{Metal Hydroxide} \quad (6.11)$$

$$CaO(s) + H_2O(l) \rightarrow Ca(OH)_2(aq) \quad (6.12)$$

- Some metal oxides do not react with water, but they are basic when they react with acid to form salt and water:

$$\text{Metal Oxide} + \text{Acid} \rightarrow \text{Salt} + \text{Water} \quad (6.13)$$

$$MgO(s) + 2HCl(aq) \rightarrow MgCl_2 + H_2O(l) \quad (6.14)$$

- Others exhibit amphoterism, that is they react with both acids and bases such as aluminum oxide that dissolves in a strong acid and a strong base:

$$Al_2O_3 + 6H^+ \rightarrow 2Al^{3+} + 3H_2O \quad (6.15)$$

$$Al_2O_3 + 6OH^- + 3H_2O \rightarrow 2Al(OH)_6^{3-} \quad (6.16)$$

- Still others are neutral and nonreactive

4. The leached metals slightly increase linearly over time throughout the test period whereas the solution pH was buffered at the same pH.
5. The leached materials are generally low because of low hydraulic conductivity of the solidified matrix (Mohamed and El Gamal, 2007). In addition, because of their hardening by solidification, metal oxides found in the fly ash are chemically bonded within the matrix because they are converted to less soluble metal sulfides and a small percentage of sulfates (Darnell et al., 1992). This property of transformation of metal oxides to less soluble sulfide forms is also reported by Mayberry et al. (1993). These reasons make the BMSC a good candidate for utilization as a matrix or binder for the immobilization of wastes. Leaching studies indicated that the stabilization process has minimized or prevented the release of the toxic elements from the solidified matrix.

Influence of Temperature

Because temperature is an important factor that greatly influences the rate of leaching of sulfur and metals from solidified matrix, specimens were tested in distilled water at temperatures of 24°, 40°, and 60°C. The results shown in Figures 6.37 and 6.38 highlight the following:

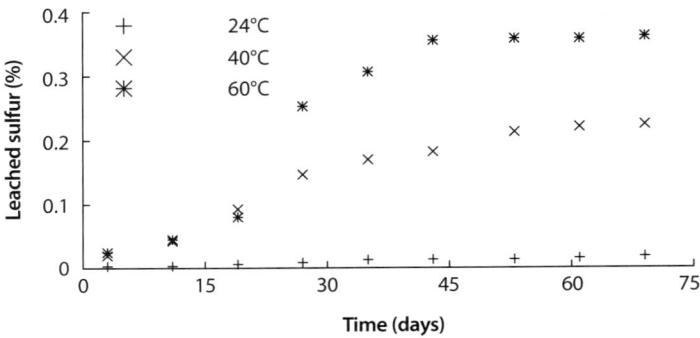

Figure 6.37. Variations of amount of leached sulfur from BMSC with time and temperature.

a. The leached sulfur for the case of distilled water at room temperature was of no consequence throughout the test period of 70 days (Figure 6.37). This means that BMSC is stable and insoluble in distilled water at room temperature.
b. The leached materials from the sulfur concrete tested in distilled water are sulfur, Ca, and Mg. Other metals such as Al and Fe were not detected in the leached products.
c. The difference in the rate of leached materials in distilled water between 24°, 40°, and 60°C was insignificant during the early test period up to 19 days and gradually increased over time; the temperature effects were small and increase slowly with time. With the further increase of immersion time, an expected increase in sulfur and metal oxides leaching into the solution was observed. This was an indication of dependence of the reaction rate of metals in the solidified matrix on temperature and time when immersed in distilled water.

The leached rate of metal oxides (Ca and Mg) was insignificant at room temperature but slightly enhanced with increased temperature as shown in Figure 6.38. High temperature has accelerated the leaching process because solubility of metals depends on temperature and increases consequently as temperature increases (Lageraaen and Kalb, 1997).

Influence of Salinity

Chemical investigations into possible leaching of sulfur from BMSC specimens immersed in de-ionized water (DIW) and a saline solution of 5% NaCl were conducted. Figures 6.39 and 6.40 show the percentage of leached sulfur and calcium

Sulfur Concrete **249**

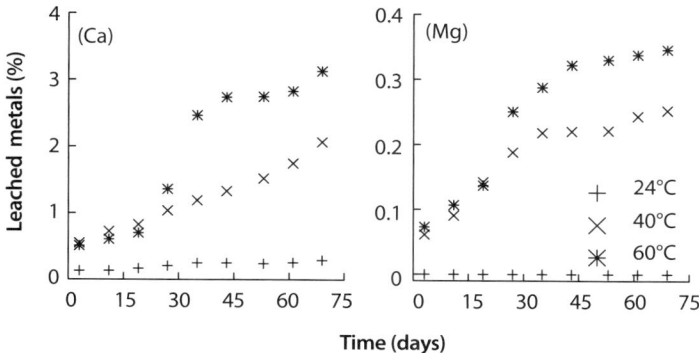

Figure 6.38. Variations of leached Ca and Mg from BMSC with time and temperature.

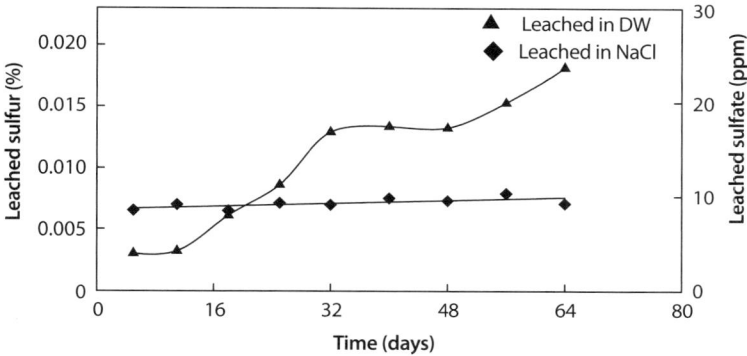

Figure 6.39. Variations of the amount of sulfur and sulfate leached from BMSC in de-ionized water (DIW) and 5% salt solution (adapted from Mohamed and El Gamal, 2009).

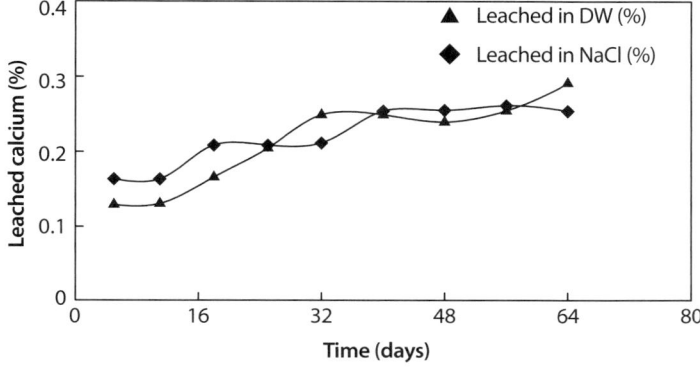

Figure 6.40. Variations of the amount of calcium leached from BMSC in de-ionized water (DIW) and 5% salt solution (adapted from Mohamed and El Gamal, 2009).

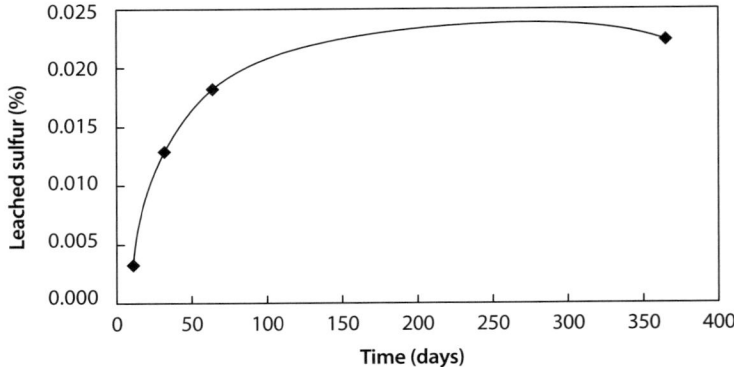

Figure 6.41. Variations of the amount of sulfur leached from BMSC in de-ionized (DIW) at 24°C as a function of time (adapted from Mohamed and El Gamal, 2009).

during 64 days in DIW and the saline solution. The results indicated that, at room temperature, BMSC was stable and insoluble in water throughout the test duration and no pH change or weight loss was observed. As the duration time of leached test increases, it was found that the total leached sulfur in DIW was nearly steady.

As discussed, calcium and sulfate are leached away by the aqueous solutions of DIW and the saline solution from BMSC. The total sulfur content leached in each solution, as sulfate, appears to be low as shown in Figure 6.39. Also, the tested solutions have extremely low concentrations of dissolved calcium sulfate as shown in Figure 6.40. Saline water contributed to calcium leaching from BMSC; approximately the same quantity as that leached in the case of DIW. Whereas the leached sulfate in saline solution is significantly lower than that in DIW, it means that there is another source of calcium in saline water rather than calcium sulfate.

Using the U.S. Environmental Protection Agency titration test method for determination of the soluble carbonates and bicarbonates after 64 days, it was found that in DIW the carbonates and bicarbonates were 0 and 24.3 mg/l, respectively, while in saline solution they were 3.6 and 112 mg/l, respectively. This indicates that the leached calcium in DIW comes mainly from the dissociation of calcium sulfate whereas, in saline solution, the major leached calcium comes from the dissociation of calcium bicarbonate. Also, it can be noted that the decomposition of sulfur in BMSC in a salt solution is significantly slower than that in water, which is compatible with previously reported studies (Lusty, 1983). The data represented in Figure 6.41 indicated that BMSC is stable in aqueous environment, which is consistent with the previous observation where there was no weight loss observed.

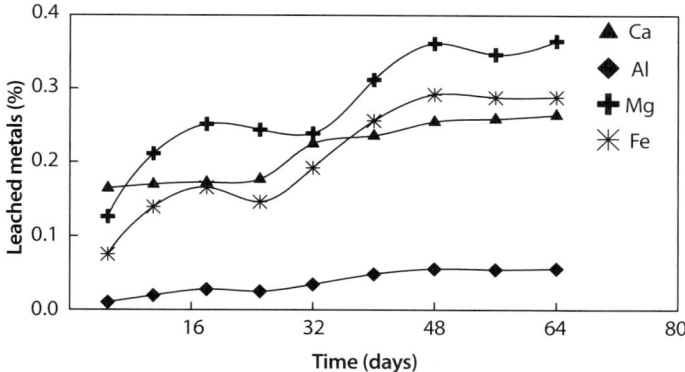

Figure 6.42. Variations of the amount of metals leached from BMSC in 70% sulfuric acid solution at 24°C (adapted from Mohamed and El Gamal, 2009).

Influence of Acidity

The attack of BMSC by 70% sulfuric acid solution resulted in the formation of calcium, aluminum, iron, and magnesium sulfates that, when they are soluble, might be leached away by the aqueous solution as indicated in Figure 6.42. Sulfuric acid reacts with most metals in a displacement reaction to produce metal sulfates. Dilute H_2SO_4 attacks calcium, aluminum, magnesium, and iron. The calcium sulfate is a major soluble salt, and the acidic environment accelerated salt solubility and the leaching process. Water acts as a solvent and a transport medium for aggressive agents and reaction products. The degree of reaction of an acid is dependent on the chemical character of anions present.

The strength of acid, its dissociation degree in solutions, and mainly the solubility of the salt formed are dependent on the chemical character of anion. The acidic attack is based on the processes of decomposition and leaching of the constituents of concrete matrix. This effect contributes to the progress of acidic attack. Figure 6.42 indicates that metals leached in an acidic solution, depending on the percentage of each metal oxide in the mixture and the level of concentrations in the leachate, are below the concentration limits set by environmental agencies for solidification of hazardous waste.

6.14 POTENTIAL ECOLOGICAL EFFECTS OF SULFUR CONCRETE

Portland cement concrete is considered toxicologically nearly inert and usually does not cause more than a physical alteration to habitats where it is used. The calcium salts in Portland cement can leach from the concrete and increase the pH of the surrounding soil. This is a localized effect and affects only acid-loving plants near a concrete installation such as a concrete block house foundation. However, significantly altering the pH of soil can impact the fate and transport of other pollutants.

The literature lacks data on leaching of sulfur from sulfur concrete structures under normal use conditions. For example, McBee et al. (1985) reported that sulfur concretes tested under actual operating conditions in 50 corrosive environments and 40 commercial facilities over a 4-year period showed no significant signs of strength loss or deterioration. These results indicate that there was little or no loss of sulfur from the structures. Elemental sulfur has a low aqueous solubility and sulfur concrete has a low water content and hydraulic conductivity, reducing the likelihood of significant leaching of sulfur (Mohamed and El Gamal, 2009; Wrzesinski et al., 1988).

If sulfur leaches from sulfur concrete, it will have the opposite effect, compared to Portland cement concrete, on the soil pH in the immediate vicinity of the concrete structure. As discussed in Chapter 3, elemental sulfur (the form of sulfur in sulfur cement and concrete) is oxidized in soil to sulfate that reacts with soil water to form sulfuric acid, reducing soil pH. Warren and Dudas (1992a) reported pH values below 3.0 in soils adjacent to an elemental sulfur stockpile. At a soil pH of approximately 2.1, carbonates, iron-containing primary minerals, chlorite, bentonite, plagioclase, and feldspars were either partially or completely dissolved (Mohamed and Antia, 1998; Yong, Mohamed, and Warkentin, 1992). Calcium carbonate and calcium-feldspars dissolved in the presence of sulfuric acid and the calcium re-precipitated as gypsum. The iron dissolved from the various minerals precipitated as iron oxyhydroxides. Clay-sized particles of mica and kaolinite were not affected by the acid environment. These changes in the mineral properties of the soil changed its texture and undoubtedly altered its suitability for local plants and soil animals. It is unlikely that soil acidification would be as extreme near sulfur concrete structures as near the sulfur pile. It is unlikely that soil in contact with sulfur concrete structures would decrease much below 5.5. At this pH, there would be little or no alteration of the mineral structure of the soil.

However, soil acidification affects metal solubility and mobility (Mohamed and Antia, 1998; Yong, Mohamed, and Warkentin, 1992; Warren and Dudas, 1992b). As a general rule, iron minerals in soils and sediments dissolve under acid conditions and precipitate under alkaline conditions (Mohamed and Antia, 1998; Yong, Mohamed, and Warkentin, 1992; Lepp, 1972). Manganese, also frequently abundant in soils and sediments, behaves in a similar manner. Large amounts of several metals may form complexes or co-precipitated with solid iron and manganese oxyhydroxides or iron sulfides. Acid dissolution of these solid phases may release dissolved metals into the soil or sediment pore water (Mohamed and Antia, 1998; Yong, Mohamed, and Warkentin, 1992).

In soils near the elemental sulfur stockpile examined by Warren and Dudas (1992b), arsenic, cobalt, and gallium were mobilized by the acid and re-precipitated with the iron oxyhydroxides. Barium and hafnium were enriched in the most acidified layers of soil. Soil acidification caused volatilization of mercury to the atmosphere, probably by converting it to volatile elemental form. Other metals were not affected by the sulfur-induced soil acidification. All the metals originated in the soil; none came from the pile leached solution. Because soil pH near sulfur concrete structures will not be as low as the pH values measured near the sulfur stockpile, it is doubtful that there will be substantial mobilization and redistribution of metals in the soil (Beall and Neff, 2005). Kalb et al. (1991a, b) showed by TCLP analysis that leaching of metals from incinerator fly ash was reduced to acceptable levels by encapsulating the fly ash in modified sulfur cement. Thus, sulfur concrete is unlikely to leach metals even under severe environmental conditions (Beall and Neff, 2005). Mohamed and El Gamal (2009) reported that level of metal leachability from modified sulfur concrete is below concentration limits set by environmental agencies (Figure 6.42).

Ferenbaugh et al., (1992) have shown that when a commercial sulfur binder known as *Sulflex* was added to soils, the sulfur apparently was oxidized to sulfate by soil-dwelling *Thiobacillus thiooxidans* with simultaneous acidification of the soil. It is worth noting that Sulflex is a mixture of elemental sulfur and plasticizers and has been marketed as an alternative binder for asphalt or Portland cement in concrete. Plants grown on Sulflex-amended soils accumulated sulfur and their productivity was reduced. It is uncertain if sulfur concrete or sulfur modified asphalt would behave in the same way (Beall and Neff, 2005). Lee (1998) reported that sulfur concrete sewer pipe was resistant to acid and bacterial corrosion. Further work is needed on the susceptibility of sulfur concrete and sulfur asphalt of the types used for roads and construction to bacterial oxidation and reduction under realistic environmental conditions.

As previously discussed in Chapters 5 and 6, modified sulfur binder in most sulfur cement and concretes contains 5 to 20% of polymer modifiers. As an example, the polymer modifiers used most frequently are mixed dicyclopentadiene and cyclopentadiene oligomers. These oligomers apparently polymerize and bind tightly to the sulfur phase during the cooling of the sulfur concrete. They are considered to be insoluble in water and are unlikely to leach from the concrete. The monomers probably are slightly soluble. According to the Agency for Toxic Substances and Disease Registry, dicyclopentadiene has a low toxicity to mammals (http://ntp-server.niehs.nih.bov/htdocs/CHEM_H&S/NTP_CHEM7/). The median lethal dose (LD_{50}) by oral ingestion or intra-peritoneal injection in rats and mice ranges from 200 to 353 mg/kg body weight. The lowest inhalation dose causing adverse effects in rats during four hours of exposure is 500 ppm. The LD_{50}, when applied to rabbits by skin painting, is 5,080 mg/kg body wt. Exposure to 500 ppm causes eye irritation. There is no evidence that either cyclopentadiene or dicyclopentadiene monomers are mutagenic or carcinogenic in mammals. It is toxic to aquatic animals at concentrations of 1 to 10 ppm in water. These chemicals are classified as hazardous, primarily because of their flammability.

The only hazard of these chemicals may occur during the manufacture of sulfur concrete products when the binder modifier is handled neat. Inhalation of fumes of these chemicals could be harmful. The recommended maximum concentration in indoor air is 0.1 ppm. However, most people can smell dicyclopentadiene at levels in the atmosphere of 0.003 ppm. Once incorporated into the cement, these chemicals are immobile and inert and are unlikely to leach from the cement into the environment (Beall and Neff, 2005).

Solid Portland cement concrete is considered nontoxic. However, as with sulfur concrete, there may be slight health hazards associated with its manufacture. Inhalation of cement dust may cause nose, throat, or lung irritation and choking. Because the cement may contain crystalline silica, prolonged inhalation may cause silicosis. Silica dust is considered a human carcinogen when inhaled. Mainly because of the silica in Portland cement, the American Conference of Governmental Industrial Hygienists has set a threshold limit value for respirable Portland cement dust of 10 mg/m^3 of air.

In conclusion, the potential environmental effects of sulfur concrete and Portland cement concrete appear to be similar. The main environmental effects of both products are on soil pH and resulting changes in metals mobility and the suitability of the soils for various types of plants. Both products are not hazardous in their finished states, but the ingredients may be harmful to people manufacturing products with the two types of concrete. The health hazards are not great and methods of handling and working with these raw materials are available to minimize exposure to harmful dusts and chemicals.

6.15 SUMMARY AND CONCLUDING REMARKS

In this chapter, we have discussed various types of sulfur concrete products in view of the type of chemical additives and the overall performance. Each type behaves differently, depending on the chemical additives and the material composition. Overall, sulfur concrete products are highly durable and could be used as construction materials in locations within industrial plants or other locations in which acid and salt environments result in premature deterioration and failure of Portland cement concrete. While ultimate life or durability of sulfur concrete has not been completely established in many end-user applications, enough evidence of its corrosion resistance and durability has been accumulated to show that it has several times the life of other construction materials now being used in corrosive environments.

Generally, sulfur concrete has the following useful characteristics: (a) low hydraulic conductivity and porosity, (b) mechanical properties, including tensile, compressive, and flexural strengths, and fatigue life are greater than those obtained with normal Portland cement concrete, (c) comparable density with that of Portland cement concrete, (d) excellent resistance to attack by most acids and salts, some at high concentrations, also resists corrosive electrolyte attack, (e) can be placed year-round, (f) compositional materials could be recycled and used again, (g) impurity of used materials does not have any effect on the final strength properties, (h) no need of water for production, (i) possibility of color coding without any problem, (j) can be utilized as construction materials for floors, walls, and sump pits in the chemical, metallurgical, battery, fertilizer, food, pulp, and paper industries, and (k) can be utilized as road paving and bridge decking where salt corrosion problems are encountered.

Special attention should be given to the leachability of sulfur and other metal ions due to exposure to various environmental conditions. Therefore, material selection is of prime importance. The main environmental effects of sulfur concrete products are on soil pH and resulting changes in metals mobility and the suitability of the soils for various plants. Sulfur concrete products are not hazardous in their finished states, but the ingredients may be harmful to people who manufacture the products. The health hazards are not great and methods of handling and working with these raw materials are available to minimize exposure to harmful dusts and chemicals.

7

TECHNOLOGICAL ASPECTS OF SULFUR CONCRETE PRODUCTION

7.1 INTRODUCTION

Sulfur concrete is a thermoplastic material containing minerals with a modified sulfur binder. In principle, it is recoverable without limit; in practical experience, a fivefold recovery has been reported. The recovery process is energy efficient due to the low melting point (120°C) of the modified sulfur. Also, it leaves the aggregates intact from one application to the next. It is true that sulfur-modified concrete and Portland cement concrete contain the same amount of binder. Energy consumption is the biggest environmental concern with cement and concrete production. Melting elemental sulfur requires about one tenth of the energy it takes to produce Portland cement. However, taking into consideration that all aggregates must be heated to above the melting point of elemental sulfur, which is lost in the cooling process, sulfur-modified concrete consumes only about one third of the the energy of Portland cement concrete.

The durability of sulfur concrete relates to a number of conditions: mix design; aggregate composition and amount of binder; choice of aggregates that resists harsh environmental conditions; choice of filler that influences workability and thermal stress; composite (sulfur binder, fillers, and aggregates) resistance to acidic, alkaline, and salt environments; water permeability and absorption; casting procedures and binder compensation; frost resistance; service temperature; fire load; creep; fatigue load; steel and other reinforcements; and abrasion resistance.

This chapter discusses the preceding issues as well as other relevant ones that contribute to the development of a sustainable manufacturing technology for sulfur-modified concrete.

7.2 SULFUR CONCRETE PRODUCTION

As discussed in Chapter 2, the worldwide production of sulfur, especially where fuels were de-sulfurized due to concerns for the atmospheric environment, escalated in the late 1960s. This occurred in the United States and Canada with an increased recovery from sour natural gas and the refining of higher sulfur crude oils. Production increased along with projected sulfur recovery from conversion sources such as power plants and smelters, as well as recovery from an increased use of coal for energy production, and led to forecasts of a large sulfur surplus in the late 1980s.

7.2.1 The 1970s

The United States Bureau of Mines and The Sulphur Institute (TSI) (Washington D.C.) launched a cooperative program in 1971 to investigate and develop new uses of sulfur. After examining many potential uses, the conclusion was that the most promising use for engaging large amounts of this natural resource was in construction materials. Research was initiated at the Bureau of Mines in 1972. The intention was to go commercial no later than 1976.

In 1972, the Canada Center for Mineral and Energy Technology (CANMET) and the National Research Council (NRC) of Canada initiated a research program for the development of sulfur concrete. This was followed by work at the University of Calgary, Alberta, Canada.

In 1972, Vroom, with assistance from the NRC of Canada and Ortega of McGill University, Montreal, Canada, developed a process based on sulfur modification with olefinic hydrocarbon polymers called SRX polymer. The resulting sulfur concrete was first produced for commercial use in Calgary, Alberta, Canada, in 1975. Starcrete Technologies has been operating with this technology in Calgary since 1977. The properties and behavior of olefinic hydrocarbon polymers are discussed in Chapters 5 and 6.

The SRX modifier allows the sulfur to adopt a thermodynamically stable orthorhombic form, while preventing the growth of microcrystals. The stressing contraction upon phase transition does not occur when the crystals are prevented from growing any larger than approximately one micron in size (Vroon and Hyne, 1995). The choice of aggregates is apparently less particular than with Chempruf-modified sulfur concrete. Small amounts of organic matter, salts, and other impurities may be tolerated, but the aggregates must be essentially free (<1%) of swelling clays.

Observed failures in sulfur concrete is generally due to a growth of macro-sulfur crystals as the amorphous sulfur gradually reverts to the more stable orthorhombic form. The reversion is wholly dependent upon the plasticizers used and the individual processes.

In 1973, the Sulfur Development Institute of Canada (SUDIC), jointly founded by the Canadian Federal Government, Alberta Provincial Government, and the Canadian sulfur producers, was established to develop new markets for the increasing Canadian sulfur stockpile. By 1988, the activities of SUDIC were ceased because of product failures.

In 1977, U.S. Bureau of Mines and TSI initiated a cooperative test program on sulfur concrete. The objective was to evaluate the performance of the material under operating conditions in corrosive, industrial environments. Under the program, sulfur concrete was tested in 40 commercial plants in 56 corrosive environments. The plants used in the corrosion testing program included metal production and refining operations for aluminum, copper, nickel, lead, manganese, magnesium, titanium, uranium, vanadium, zinc, and precious metals, as well as chemical and fertilizer production plants for phosphoric, sulfuric, chromic, and nitric acids and sodium and potassium salts. The main objectives of the program were to: (a) establish the feasibility of using sulfur concrete in large-scale applications and (b) determine the ultimate performance of sulfur concrete under operating conditions in harsh environments.

Precast components such as tiles, slabs, tanks, and pump foundations were cast at a laboratory and then transported and placed in industrial plants. An example of the industrial evaluation is a sulfur concrete support pier that was installed to replace a deteriorated Portland cement concrete support pier in a potash storage plant. The performance was good enough to overcome the previously experienced degradation problems (McBee et al., 1985). The average test results when exposing sulfur concrete to various chemical environments ranged between three to five years of performance without showing any sign of corrosion or deterioration. Under the same conditions, conventional concrete materials were attacked and, in some cases, completely destroyed (McBee et al., 1985).

The test results indicated that sulfur concrete performed well in the majority of corrosive environments. However, sulfur concrete deteriorated in hot chromic acid solutions, sodium chlorate-hypochlorite, copper slimes, and hot organic solvent solutions. Failure occurs also in environments in which sulfur concrete was exposed to a temperature above 110°C. The final evaluation concluded that and Hyne, 1995). The choice of aggregates is apparently less particular than with Chempruf-modified sulfur concrete. Small amounts of organic matter, salts, and other impurities may be tolerated, but the aggregates must be essentially free (<1%) of swelling clays.

Observed failures in sulfur concrete is generally due to a growth of macro-sulfur crystals as the amorphous sulfur gradually reverts to the more stable orthorhombic form. The reversion is wholly dependent upon the plasticizers used and the individual processes.

In 1973, the Sulfur Development Institute of Canada (SUDIC), jointly founded by the Canadian Federal Government, Alberta Provincial Government, and the Canadian sulfur producers, was established to develop new markets for the increasing Canadian sulfur stockpile. By 1988, the activities of SUDIC were ceased because of product failures.

In 1977, U.S. Bureau of Mines and TSI initiated a cooperative test program on sulfur concrete. The objective was to evaluate the performance of the material under operating conditions in corrosive, industrial environments. Under the program, sulfur concrete was tested in 40 commercial plants in 56 corrosive environments. The plants used in the corrosion testing program included metal production and refining operations for aluminum, copper, nickel, lead, manganese, magnesium, titanium, uranium, vanadium, zinc, and precious metals, as well as chemical and fertilizer production plants for phosphoric, sulfuric, chromic, and nitric acids and sodium and potassium salts. The main objectives of the program were to: (a) establish the feasibility of using sulfur concrete in large-scale applications and (b) determine the ultimate performance of sulfur concrete under operating conditions in harsh environments.

Precast components such as tiles, slabs, tanks, and pump foundations were cast at a laboratory and then transported and placed in industrial plants. An example of the industrial evaluation is a sulfur concrete support pier that was installed to replace a deteriorated Portland cement concrete support pier in a potash storage plant. The performance was good enough to overcome the previously experienced degradation problems (McBee et al., 1985). The average test results when exposing sulfur concrete to various chemical environments ranged between three to five years of performance without showing any sign of corrosion or deterioration. Under the same conditions, conventional concrete materials were attacked and, in some cases, completely destroyed (McBee et al., 1985).

The test results indicated that sulfur concrete performed well in the majority of corrosive environments. However, sulfur concrete deteriorated in hot chromic acid solutions, sodium chlorate-hypochlorite, copper slimes, and hot organic solvent solutions. Failure occurs also in environments in which sulfur concrete was exposed to a temperature above 110°C. The final evaluation concluded that sulfur concrete has the potential for future expanded applications, for example, in the metallurgical, chemical, and fertilizer industries.

In 1978, a number of field test coupons prepared by the U.S. Bureau of Mines were placed in appropriate locations in several of the American Smelting and Refining Company (ASARCO) plants at El Paso, Texas. ASARCO is a major producer of nonferrous metals and minerals. The experience with sulfur concrete at ASARCO indicated complete satisfaction with the precast components. However, results were unsatisfactory in the case of insitu work because of the lack of experience of contractors. The failure was associated with procedures rather than the sulfur concrete material itself.

In the winter of 1979, 750 pipeline swamp weights, each weighing 1730 kg, were poured with Starcrete technology in a field in northern Alberta, Canada. The ambient temperature during the production period was consistently below −40°C. A relatively simple plant for producing Starcrete technology was set up specifically for this job. The equipment consisted of an aggregate drier with a feed conveyor and a bucket elevator to lift the heated aggregate to a height sufficient to drop it into conventional concrete mixer trucks. A wheeled loader was used to place aggregates in the feed hopper of the conveyor. Two unmodified concrete mixer trucks were used for mixing crushed sulfur and STX™ polymer with the heated aggregates. While one truck was mixing and discharging its load into steel molds, the second truck was being loaded with the raw materials. In spite of the low ambient temperature, there was no need to heat the mixer trucks because they could be emptied into the molds before any significant cooling of the hot mix occurred.

7.2.2 The 1980s

In 1980, 1100 sulfur polymer concrete pipeline weights were used to provide negative buoyancy to a 107 cm (42 in) diameter natural gas pipeline—the largest ever installed in Canada—using Starcrete technology. Sulfur polymer concrete was the preferred material because there was concern about the effect of sulfates and organic acids (i.e., muskage) that were produced from the substrate on sulfate-resistant Portland cement concrete pipes that will be placed in the substrate. The price was only slightly higher than that of sulfate-resistant concrete weights because the fast setting permitted a substantial savings in mold costs. Two different types of weights were produced—saddle weights for use in muskeg areas and bolt-on weights for river and lake crossings. The saddle weights weighed approximately 5600 kg each and the bolt-on weights, made in 2 halves, had a combined weight of about 6200 kg. The sulfur polymer concrete hot mix was produced in a mobile asphalt plant by mixing liquid sulfur and STX™ polymer with the hot, dry aggregates in the pug mill. This hot mix was then fed to concrete mixer trucks that had been modified with an insulating shroud over the mixer section. Mix temperatures were maintained by propane-fired heaters placed under the mixer. The molds were made of rolled, mild steel plate. Felt cushioning material was placed over the convex surface of the molds to prevent damage to the plastic wrap on the pipeline when the weights were placed on it. Pre-assembled cages of uncoated reinforcing steel, with lifting hooks attached, were positioned in the molds that were then filled from the mixer trucks. Both internal and external vibration were used during pouring. Setting of the hot mix was accelerated by occasional spraying on the outside of the molds with water. In this way, the weights could be lifted out of the molds with their own lifting hooks in 2.5 to 3 hr and immediately loaded onto flat deck trailers for transportation to the field. When running 24 hr a day, this precasting operation produced 240 tons of product per day.

After ten years of development at the U.S. Bureau of Mines, the new Chempruf modifier was patented in 1982 (McBee and Sullivan, U.S. Patent No. 4348313). In 1984, the U.S. Department of Commerce gave a license to exploit the invention to National Chempruf Concrete and operated in Illinois (Sulcon), Kansas (Reece Construction), Tennessee (GRC, Chempruf), Pennsylvania (Wagman Construction), Washington (F. E. Ward Contractors), and Copenhagen, Denmark (4KA/S). The Danish company acquired the license to carry out Chempruf concrete activities in most of Europe, the Middle East, and India. Recently, the Chempruf-modified sulfur is produced under the name of Chement 2000 (Chemically Resistant Cement) by Chemical Enterprises, Houston, TX.

The binder of sulfur concrete used in the Chempruf is elemental sulfur modified on a 50-50 basis, i.e., with a 5% modifying agent containing 2.5% dicyclopentadiene (DCPD) and 2.5% cyclopentadiene (CPD) oligomer. The properties and behavior of DCPD-CPD oligomer are discussed in Chapters 5 and 6.

As discussed in Chapter 5, Chempruf modified does not undergo a rapid phase transition from $S(\beta)$ to $S(\alpha)$ after solidification. This means that the transition is retarded or delayed, which may be sufficient for practical purposes. The durability was improved by eliminating the internal stresses caused by the phase transition. By the addition of 5% Chempruf modifier, volume contraction is reduced from 13% of elemental sulfur to only 4% (McBee et al., 1983). The corresponding thermally induced shrinkage of sulfur concrete is 0.7% (Aarsleff, 1989), which may be subdivided into 0.1 to 0.15% in the solid state (Weber, 1993) and 0.55 to 0.6% at solidification.

In 1982, SULCON applied the sulfur concrete technology developed by TSI in Washington, D.C. and the Bureau of Mines in Boulder City, NV to rehabilitate a large basement area under a copper electrowinning operation at AMAX Nickel, Braithwaite, LA. The project demonstrated the commercial potential of sulfur concrete because of the superior characteristics of sulfur concrete, including acid resistance, strength achieved in relatively few hours, low permeability, reusable, and the possibility to be used all year round.

Between 1983 and 1986, Brown and Root, Houston, TX, completed two sulfur concrete floor installations in its electrowinning facilities (Miele, 1986). The first installation was in New Mexico, and the sulfur concrete was prepared applying a Canadian patent. The second installation was an expansion of the same industry in New Mexico. The results indicated that sulfur concrete is the proper material for use in tank-house flooring. If good construction techniques, proper equipment, and well-qualified craftsman are utilized for the installation, sulfur concrete can provide an excellent floor system.

In 1983 to 84, the James River Pier at the Hopewell Plant was rehabilitated using sulfur concrete (Allied Signal Corporation, Collins, 1986). The pier was constructed in 1920 and deterioration started in 1954. By 1974, the sulfate attack was so severe that a major repair was necessary. The repair was completed using aluminous cement with a higher resistance than Portland cement concrete. Despite all efforts, deterioration continued. Sulfur concrete was used as a construction material protection cap of the structural concrete.

In Southern British Columbia, an outdoor vat—4 m wide, 11 m long, 2.5 m deep—was cast in 7 sections using Starcrete technology. U-shaped, plywood forms were used for each section. Epoxy-coated rebar was used in this case, although uncoated rebar is normally used. Tubing to carry post-tensioning tendons was tied to the rebar. This tubing extended 0.10 m from one side of each form. The forms were set on their sides for filling with the sulfur polymer concrete mix. Internal, plus some external, vibration was used during pouring. The floor and all walls were 0.46 m (18 in) thick. Forms were removed the day after pouring. The individual sections were placed in the excavation, using the tubing extensions to provide a uniform 0.10 mm space between each section. These spaces were then enclosed by additional form work and filled with more sulfur polymer concrete mix in which finer aggregate was used. After setting overnight, longitudinal post-tensioning was carried out. After backfilling, the vat was filled with 40% sulfuric acid and used for dissolving scrap zinc. The resulting zinc sulfate solution, containing about 20% sulfuric acid, is fed to electrolytic cells of an electrowinning plant for recovering pure zinc metal. The spent electrolyte, still containing 20% sulfuric acid and a small amount of zinc sulfate, is then returned to the vat where more acid is added for dissolving more zinc. The vat has performed satisfactorily in this service since 1985.

7.2.3 The 1990s

Sulfur polymer concrete (with STX™ polymer) was selected for the expansion of a large copper electrowinning plant in Arizona early in 1991. In such a plant, copper metal is electrolytically deposited from a solution of copper sulfate in sulfuric acid. During the electrolysis, hydrogen bubbles rise to the surface of the cell tank and burst, emitting a fine spray of sulfuric acid that collects on the floor below and that must be drained off. Due to the magnitude of the job and the owner's desire to have it completed in a timely manner, two types of equipment were used to supply the hot mix. One was a custom-made, modified asphalt plant in which the mix was produced and delivered from a pug mill to motorized buggies in 400-kg batches every 2.5 min. The other was a modified concrete mixer truck designed to heat and dry the aggregates in the truck. Sulfur and STX™ polymer were then introduced and the mixing was completed. This system was able to deliver 9-ton batches directly to the pouring site. Screened smelter tailings were used as acid-resistant aggregates for the sulfur concrete mix. Due to the high density of these tailings, the sulfur concrete had a density of approximately 2.9 g/cc (180 lb/cu ft) after placement. Sections of floors 3.6 m wide and approximately 4.6 m long were poured in a checkerboard fashion. The finishing of the floors was done with vibratory screeds consisting of electric vibrators mounted on two 50 by 150 mm boards separated laterally by 320 mm. Conventional concrete trowels were used for touching up where necessary. These trowels were left on top of the hot mix to keep the material from solidifying on them. Even with exposure to hot sun and ambient temperatures of approximately 30°C (86°F), bulkhead forms could be removed in approximately 2 hr after pouring. Under somewhat cooler conditions, this might be completed within 1 hr. To pour the sulfur concrete mix into areas that were not readily accessible by motorized buggies or the mixer truck, a crane and conventional concrete buckets were used. The bottom-opening mechanism of the buckets was preheated with a propane torch before the buckets were filled. A rubberized asphalt, hot-melt sealant was used in all construction joints, and vinyl water-stop was used at all joints where leakage of acid might possibly occur. During the operation of this plant, a fine spray of 15 to 20% sulfuric acid is emitted from the electrolytic cells mounted over the floor. Consequently, the floor, drainage canals, and sump are continuously bathed with this acid.

In 1992, a project was implemented by Construction Equipment Services in South Africa for the replacement of existing effluent drains that were used to discharge an effluent of 45% sulfuric acid concentration. The installed system using Starcrete technology was successful and the drains served the purpose with no sign of degradation or failure. Until 1996, there was no evidence of uncontrolled cracking, joint failure, or degradation of the sulfur concrete.

Vorwerk (1996) reported on the engineering experience with the use of sulfur concrete in Chile. The results indicated that sulfur concrete possesses the desired mechanical properties, including thermal cycling between 10°C and 30°C and the excellent performance in alkaline environment. These properties remained stable over a period of eight years.

Nevin (1996) reported a case study involving two separate contracts for the supply and installation of acid-resistant drainage systems on a zinc factory undertaken three years apart. Starcrete technology that was developed in Canada was selected as a construction material. The first project was completed in 1992 and the second began in 1995. Two major problems were encountered in the first project that were related to shrinkage and cracking, and joint failure. Since the reported failure was related to installation procedures rather than material performance, these problems were resolved during the execution of the second project.

Okumura (1996) reported the case study of the Cominco Ltd. zinc factory in Trail, British Columbia, Canada. As part of the sulfur dioxide recovery process at the factory, an ammonium sulfate fertilizer plant was in operation. The process involved the evaporation of a 40% ammonium sulfate solution at approximately 100°C. It regularly came in contact with the floor at this temperature, and the regular steel and concrete at the plant showed deterioration. In 1992, there was a reconstruction at the plant and most of the floors were replaced with sulfur concrete. Visual inspection of the area showed that floors, walls, trenches, and sumps were in excellent condition. Sulfur concrete coatings around the footings showed no corrosion, whereas behind most of the footings there was a visible sulfate attack on Portland cement concrete.

7.2.4 The 2000s

According to Lee (2000), Brimstone Pipe Company conducted a research project at the Spangler Geotechnical Laboratory at Iowa State University from 1977 to 2000. The purpose was to develop the equipment, technology, and material for manufacturing a low-cost, acid-resistant lining and watertight joint for concrete piping. The first phase was intended to evaluate a feasibility study of manufacturing low-cost, acid resistant sanitary pipes manufactured from sulfur concrete. The results demonstrated the viability of the lining process. The second phase concentrated on manufacturing sulfur concrete-lined Portland cement concrete pipe and manhole components. The concentration was mainly on quality control during the manufacturing process through a demonstration project.

Nippon Oil Corporation, Research and Development Dept., RECOSUL Business Development Group, Japan, certified the developed sulfur concrete technology that was based on sulfur modification using ENB as discussed in

Chapters 5 and 6. The technology was further implemented in a number of projects in Japan and in the United Arab Emirates (UAE) whereby 80 meters of pipe and one manhole were installed in Al Ain, UAE (Jan. 2009). The project was executed in association with the United Arab Emirates University and Abu Dhabi Sewerage Services. The project was financially supported by Japan Cooperation Centre, Petroleum (http://www.jase-w.eccj.or.jp/technologies/pdf/petrochemicals/P-4.pdf).

The ENB-modified sulfur concrete technology was certified in Japan by:

- Japan Sewage Works Association's Product Certificate (Class II): sewage pipes and assembled manholes
- Japan Institute of Wastewater Engineering Technology: construction technology evaluation
- Certificate (sewerage technology): eco-sulfur anti-corrosion method No. 0537; a member of the Sheet Lining Method Association (March 2006)
  - Adopted at Ariake Water Reclamation Center operated by the Tokyo Metropolitan government and other facilities
- Hokkaido government's recycle brand certified product: eco-sulfur concrete civil engineering secondary product (March 2007)
  - Adopted at the Hakodate Harbor Basin Sewage Purge Center of the Hakodate Bay Sewerage Treatment Plant
  - Hokkaido government's algal reef qualification: eco-sulfur concrete algal reef product (March 2008)

7.3 MIX DESIGN

In the hypothetical model proposed by Makenya (1997), the minerals are tied into a thermoplastic composite sulfur concrete of good durability. A maximum aggregate size should be used that complies with the product geometry. The maximum weight density should be attained that has the lowest porosity and fulfills the conditions of the best durability under fluctuating humidity and temperature loads. As discussed in Chapter 4, the aggregates above the filler level (0.074 mm) should be dense-graded, tough, and free from deleterious materials such as swelling clays.

The sulfur content of the mix must be designed to achieve an optimum level of viscosity and a strictly limited absorption of moisture. This is the critical sulfur content value (Figure 7.1). If the amount of sulfur is increased above that value, the thermal expansion of the composite goes up, causing microcracking (Makenya, 1977, 2001). If the sulfur content is decreased from the critical value,

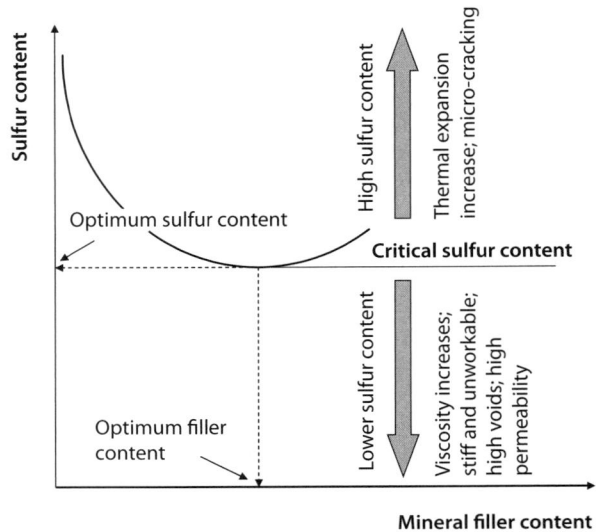

Figure 7.1. Hypothesis of the concept of optimum sulfur content in sulfur concrete design mix.

the viscosity of the mix goes up and the sulfur concrete may become stiff and unworkable, causing a final void content that is higher than desirable. This creates an interconnection between voids and providing permeability throughout the sulfur concrete body. Additionally, it leads to a durability problem in conditions such as freeze/thaw, fluctuating moisture, and temperature.

As discussed in Chapter 4, the optimum range of sulfur should be just less than the amount necessary to fill the voids in the aggregate to 100% saturation, yet, high enough to keep the final void content at less than 8% (Pickard, 1984). We are considering the sulfur cement as the sum of sulfur, modifier, and filler.

Furthermore, as discussed in Chapter 4, mineral fillers are the most important component in the mix because they control the plasticity of the hot mix. The aggregates above 0.074 mm (mesh 200) diameter should follow a Fuller distribution with a low coefficient n of 0.3 for crushed aggregates. The amount of sulfur cement should be adjusted to fill up the voids of the matrix. This means that the grain size distribution of sulfur cement should satisfy the condition of dense-graded distribution at the level of transition to filler (0.074 mm) (Makenya, 1997).

For aggregates, the grain size distribution at the filler level should be a gap-grading (Soderberg, 1983; Dohner, 1996). However, McBee (1993) indicated that gap-grading does not satisfy the low-absorption capacity especially for frost resistance. McBee et al. (1983) indicated that the suitable gradation for sulfur concrete was achieved based on the Fuller maximum density relationship discussed in Chapter 4. It was shown by Gregor and Hackl (1978) that Fuller parabola (n =

Table 7.1. Calculated percent passing for $D = 20$ mm and $n = 0.3$ for crushed aggregates.

Sieve diameter (mm)	0.075	0.15	0.312	0.625	1.25	2.5	5	10	20
Percent passing (P)	18.7	23	28.7	35.3	43.5	53.6	65.9	81.2	100

Table 7.2. Mineral fillers and sulfur contents for crushed aggregates with $n = 0.3$

D (mm)	6	10	12	20	25
Mineral filler (%) (minus mesh 200)	8	7	6	5	4
Modified sulfur (% wt)	18.5	16.5	15.0	14.0	13.5

Table 7.3. Sulfur concrete composition for crushed aggregates with $n = 0.3$

D (mm)	6	10	12	20	25
Coarse aggregates (> 2 mm)	29.5	37.5	42.5	49.0	53.5
Fine aggregates	44.0	39.0	36.5	32.0	29.0
Mineral filler (%) (minus mesh 200)	8	7	6	5	4
Modified sulfur (% wt)	18.5	16.5	15.0	14.0	13.5

0.5) does not furnish the optimum packing densities. In the case of spherical aggregates, $n = 0.4$ is closer to the greatest density and, if coarse aggregates are crushed, $n = 0.3$ is sufficient.

Makenya (2001) used the Fuller relationship with $n = 0.3$ for crushed aggregates with $D = 20$ mm and calculated the filler and sulfur contents of a proper mix as shown in Tables 7.1 to 7.3. Figure 7.2 shows a typical mix proportions for sulfur concrete for construction industry.

7.4 MIXING PROCESS

The mixture of sulfur concrete must be in proportion to achieve the best balance between mechanical strength, density (limited voids), and good workability (Chempruf, 1988). The flowing product should be a well-coated homogenous mixture. A limited absorption into the solidified product will lead to good durability under severe conditions. Binder levels will vary, depending on aggregate size and shape as discussed.

Sulfur concrete is prepared by hot-mixing coarse and fine aggregate, filler, and binder at 130 to 150°C (McBee et al., 1983). Any type of mechanical mixer may be used that can maintain the mix at the required temperature or keep it within a limited temperature range. It is preferable to mix at the higher end of the temperature range to allow more time for transporting and casting the mixture

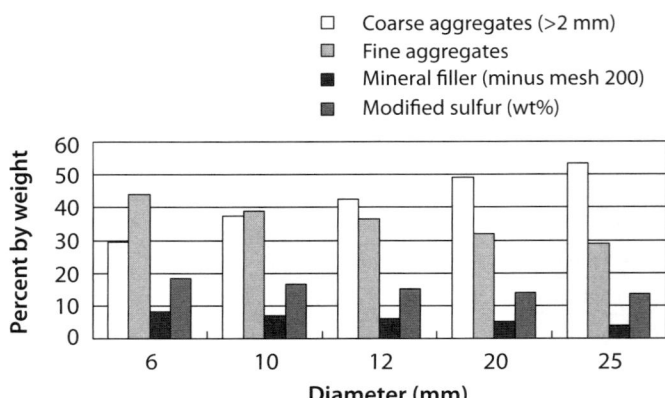

Figure 7.2. Typical mix proportions for sulfur concrete for construction industry.

before solidification. The modified cement may be added either in the liquid or in the solid form. When using liquid sulfur cement, it is added at 140°C to aggregates at approximately 160°C to obtain a mixture in the 140 to 150°C range. Solid flake binder (cold) can be added to aggregates at a somewhat higher temperature level (170 to 190°C) to melt the binder and obtain the desired mixing temperature. Adding binder and filler simultaneously in solid form requires a higher initial temperature of the fine and coarse aggregates.

The most recent development is linked with the equivalent asphalt-mixing process in which the filler is mixed into the slurry after the binder. An early attempt in this direction has been described by Malhotra (1983) for elemental sulfur concrete. This more complex procedure coats the aggregates after the distribution of the sulfur and adds extra sulfur with the sand and filler. Mixing time, to obtain homogenous material, depends on the equipment being used. Small batch mixtures require 1 to 2 minutes. The batching procedure may follow different schemes as explained next (Makenya, 1997).

First Scheme

All aggregates (coarse, fine, and filler) are preheated and the modified sulfur is added in melted form (Chempruf, 1982, 1988) as shown in Figure 7.3. An alternative is to mix all components at ambient temperature then stir and heat until everything is in liquid form (Al-Tayyib and Khan, 1988) as shown in Figure 7.4. This procedure was also followed in the manufacturing of Sulkret (Ecker and Steidl, 1986).

Technological Aspects of Sulfur Concrete Production

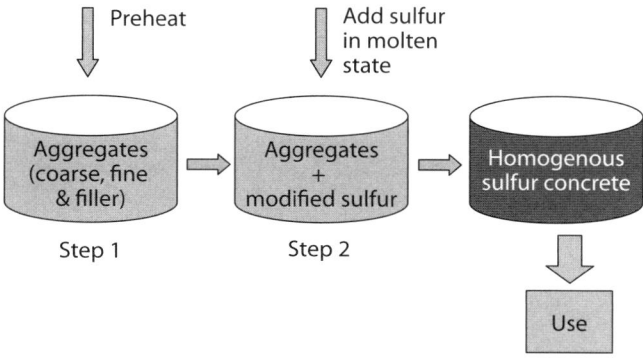

Figure 7.3. Early Chempruf mixing scheme.

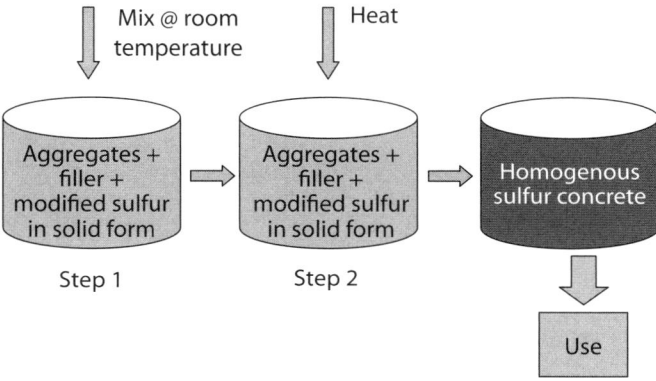

Figure 7.4. Sulkret mixing scheme.

Second Scheme:

All aggregates (coarse, fine, and mineral filler) are preheated to a somewhat higher temperature level and the modified sulfur is added in solid form (Ekblad, 1992) as shown in Figure 7.5. This method is being used by 4K A/S.

Third Scheme:

Coarse and fine aggregates are preheated and mixed, simultaneously adding filler and modified sulfur (Chempruf, 1994) as shown in Figure 7.6.

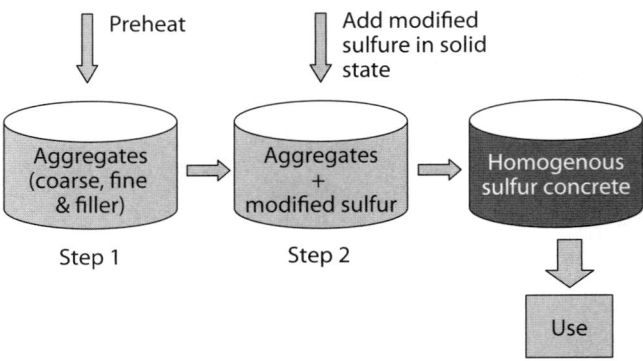

Figure 7.5. 4K A/S mixing scheme.

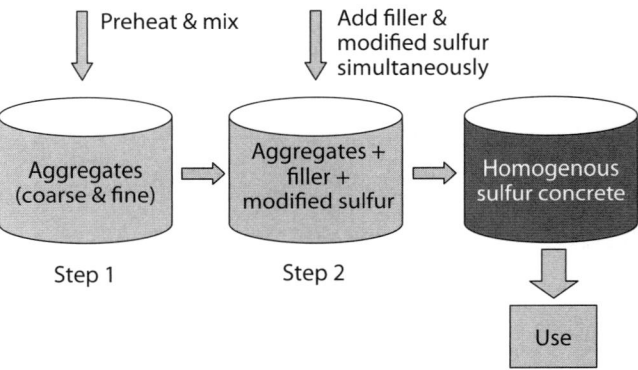

Figure 7.6. Later Chempruf mixing scheme.

Fourth Scheme:

Coarse and fine aggregates are preheated. Modified sulfur is added in solid form, stirring until melted and all aggregates are covered with sulfur. The cold filler is added in the flowing mix under continued stirring and heating until the sulfur concrete is homogeneous and ready to pour. The procedure is explained in the ACI (1988) that indicates that adding the filler after the sulfur prevents problems from dusting and balling of the filler. It is believed that this procedure is also favorable in the later process of solidification when ductility helps to prevent microcracking. This procedure also produces a good surface as reported by Thylén (1991). A schematic representation of this process is shown in Figure 7.7.

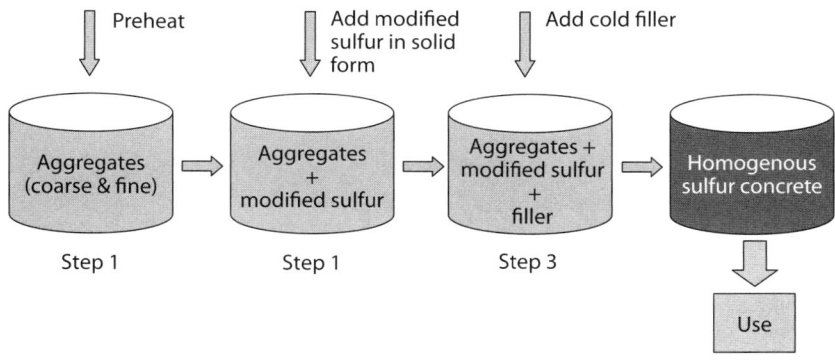

Figure 7.7. ACI mixing scheme.

Fifth Scheme:

Elemental sulfur (step 1) and modifiers (step 2) are heated to 130 to 150°C (Mohamed and El Gamal, 2006). At a controlled rate, molten sulfur is mixed with the modifier (step 3) to from modified sulfur. Mineral fillers (step 4) are preheated to higher temperatures than that of step 3 and transferred at a controlled rate to form sulfur cement (step 5) after mixing it with modified sulfur. Fine aggregates (step 6) are preheated to a temperature similar to that in step 4 and mixed with sulfur cement (step 5) to form homogenous sulfur concrete (step 7). A schematic representation of this process is shown in Figure 7.8.

7.5 EQUIPMENT

7.5.1 Development

The initial development and testing of sulfur concrete technology is generally accomplished in the laboratory with small batches, using crock pots, ovens, and small concrete mixers with an external heat application. The first generation of commercial scale sulfur concrete equipment was largely borrowed from the Portland cement concrete and asphalt industries. The aggregates were measured by flow rate or loader bucket and then conveyed into a standard concrete mixer truck in which propane torches were held on the drum to maintain material temperature.

The next generation of commercial scale sulfur concrete manufacturing equipment featured better flow rate controls for aggregate measurement, more automated methods for measuring sulfur and mineral filler, and automatic temperature control on the aggregate dryer. Mixer trucks with direct-fired burners

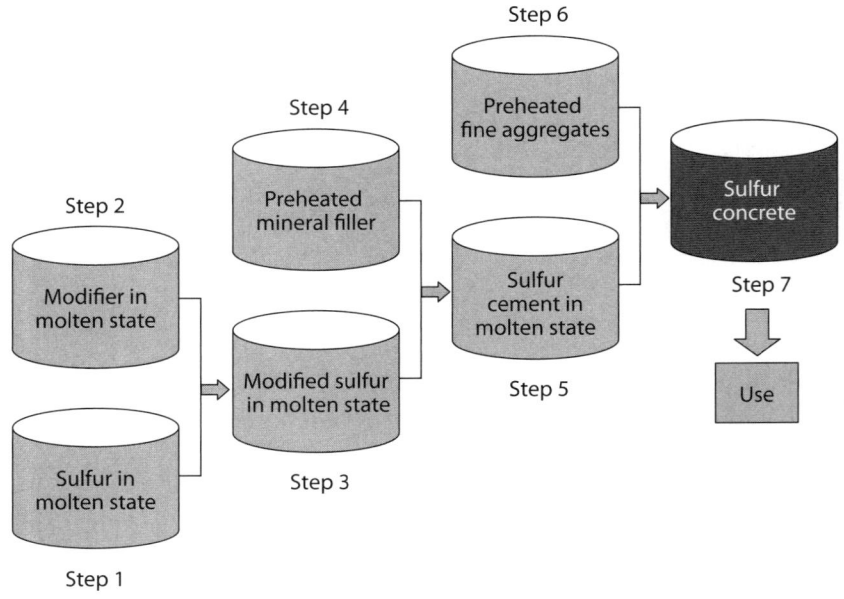

Figure 7.8. Mohamed and El Gamal mixing scheme.

were developed that enabled the trucks also to be used for heating and drying the aggregates.

Modern sulfur concrete batching equipment has addressed the need for better control of the aggregate matrix by screening and sorting the hot dry aggregates and then batching the segregated aggregates by weight. Newer mixer trucks feature an insulated shroud over the mixer drum and an indirect-fired thermostat-controlled burners. This type of truck is capable of maintaining the sulfur concrete mix at the desired temperature level for an indefinite period without the risk of localized overheating.

With current equipment, typical production is 19 to 54 m³ of sulfur concrete during a 10-hour construction workday. Although this amount pales in comparison to the high output of Portland cement concrete batch plants, one must keep in mind that sulfur concrete is different from Portland cement concrete both in performance and in the placing and finishing techniques. As such, specific installation adjustments must be made. Effective sulfur concrete placement is accomplished by preparing smaller areas at a time and then by placing these areas over consecutive days. As a result, the actual time required to complete the entire sulfur concrete placement is only marginally greater than the time required to complete Portland cement concrete placement as shown in Table 7.4 (Crick and Whitmore, 1998).

Table 7.4. Construction time of sulfur concrete versus Portland cement concrete with protective lining for 900 m², 0.15 m slab (Crick and Whitmore, 1998)

ID	Task Name	Duration	Week No. 1	2	3	4	5	6	7
1		days							
2	**Portland cement concrete**	41	▬	▬	▬	▬	▬	▬	
3	Base preparation	3	▬						
4	Form & rebar	6	▬						
5	Place & finish	1		▬					
6	Curing time	28			▬	▬	▬	▬	
7	Lining installation	3							▬
8									
9	**Sulfur concrete**	14	▬	▬					
10	Base preparation	3	▬						
11	Form & rebar (125 sq m/day)	9	▬						
12	Place & finish (19 cu m/day)	8	▬						
13	Seal joints	1		▬					

7.5.2 Commercial Scale Application

When evaluating the suitability of sulfur concrete for a particular project, the following should be considered (Crick and Whitmore, 1998):

1. *Environment*: Sulfur concrete should be considered for acid and mineral salt environments in which Portland cement concrete would be attacked or destroyed without additional chemical protection.
2. *Project size*: To be commercially competitive, most project applications need a minimum of 23 m^3 to offset the cost of mobilizing the equipment to the site. For projects smaller than 23 m^3, alternative materials such as polymer concrete become more viable because their relatively high material cost per m^3 is frequently offset by the cost of mobilizing equipment for sulfur concrete.
3. *Feasibility*: There are substantial transportation costs associated with securing suitable, corrosion-resistant aggregates and mineral fillers. In light of the higher cost of manufacturing the equipment and raw materials, as well as the expense of transporting the materials and equipments, a sulfur concrete slab is more expensive than an unprotected Portland cement concrete slab when compared on a price per m^3 basis. However, when the price of an in-place chemi-

cally resistant sulfur concrete slab is compared with an in-place Portland cement concrete slab complete with protective lining systems, the cost of the sulfur concrete becomes competitive.

4. *Schedule*: Once placed, sulfur concrete can be put in service within 24 hr. During the winter, using sulfur concrete for outdoor work would eliminate the need for heating and hoarding during curing and would allow for substantial savings. These scheduling advantages serve to reduce installation costs and could give sulfur concrete a significant cost advantage over Portland cement concrete with a protective lining.

5. *Scope*: Sulfur concrete is best suited for construction of slab-on-grade floors, sumps, trenches, pump and machine, containment curbs, and pony walls. Although it would be possible to place structural beams, columns, and suspended slabs using sulfur concrete, engineering applications for these areas are currently limited. An additional consideration in using sulfur concrete for structural members is that sulfur concrete is thermoplastic and would require suitable fire protection measures if used for the construction of structural elements. Use of fire retardant additives with sulfur concrete is discussed later in this chapter.

7.6 MANUFACTURING

Various equipment has been used to prepare sulfur cement mixes, including standard laboratory equipment, concrete mixers and heated transit mix trucks. Uniform heat distribution in the mixer is important to avoid hot spots that can cause the sulfur to ignite. A cast iron bottom in the mixer would help to distribute the heat evenly. The basic equipment and processes required to produce sulfur concrete are: (a) butane gas or electricity to melt sulfur (120° to 140°C); (b) a regular concrete mixer to mix molten sulfur with aggregates, and (c) metals or woods to fabricate molds for casting sulfur concrete. Construction with sulfur involves precautions similar to those encountered in hot-mixed material such as in asphalt paving operations (ACI, 1988).

7.6.1 Pre-cast Mixing and Production

Nippon Oil Corporation and Japan Petroleum Energy (EP 1886 781 A1, 2008) reported a system for the production of sulfur concrete products. The basic components of a pre-cast manufacturing system are shown in Figure 7.9 and are also discussed.

Technological Aspects of Sulfur Concrete Production

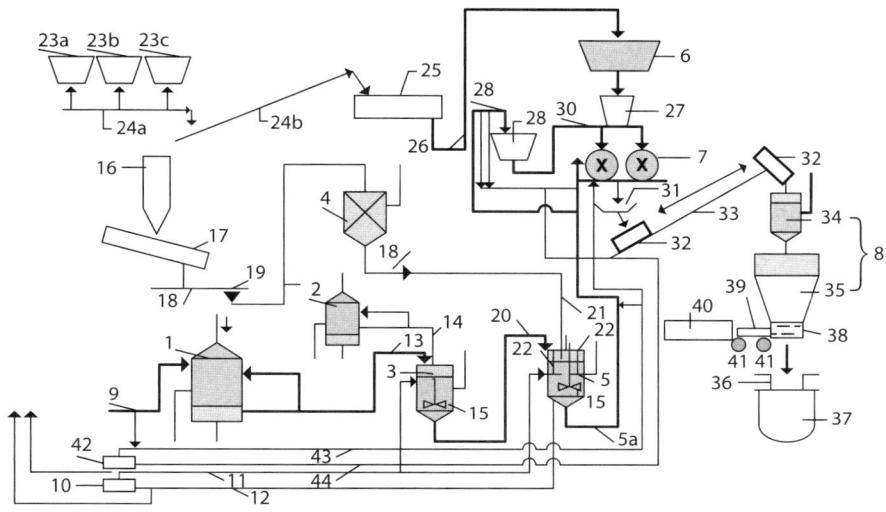

1	Molten sulfur tank	26	Hot elevator
2	Sulfur modifying agent tank	27	Aggregate scale hopper
3	Modified sulfur producing tank	30	Injection pipe
4	Fine aggregate hot bin	31	Cushion hopper
5	Modified-sulfur intermediate material tank	32	Skip bucket
6	Coarse aggregate hot bin	33	Transfer skip device
7	Mulling device	34	Surge bin
8	Pouring device	35	Scale hopper
10	First heating medium heater	36	Mold form
11	Supply line of heating medium	37	Vibrator
12	Reflux line of heating medium	38	Outlet opening of raw material
15	Anchor-shaped rotating blade	39	Flat plate
16	Coal ash silo	40	Drive cylinder
17	Drying kiln	42	Second heating medium heater
18	Transfer device	43	Supply line of heating medium
19	Transfer pump	44	Reflux line of heating medium
22	Rotating blades		

Figure 7.9. An explanatory view of a modified sulfur solidified concrete production system (Nippon Oil Corporation and Japan Petroleum Energy, EP 1886 781 A1, 2008).

From Figure 7.9, the modified sulfur-solidified substance-producing system includes a molten sulfur tank (1), a sulfur modifying agent tank (2), a modified sulfur producing tank (3), a fine aggregate hot bin (4), a modified sulfur intermediate material tank (5), a coarse aggregate hot bin (6), a mulling device (7) and a pouring device (8).

Molten Sulfur Handling Process

The molten sulfur tank accommodates molten sulfur and heats the inside substance (molten sulfur) to a predetermined temperature (e.g., 119°C) or higher. The molten sulfur is melted at 150°C, for example, and is supplied from a tank truck (not shown) into the molten sulfur tank (1) through the supply pipe (9). The molten sulfur tank (1) has a jacket-type structure and the heating element is added to the molten sulfur tank (1). That is, a first heating medium heater (10) is located on an outer portion of the molten sulfur tank. An electric heater warms the water and other heating medium and this medium is supplied to the jacket-type structure of the molten sulfur tank through a supply (11) and a reflux (12) line.

Modifying Agent Handling Process

The sulfur modifying agent tank (2) accommodates the sulfur modifying agent that denatures the molten sulfur. Denaturing the molten sulfur is equivalent to polymerizing sulfur, for example, by a sulfur modifying agent.

Modified Sulfur Production Process

The modified sulfur producing tank (3) receives and mixes molten sulfur from the molten sulfur tank (1) and the sulfur modifying agent from the sulfur modifying agent tank (2) to produce modified sulfur. In addition, it heats the inside accommodated substance (modified sulfur) to a predetermined temperature (e.g., 119°C) or higher. A supply pipe (13) from the molten sulfur tank (1) and a supply pipe (14) from the sulfur modifying agent tank (2) are connected to a ceiling portion of the modified sulfur producing tank (3). An anchor-shaped rotating blade (15) is in the tank for mixing and stirring. The modified sulfur produced is obtained by polymerizing the sulfur by the sulfur modifying agent. The modified sulfur producing tank (3) has a jacket-type structure and a warming mechanism to heat the modified sulfur. The supply line (11) and reflux line (12) from the first heating medium heater (10) are, respectively, connected to the modified sulfur producing tank (3). The material heated by the first heating medium heater (10) is supplied to the jacket-type structure of the modified sulfur producing tank (3) through the supply (11) and reflux (12) lines.

Fine Aggregates Handling Process

The fine aggregate hot bin (4) accommodates fine aggregate and heats the accommodated substance (fine aggregate) to a predetermined temperature (e.g., 150°C) or higher. The fine aggregate is usually comprised of aggregate having a particle diameter of 1 mm or less. Examples of fine aggregates are one or more of natural stone, sand, quartz sand, steel slag, ferronickel slag, copper slag, a byproduct produced when metal is produced, coal ash, fuel crematory ash, electric dust collection ash, and molten slag. When coal ash is used as the fine aggregate, coal ash stored in a coal ash silo (16) is put into a drying kiln (17) such as a rotary kiln and is then dried at 150°C, for example. Then, it is conveyed to a transfer pump (19) by a transfer device (18) such as a conveyor belt that is locally heated by an electric heater. It is then pumped and sent to the fine aggregate hot bin (4) by the transfer pump (19). A sidewall of the fine aggregate hot bin (4) is also heated by an electric heater with the fine aggregate reaching 150°C.

Sulfur Cement Production Process

The modified sulfur intermediate material tank (5) receives modified sulfur from the modified sulfur producing tank (3) and fine aggregate from the fine aggregate hot bin (4) and mixes them to produce modified sulfur intermediate material. It then heats this material to a predetermined temperature (e.g., 119°C) or higher. A supply pipe (20) from the modified sulfur producing tank (3) and a supply pipe (21) from the fine aggregate hot bin (4) are, respectively, connected to a ceiling portion of the modified sulfur intermediate material tank (5). The anchor-shaped rotating blade (15) is used for mixing and stirring in the tank. Two rotating blades (22) move at a high speed in an upper portion of the tank. This is to dry to approximately 150°C fine aggregate that enters the tank from the upper portion to be dispersed and easily mixed with previously supplied modified sulfur. This allows for the modified sulfur and fine aggregate to be immediately mixed uniformly within the tank.

The modified sulfur intermediate material tank (5) has a jacket-type structure and provides a means for heating the modified sulfur intermediate material. That is, the supply line (11) and the reflux line (12) from the first heating medium heater (10) are, respectively, connected to the modified sulfur intermediate material tank (5). The heating is performed by the first medium heater (10) that is supplied to the jacket-type structure portion of the reformed sulfur intermediate material tank (5) through the supply (11) and reflux (12) lines. Further, the material from the fine aggregate hot bin (4) is heated by a transfer device (18) and, in this state, the fine aggregate is supplied through the supply pipe (21).

Coarse Aggregates Handling Process

A coarse aggregate hot bin (6) accommodates a coarse aggregate and, additionally, heats it to a predetermined temperature (e.g., 120 to 130°C). Wide-range coarse aggregates can be used and examples of the fine aggregate have been described previously. The particle diameter of the coarse aggregate ranges from 1 to 50 mm. In Figure 7.9, the coarse aggregates are put into a first aggregate cold hopper (23a), a second aggregate cold hopper (23b) and a third aggregate cold hopper (23c). They are transferred into a drying kiln (25) such as a rotary kiln by a first conveyor belt (24a) and a second conveyor belt (24b), and then are heated to 150°C and dried. They are placed on a hot elevator (26) that is locally heated by an electric heater, lifted up, and then sent to the coarse aggregate hot bin (6). A sidewall of the coarse aggregate hot bin (6) is also heated to approximately 120 to 130°C.

Modified-Sulfur Concrete Production Process

The mulling device (7) receives and mixes the modified sulfur intermediate material from the modified sulfur intermediate material tank (5) and a coarse aggregate from the coarse aggregate hot bin (6) and heats the mulled substance to a predetermined temperature (e.g., 119°C) or higher. The coarse aggregate discharged from the coarse aggregate hot bin (6) is put into the aggregate scale hopper (27), and a predetermined amount of coarse aggregate is put into the mulling device (7). At the same time, the modified sulfur intermediate material from the modified sulfur intermediate material tank (5) is moved into the intermediate material scale hopper (29) through a supply pipe (28) which is continuously connected to the outlet pipe (5a). A predetermined amount of modified sulfur intermediate material is pumped through an injection pipe (30) by a spray pump (not shown), and is injected into the mulling device (7) from an injection opening formed in a tip end of the injection pipe (30). Then, the coarse aggregate and the modified sulfur intermediate material are mulled by a rotating blade in the mulling device (7). The mulling device (7) also has a jacket-type structure and a means for heating a mulled substance therein is added.

The pouring device (8) receives and stores a raw material (mulled substance), which is mulled by the mulling device (7), heats the same to a predetermined temperature (e.g., 120 to 130°C) or higher and, furthermore, measures the inside raw material to subsequently pour it into a mold form. The raw material discharged from the mulling device (7) is received by a cushion hopper (31) placed below the mulling device (7) and is put into a skip bucket (32) found below the cushion hopper (31). The skip bucket (32) rises on an inclined transfer skip device (33) and deposits the raw material into the surge bin (34). A scale hopper (35) is connected to a lower portion of the surge bin (34) and measures a predetermined amount of raw material to be poured into the mold form (36). The surge bin (34) and the scale hopper (35) constitute the pouring device (8). The sidewalls of the cushion hopper (31), the skip bucket (32), the surge bin (34) and the scale hopper (35) are heated by an electric heater or the like with the inside raw material reaching a temperature of approximately 120 to 130°C.

The mold form (36) below the scale hopper (35) is disposed on an upper surface of a vibrator (37), which vibrates a raw material that has been put into the mold form (36), to remove air found in the raw material. As a result, the strength of the modified sulfur substance after it is cooled and solidified is enhanced to improve the finish. Examples of the mold form (36) include, but are not limited to, a panel shape, a tile shape, and a block shape. Additionally, ultrasonic vibration may be applied to carry out the forming process.

Piping and Transfer Systems

In the structure of the entire system, all of the piping system that connects the devices from the molten sulfur tank (1) to the mulling device (7) have a double structure. A means for heating fluid in the piping system to a predetermined temperature (e.g., 119°C) or higher is added to the piping system. That is, a second medium heater (42) is provided outside of the molten sulfur tank (1). Water and other heating medium are heated and this heated medium is supplied through the supply line (43) and the reflux line (44) to the supply pipe (9) for transferring molten sulfur to the molten sulfur tank (1). The supply pipe (13) for molten sulfur from the molten sulfur tank (1) to the modified sulfur producing tank (3), the supply pipe (20) for modified sulfur from the modified sulfur producing tank (3) to the modified sulfur intermediate material tank (5), the supply pipe (28) for modified sulfur intermediate material from the modified sulfur intermediate material tank (5) to the intermediate material scale hopper (29), and the injection pipe (30) through which modified sulfur intermediate material is sent from the intermediate material scale hopper (29) to the mulling device (7) is *then* sent to the jacket-type structure portion of the mulling device (7). Therefore, in the producing system of the modified sulfur solidified substance, sulfur material that is being produced is always heated to the solidification point of sulfur or higher, and then the modified sulfur solidified substance can be safely and easily produced.

In the modified sulfur solidified substance producing system, a generating device that supplies nitrogen gas to the molten sulfur, the modified sulfur, or the modified sulfur intermediate material may be added (not shown). In this case, sealing or purging is carried out using nitrogen gas, which is inert, so that oxidation of sulfur can be prevented and, at the same time, the generation of hydrogen sulfide can be prevented.

Final Modified-Sulfur Concrete Product

The state of the produced modified sulfur solidified substance using the modified sulfur solidified substance producing system having the described configuration is now discussed.

As shown in Figure 7.9, molten sulfur is put in the molten sulfur tank (1) and is heated to a predetermined temperature (e.g., 119°C) or higher. A sulfur-modifying agent for denaturing the molten sulfur is put in the sulfur reforming agent tank (2). In this state, molten sulfur delivered from the molten sulfur tank (1) through the supply pipe (13) and sulfur modifying agent delivered from the sulfur reforming agent tank (2) through the supply pipe (14) are received and mixed in the modified sulfur producing tank (3) so as to produce modified sulfur. The inside modified sulfur is heated to a predetermined temperature (e.g., 119°C) or higher. At the same time, a fine aggregate is placed within the hot bin (4) and heated to a predetermined temperature (e.g., 150°C) or higher. Next, modified sulfur delivered from the modified sulfur producing tank (3) through the supply pipe (20) and a fine aggregate delivered from the hot bin (4) through the supply pipe (21) are received and mixed in the reformed sulfur intermediate material tank (5) to produce the modified sulfur intermediate material. The inside reformed sulfur intermediate materials are then heated to a predetermined temperature (e.g., 119°C) or higher. The modified sulfur intermediate material produced in this manner is handled as a nondangerous substance, and the piping equipment after the outlet pipe (5a) of the reformed sulfur intermediate material tank (5) can avoid specifying the inside fluid being handled as a dangerous substance.

The modified sulfur intermediate material delivered from the modified sulfur intermediate material tank (5) through the outlet pipe (5a) and the supply pipe (28) is measured by the intermediate material scale hopper (29), and then injected into the mulling device (7) through the injection pipe (30). Prior to this, a coarse aggregate discharged from the coarse aggregate hot bin (6) is put into the aggregate scale hopper (27), and a predetermined amount of coarse aggregate is put into the mulling device (7). In this state, modified sulfur intermediate material from the modified sulfur intermediate material tank (5) and a coarse aggregate from the hot bin (6) are mulled by the mulling device (7). The inside mulled substance is heated to a predetermined temperature (e.g., 119°C) or higher.

A raw material mulled by the mulling device (7) is sent by the skip bucket (32) and the transfer skip device (33) and put into the surge bin (34). A predetermined amount of raw material is measured by the scale hopper (35) located on a lower portion of the surge bin (34), and the raw material is poured into the mold form (36) located below the scale hopper (35). Then, the raw material is cooled and solidified in the mold form (36) and a modified sulfur solidified substance having a desired shape defined by the mold is produced.

In 2008, Nippon Oil Corporation, Japan, utilized its pre-cast manufacturing process plant described to manufacture pre-cast pipes and manholes for a demonstration site in Al Ain, United Arab Emirates, as shown in Figure 7.10.

Figure 7.10. Placed modified-sulfur concrete pipes and manhole within the sewerage distribution system in Al Ain, United Arab Emirates, by Nippon Oil Corporation, Japan, in January 2009.

7.6.2 *In situ* Construction Mixing and Placing Techniques

Equipment and techniques from both the concrete and asphalt technologies are used to batch, transport, place, and finish sulfur concrete. The typical equipment for a cast in-place sulfur concrete operation would include: (a) an aggregate drying system with an efficient means of drying and heating aggregates is a rotary kiln system (manufactured for the asphalt industry). Specially heated concrete transit mixer trucks can be modified for the preparation of sulfur concrete (Figure 7.11a); (b) weight hopper/scale for proportioning mixture materials (Figure 7.11b); (c) mixing/transporting equipment (Figure 7.11c); and (d) typical concrete hand-placing and finishing tools.

Proper techniques are necessary when mixing and placing sulfur concrete. Many methods of manufacturing sulfur concrete and various types of equipment are currently being used to produce, transport, and place the materials.

Generally, sulfur concrete can be prepared in nearly any mixer used for preparing Portland cement concrete or asphalt cement concrete provided that it has been modified to monitor and loosely control the optimum temperature range (135 to 141°C) of the mixture. Essential to the preparation of sulfur concrete is a means of drying and heating the aggregate to the desired temperature for mixing

Figure 7.11a. Special heated concrete mixer (left) and aggregate dryer/superheated (right) for sulfur concrete application (Pickard, 1985).

and maintaining the sulfur concrete mixture with a relatively limited temperature range during transport and placement. In one method of preparing sulfur concrete, divided bins are used to proportion coarse and fine aggregates into a fuel oil-, propane-, or gas-heated rotary kiln. The kiln discharges a weighed batch of heated aggregate into a heat-jacketed mixer followed by a proportion of an ambient temperature flaked, modified sulfur and mineral filler. The heated aggregates transfer heat to the mineral filler and melt the modified sulfur producing a homogenous sulfur concrete hot mix.

Heat-jacketed Portland cement concrete transit mixers with capacities up to 9 m^3 are used to mix, transport, and place sulfur concrete. The mixer can be equipped with an onboard microprocessor that controls four propane infrared catalytic heaters used to maintain the sulfur concrete within its optimum temperature range for placement. Flaked, modified sulfur and mineral filler at an ambient temperature are then added to the hot aggregates in the transit mixer to produce a homogenous sulfur concrete mixture. The heat-jacketed Portland cement concrete transit mixer maintains the sulfur concrete within a limited temperature range for long periods of time, keeping the sulfur concrete plastic during transport and workable until placement at the job site. Newer sulfur concrete transit mixers also have onboard capability to dry and heat the aggregates efficiently, thereby eliminating the need for a separate rotary kiln.

A new generation, self-contained machine that is capable of mixing, in succession, a series of batches of sulfur concrete at a production rate up to 8 m^3/h has been designed and built. The portable machine weigh all mix components immediately before discharge into the mixer for enhanced quality control. The

Technological Aspects of Sulfur Concrete Production 283

Figure 7.11b. Weigh-hopper, scale and conveyor for the preparation of sulfur concrete mix (Pickard, 1985).

Figure 7.11c. Heated concrete mixer for the preparation of sulfur concrete (Pickard, 1985).

batching system is fully automated, allowing its operator to preset the desired mixture composition for continuous batch duplication. Each subsequent batch of sulfur concrete is produced in approximately 1 min intervals (Weber et al., 1990).

The requirements for mixing/transporting equipment are defined by the unique thermoplastic characteristics of sulfur concrete as recommended by ACI Committee 548: (a) sulfur concrete must be maintained in molten form within a narrow temperature range, (b) the concrete mixture must be thoroughly mixed so that the molten sulfur cement adequately coats the fine and coarse aggregate and mineral filler, and (c) the temperature of the mixture should be continuously monitored to prevent under- or over-heating.

7.6.3 Placing and Finishing

Once the section is filled with sulfur concrete, it should be struck off with a simple screed (a leveling device drawn over freshly poured concrete—a board can be used). A vibrating screed can be used to achieve a relatively smooth, sealed surface. After striking off, there are only minutes left to finish the surface. With a 2-in overlay, there are 2 to 3 min to make a finishing pass before the surface begins to crust, with a 6 to 8 in. slab, there are 10 to 15 min. High-quality metal floats are recommended for finishing. One pass over the slab is enough to smooth and seal it. If the surface is accidentally worked after it has crusted, a propane torch can be used to re-melt the area so that it can be finished properly. In an acid environment, if acid drips on sulfur concrete that has already solidification, the slab will not be damaged. Sulfur concrete can be placed on fairly steep grades without flow problems. This requires silica flour (not fly ash) in the mix; the silica yields a stiffer, less flowable mixture. Also, excessive flowing can be minimized by employing sulfur cement at the subsaturation level in the mix (Pickard, 1985).

Typical placing and finishing equipment includes steel shovels and rakes, wood screed and floats, and standard concrete vibrators. Wood floats should be treated with oil to make cleaning easier and to avoid buildup of sulfur concrete. Metal floats and screeds are not recommended due to their tendency to conduct heat away from the sulfur concrete surface, causing it to freeze on the surface and making finishing difficult. The need to place and finish the sulfur concrete simultaneously lends itself to a phased approach that will optimize the production capability of the sulfur concrete batch plant and formwork reuse, taking advantage of sulfur concrete's rapid hardening. Additional forming costs can be kept to a minimum by using previously placed material as part of the formwork. An important placing consideration is that sulfur concrete shrinks as it transforms from liquid to solid, not as a result of the curing and hydration processes that cause Portland cement concrete to shrink. As such, the Portland cement concrete practice of saw-cutting control joints the day after the placement is of limited value in most sulfur concrete placements because most sulfur concrete shrinkage will occur within 12 to 24 hr of placement. Formwork for sulfur concrete should, therefore, be designed to minimize the risk of shrinkage cracks. Standard rules for length to width to depth ratios can be used to achieve this purpose (Crick and Whitmore, 1998).

7.6.4 Cold Weather Placements

Sulfur concrete can effectively be placed at temperatures as low as −34°C without adversely affecting in-place material strength or performance. For cold weather placement, sulfur concrete freezes more quickly and, as a result, special care should be taken to ensure that the finishing crew is large enough to place and finish the material prior to freezing (Crick and Whitmore, 1998). Large open placement should be avoided due to the difficulty of finishing these quickly. Likewise, small placements or thin slabs should also be avoided during cold weather because the small amount of material being placed does not have sufficient mass to maintain its heat through placing and finishing operations.

For better placing and finishing during cold weather, the use of a temperature-controlled batch plant and a heated and insulated sulfur concrete truck is essential to allow for batching of hot mixes at the high end of the allowable temperature range and for maintaining the material at that temperature until placement. In cold weather, it is especially important to keep material handling to a minimum and the mass of individual placements to a maximum.

Failure to take appropriate measures during cold weather placement can result in a rough or undesirable finish, cold joints, excessive honeycombing, cracking, and flash freezing of sulfur concrete to the reinforcing steel and framework. Despite the difficulties associated with cold weather sulfur concrete work, sulfur concrete can offer considerable cost advantages over Portland cement concrete because heating and hoarding is not required during curing.

7.6.5 Wind and Moisture

Wind also poses a concern for the proper placement of sulfur concrete. For work in open areas, care should be taken to ensure that the placement is adequately protected from the wind during placement. Wind causes the sulfur concrete to flash freeze on the surface even if the material temperature (118°C) is well above the freezing point (Crick and Whitmore, 1998). Propane torches can be used to help finish the slab in these conditions, but the final appearance will be fairly rough.

The presence of water or snow in the substrate or formwork poses a concern for proper placing and finishing. When the hot sulfur concrete comes in contact with moisture in the substrate, the moisture is vaporized with the resulting steam percolating up through the hot mix. The steam passing through the sulfur concrete is approximately at a temperature of 100°C and freezes a capillary into the still molten sulfur concrete. Then, once the surface of the concrete starts to freeze, the escaping steam becomes trapped and starts to form a small air pocket beneath the surface of the sulfur concrete, often pushing up on the finished surface creating a bump. If these bumps are later broken, small air pockets will become evident. Therefore, care should be taken to line the bottom of the slab-on-grade forms with an impermeable material such as heavy plastic sheeting. For work outdoors, finished formwork should be covered to protect it from rain prior to sulfur concrete placement.

To minimize the effects of adverse weather conditions, placements should be designed for the capacity of the sulfur concrete delivery system. With a batch plant concrete truck system, the total placement volume (plus a small allowance for wastage) should not exceed the capacity of the truck. In this way, there will not be a cold joint in the middle of the placement where the material froze while waiting for additional material to be batched.

Some sulfur concrete delivery systems work on the basis of continuous small-batch deliveries. This type of system is effective at eliminating the need to design a placement around a truck-sized batch. However, in cold ambient conditions, there could be a problem with cold joints if the time between batches is too long or if the batch plant is far removed from the placement area.

7.6.6 Repairing Damages

Sulfur concrete surfaces that were not properly finished, or that have become damaged in use, can be reheated and refinished. The area can be re-melted by using indirect heating such as with an infrared heater and then quickly refinished or repaired as needed. It is not advisable to use a concentrated source of heat, because concentrated heat on a cold concrete slab can induce stress cracks on thin sections. Holes, spalls, voids, and other minor surface blemishes can be patched easily by reheating the damaged area and filling it with sulfur mortar.

7.6.7 Joints and Joint Sealing

Joints are formed to control cracking, to terminate pours, and to allow for thermal expansion. Location is on a job-by-job basis, but is limited to maximum width of pours (3 m).

A flexible joint sealant compatible with the environmental conditions and ACI 504R-77 *Guide to joint sealants for concrete structures* should be applied to all joints, extending the full width of the joint with a depth no greater than the width. A bond-breaker should be put in below the sealant. If not, the sealant will adhere to the sulfur concrete and crack during thermal expansion. A pre-formed backer-rod made of closed-cell polypropylene is often used as a bond-breaker and will keep uncured sealant from seeping through the joint.

Typically, contraction joints should be constructed at a distance in feet of not more than three times the slab thickness in inches. Contraction joints are usually 6 mm to 9 mm wide and extend 20 to 25 % into the depth of the slab.

Expansion joints are usually 10 to 13 mm wide. They are constructed to relieve thermal stresses and to separate slabs from the vertical surfaces of the structure members. Maximum spacing of the expansion joints should be limited to 18 m for a normal design mixture. If a higher sulfur content is used, the maximum distance should be reduced accordingly.

$$Q_{S,20-160°C} = \underset{\text{Monoclinic}}{\Delta T_\beta \times c_\beta} + \underset{\text{Melting}}{Q(\beta \to l)} + \underset{\text{Liquid}}{\Delta T_l \times c_l} \qquad (7.2)$$

Considering $c_\beta = 0.8$ (J/g°C) for monoclinic form and $c_l = 1.05$ (J/g°C) for liquid form as average values, the quantity of heat is:

$$Q_{S,20-160°C} = (120 - 20) \times 0.8 + 50 + (160-120) \times 1.05 = 172 \quad J/g$$

Since 1 J/s = 1 W, 1 Kg of binder heated from 20 to 160°C requires 172,000 J, or 172000/3600 Wh = 0.05 kWh, which is indeed a small input of energy.

The reason for limiting the temperature range for sulfur to 160°C is the rapid increase of the dynamic viscosity that occurs above this temperature. Sulfur polymerizes and turns dark red. In fact, such temperatures would make the binder useless in practice, even considering that small amounts of organic matter is included with the modifying agent that will dampen the rate of viscosity change (Scientific American, 1970). The narrow operational temperature span 120 to 160°C is related to liquid sulfur of viscosity less than 10 cP, still more than water, that at 20°C is a little less than one cP, but quite feasible in a mixing process with mineral aggregates and in the placing of sulfur concrete (Makenya, 2001).

As discussed in Chapter 3, sulfur is a poor heat conductor, $\lambda_\beta = 0.16$ W/m°C, and has a high resistivity, ρ, of 2×10^{10} Ohm.m (Ecker and Steidl, 1986), a hundred times better than carbon.

Spontaneous ignition in open air happens at some 250°C (Thieler, 1936; Ecker and Steidl, 1986). Sulfur burns with a bluish flame under the formation of sulfur dioxide, which has a pungent smell and is highly corrosive. If the room containing the fire is unventilated, it raises the ignition point. The released heat in the combustion of sulfur is rather low, 11 kJ/g (Ecker and Steidl, 1986). In comparison with bituminous products and plastics, this is only one-fourth of their energy release (Anderberg and Pettersson, 1992), which delays the fire spreading process.

7.7.2 Recovery Process

In principle, sulfur concrete can be recovered fully an infinite number of times by raising the temperature and cooling without any impact on the aggregates (Makenya, 2001). However, in practice, there are limits to every such activity, depending on the gradual contamination of the binder. Pickard (1984) demonstrated that the compressive strength is recovered in a growing sequence up to five times, but then dropping to less than the initial strength after ten times as shown in Figure 7.13. Each cycle consists of heating the sample to 160°C, taking out the raw materials and cooling, disregarding the heat loss and using the raw materials again for creating sulfur concrete.

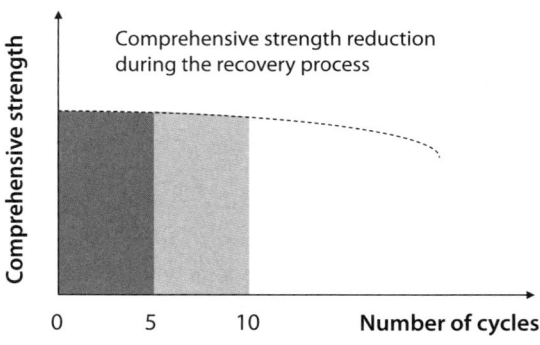

Figure 7.13. Compressive strength reduction with number of recovery cycles.

Example:
Calculate the energy requirement during the recovery process for sulfur concrete of density 2.25 Mg/m³, D = 10 mm.

Solution:
For $D = 10$ mm, the potential composition of sulfur concrete by weight is 37.5% coarse aggregates, 39.0% fine aggregates, 7.0% filler, and 16.5% modified sulfur.

Taking 1 m³ of sulfur concrete and for sulfur concrete of density 2.25 Mg/m³, the weight of each component is 843.75 kg of coarse aggregates, 877.50 kg of fine aggregates, 15.50 kg of mineral filler, and 371.25 kg of modified sulfur. The total aggregates plus mineral filler is 1878.75 kg.

Assuming the heat absorption capacity of aggregates and filler is 1.4 J/g°C, the quantity of heat required for raising the temperature from 20 to 160°C is:

$$Q_{Aggregates\ and\ Filler,\ 20-160°C} = (160-20) \times 1.4 + 1000 \times 1878.75 = 368.24 \quad MJ \quad (7.3)$$

$$Q_{Modified\ Sulfur,\ 20-160°C} = 172 \times 1000 \times 371.25 = 63.86 \quad MJ \quad (7.4)$$

$$Q_{Sulfur\ Concrete,\ 20-160°C} = 368.24 + 63.86 = 432.10 \quad MJ \quad (7.5)$$

$$= 120 \text{ kWh/m}^3$$

To add the second half of the cycle, we would require 120kWh/m³, without including any form of heat recovery, for the reuse of the raw materials for producing sulfur concrete. A reasonable heat recovery process might reduce the required energy by half (Makenya, 2001). Since Portland cement requires about 1.1 kWh/kg in the dry production (Wilck, 1991), the 370 kg of binder in Portland

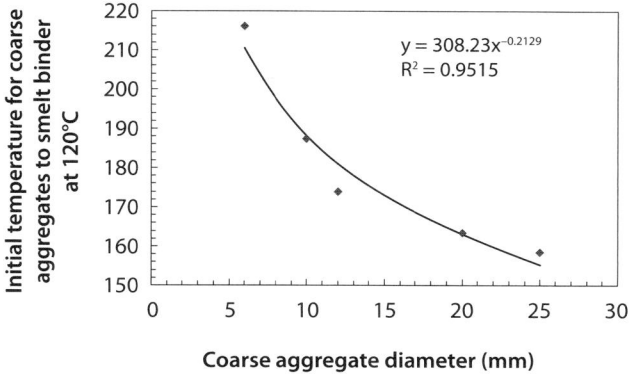

Figure 7.14. Initial temperature for coarse aggregates to smelt modified sulfur at 120°C (data obtained from Makenya, 2001).

cement concrete would require 410 kWh/m³, or 70% more than the equivalent amount for sulfur concrete.

It is of interest to know the initial heating temperature of coarse aggregates to melt the modified sulfur when producing sulfur concrete. Makenya (2001) has shown that as the coarse aggregate diameter decreases the initial temperature (for coarse aggregates to smelt a binder at 120°C) increase as shown in Figure 7.14, reaching a value of 216°C for 6 mm coarse grain diameter. For finer aggregates, the required temperature will increase. However, since the heating temperature of sulfur should not exceed 160°C, it is obvious for smaller dimension aggregates, that modified sulfur should be added to the fine aggregates at 160°C under a continuous supply of heat.

7.7.3 Cooling Process

The time duration between a pour at 140°C, through solidification, and down to an ambient temperature around 20°C is significant. In order to speed up the cooling, it is, therefore, advisable to maintain a high surface to depth ratio of any product made of sulfur concrete.

Makenya (2001) has demonstrated that the velocity of heat propagation into sulfur concrete is about half the equivalent velocity in Portland cement concrete. The calculation procedures follow.

After time, t, the intrusion depth of a heat wave is given by:

$$x_t = 3.6 \sqrt{\frac{\lambda_t}{\rho_c}} \qquad (7.6)$$

Table 7.6. Thermal properties of Portland cement and sulfur concretes

Type of property	Portland cement concrete	Sulfur concrete
Specific heat capacity, c (kJ/kg°C)	1.0	0.85
Density (kg/m³)	2.35	2.3
Thermal conductivity, λ, (W/m°C)	2.1	0.37

Where x_t is the intrusion depth (m) after time, t, λ is thermal conductivity (W/m°C), ρ is density (kg/m³), and c is specific heat capacity (kJ/kg°C).

Considering the thermal properties of Portland cement and sulfur concretes shown in Table 7.6, it is possible to establish an intrusion depth relation between Portland cement concrete and sulfur concrete as:

$$\frac{x_t^{PCC}}{x_t^{SC}} = 2.16 \qquad (7.7)$$

This means that the progress of heat in sulfur concrete is only half of that of Portland cement concrete.

7.8 DURABILITY ISSUES

7.8.1 Types of Fillers and Aggregates

The durability of sulfur concrete relates to various conditions: (a) mix design, (b) aggregate composition and amount of binder, (c) choice of aggregates that resists harsh environmental conditions, (d) choice of filler that influences workability and thermal stress, (e) choice of composite (sulfur binder, fillers, and aggregates) for resistance to acidic, basic, and salt environments, (f) water permeability and moisture absorption, (g) casting procedures and binder compensation, (h) frost resistance, (i) service temperature, (j) fire load, (k) creep, (l) fatigue load, (m) steel and other reinforcements, and (n) abrasion resistance.

Sulfur mortars are generally prepared by mixing molten-modified sulfur and heated mineral aggregates. The mineral aggregates in the compositions of the sulfur mortar include fine aggregate and fines. Fine aggregate includes sand and other materials of mineral composition that have a particle size of approximately 150 μm to 4.75 mm. Fines include silica flour, stone powder, and other material of mineral composition that have a particle size less than 150 μm (Gillott, et al., 1980). Detailed aggregates descriptions are covered in Chapters 4, 5 and 6.

Bahrami et al. (2008) studied the effect of filler types and percentage on the composition and the strength of produced sulfur concrete. They found that the

Table 7.7. Composition and properties of sulfur concrete with various fillers

Sample	Wt% of sand	Wt% of modified sulfur	Wt% of filler	Density (Mg/m³)	Compressive strength (MPa)
1	58	32	10 silica flour & stone powder	2.262	20.98
2	58	32	10 silica flour, stone powder, & mica	1.152	45.21
3	58	32	10 stone powder & mica	1.146	34.42
4	58	32	10 silica flour, stone powder, & carbon	2.309	22.16

function of the filler is to stabilize the sulfur mortar and reduce the amount of modified sulfur needed. The addition of filler increases viscosity because of sufficient matrix relative to the amount of sulfur to obtain a workable mixture and decrease void spaces in the mortar. The preparation of sulfur mortar with different fillers—silica flour, carbon, and mica—are represented in Table 7.7. The results indicate that silica flour fillers produce a stiffer mix that is useful in placing sulfur mortar on a slope. Also, the tensile and flexural strength of sulfur mortars prepared by mixing 58% sand, 2% silica flour, 8% stone powder, and 32% modified sulfur are greater than 80 and 115 kg/cm², respectively.

The influence of aggregate chemical composition on the durability performance of sulfur concrete was studied by various authors.

Czarnecki and Gillott (1990) studied the durability of sulfur concrete made with seven types of rock used as aggregate; granite, syenite, diorite, basalt, greywacke, quartzite, and marble. Some sulfur concretes expanded excessively under moist conditions. The deterioration was accompanied by a decrease in dynamic modulus of elasticity over time. Treatment of the aggregates with glycerin and crude oil admixtures was improved slightly, but expansions were still unacceptably high.

Abdel Jawad and Qudah (1994) conducted an experimental program to investigate the performance of sulfur concrete made with four types of aggregates; limestone, natural river, basalt, and Sweilleh sand. They found that strength and the unit weight of sulfur concrete were affected by factors such as type of aggregate, aggregate particle shape and surface texture, and the amount of sulfur binder in the mix. The experimental results were described briefly in Chapter 4.

Mohamed and El Gamal (2009) developed cement kiln dust-based sulfur-modified concrete products. These products consisted of bitumen-modified sulfur cement, sand and cement kiln dust (CKD) generated from the manufacture

Figure 7.15. SEM image of fresh CKD (Mohamed and El Gamal, 2009).

of cement and had excellent strength properties. However, product durability was a problem especially when the sulfur-modified concrete product was exposed to humid conditions. The experimental results of the manufactured CKD-based sulfur-modified concrete are now discussed.

The major constituents of CKD are compounds of lime, iron, silica, and alumina. The pH of a CKD water mixture is generally around 12. The SEM micrograph of fresh CKD shown in Figure 7.15 indicates that the CKD particles are coarse, irregular, and random in shape and size and a mean particle size of approximately 5-7 μm. The cumulative grain size distribution of CKD is shown in Figure 7.16. The chemical analysis of CKD and sand is shown in Table 7.8.

A simple formulated sulfur-modified concrete made with elemental sulfur, modified sulfur, CKD, sand, and glass fiber has been manufactured and tested to determine its potential utilization in public works (Mohamed and El Gamal, 2009). The prepared sulfur-modified concrete is composed of sulfur binder (elemental sulfur and polymerized sulfur), aggregates (CKD and sand), and glass fiber. Depending on the type of mix, variations in the aggregate proportion, and the binder to aggregate ratio, the resultant strength of each sulfur-modified concrete mix was evaluated.

Results from the SEM have shown that the internal structure of the CKD-based sulfur-modified concrete is extremely homogeneous with a high degree of packing. Sulfur cement effectively covers and binds the aggregates and also fills

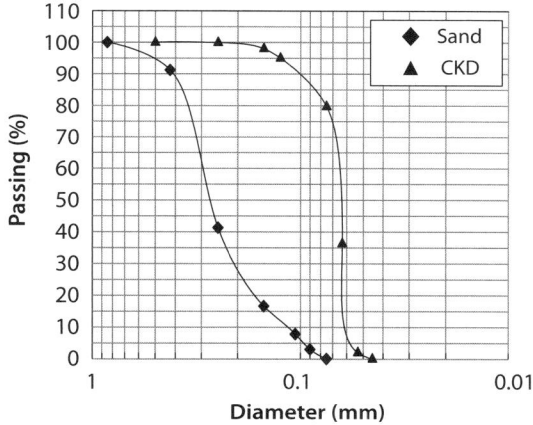

Figure 7.16. Grain size distribution of CKD and sand (Mohamed and El Gamal, 2009).

Table 7.8. Chemical composition of CKD and sand (Mohamed and El Gamal, 2009)

Waste form	Chemical composition (percent by weight)								
	SiO_2	Al_2O_3	Fe_2O_3	CaO	MgO	SO_3	K_2O	Na_2O	Cl
CKD	12.63	2.26	2.08	46.47	0.89	1.56	1.78	0.25	0.52
Sand	74.4	0.47	0.676	16.35	1.158	-----	0.13	-----	----

the inner spaces as shown in Figure 7.17. In addition, large sulfur crystals were not formed within the solidified CKD-based sulfur-modified concrete.

In a mix design, the determination of component proportions is the first step. In that regard, one aims to calculate quantities of all components and to find optimum proportions that satisfy specific requirements for workability, strength, and durability. Sulfur-modified concrete composition consisting of sulfur binder (elemental sulfur and polymerized sulfur), aggregates (CKD and sand), and glass fiber depends on various parameters as discussed below.

The effect of the bitumen-modified sulfur content on the strength of the developed mortars is shown in Figure 7.18. The results indicated that the compressive strength decreased linearly with increasing amounts of modified sulfur due to a viscosity increase that is in agreement with that previously reported by Vroom (1981). A viscosity increase due to the addition of modified sulfur has a direct impact on the crystallization of sulfur. As viscosity increases, crystal growth becomes difficult and its formation rate is reduced, causing a partial reduction in the resultant compressive strength.

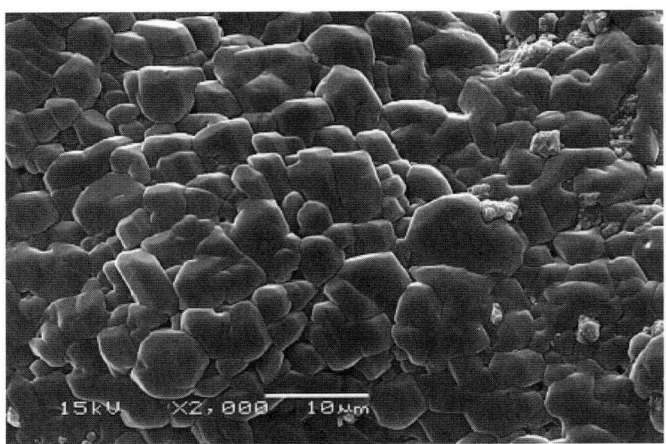

Figure 7.17. SEM photograph of CKD-based sulfur-modified concrete; the surface image shows the typical size and the arrangement (Mohamed and El Gamal, 2009).

Sulfur Binder to Aggregate Ratio

Proportioning and mixing CKD, sand, and sulfur binders would affect the final quality and serviceability of the hardened mortar. For a given quality of sulfur binder, maximum compressive strength is achieved at certain optimum binder content. Manufactured sulfur-modified concrete within this work has resulted in an increase of compressive strength as the sulfur binder/aggregate ratio increases up to 0.8 as shown in Figure 7.19 in which a thin layer of sulfur coats all particles.

Sulfur binders enhance sulfur-modified concrete corrosion resistance through the bonding of aggregate particles, filling the voids, minimizing moisture absorption, and providing sufficient fluidity in the mix to give a workable sulfur-modified concrete mixture. Sulfur binder eases interlocking between the sand and CKD particles due to its lubricating effect. However, with a large sulfur binder addition, the compressive strength is decreased as shown in Figure 7.19, because the addition of more sulfur binder increases the thickness of sulfur layers around the aggregate particles, leading to a formation of viscous bonding. Therefore, the addition of a sulfur binder has a significant influence on density, voids ratio, and the overall compressive strength.

Stability of CKD-based Sulfur-Modified Concrete in Aquatic Environments

CKD-based sulfur-modified concrete specimens were subjected to performance tests under hydrated condition to determine how well the manufactured product withstands extreme environmental conditions. High compressive strength, in the

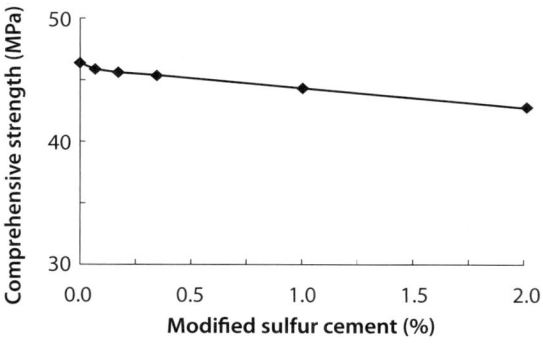

Figure 7.18. Effect of bitumen-modified sulfur content on developed compressive strength of sulfur concrete (Mohamed and El Gamal, 2009).

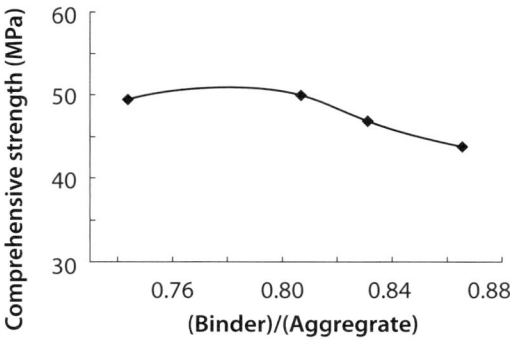

Figure 7.19. Effect of sulfur binder to aggregates ratio on developed compressive strength of sulfur concrete (Mohamed and El Gamal, 2009).

order of 50 MPa, was obtained for sulfur-modified concrete samples cured in air. However, there were adverse effects when sulfur-modified concrete samples were immersed for a period of three months in (a) water at 25 and 60°C, (b) alkaline conditions at pH 9, (c) seawater, and (d) acidic solution. Microstructure analysis by SEM showed significant fracture and damage in CKD-based sulfur-modified concrete samples as shown in Figure 7.20 due to alkali metal interaction with the aquatic environments as discussed in Chapter 6.

Effect of Cooling Rate on the Stability of Sulfur-Modified Concrete

Experimental results indicated that the cooling time of prepared specimens was relatively rapid, 15 to 30 min, due to the low specific heat capacity of the material. Prepared sulfur-modified concrete without the addition of waste solids tended to

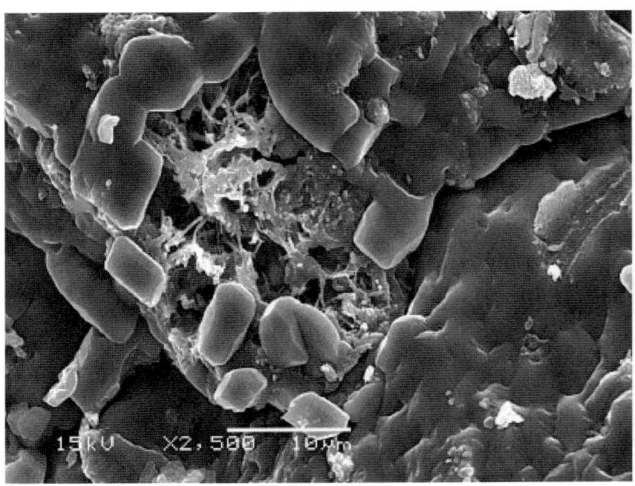

Figure 7.20. SEM photograph of CKD-based sulfur-modified concrete after immersion for seven days in distillated water at 25°C (Mohamed and El Gamal, 2009).

form a hollow void along the cylinder's axial centerline upon cooling. This phenomenon is attributed to the formation of a rigid wall on the exterior surface of the specimens due to convection. In turn, the solid exterior surface restricted the movement of the molten material at the center as it began to solidify and shrink, leaving a void in the center as shown in Figure 7.21. Voids formation decreased as the addition of waste solids increased, presumably due to high specific heat capacity and the reduction in the rate of the cooling and shrinkage potential.

Effect of Inclusion of Precipitating Agents on the Stability of Sulfur-Modified Concrete

To reduce the mobility of metals that exist in the CKD, and to comply with the U.S. EPA TCLP hazardous waste concentration limits, the potential addition of additives (i.e., precipitating agents) was examined (Mohamed and El Gamal, 2009). As a result of the scoping experiments and other considerations (such as ease of processing, cost, and availability), anhydrous sodium sulfide was selected as a preferred additive. When placed in an aqueous medium, sodium sulfide reacts with toxic metal salts to form metal sulfides of extremely low solubility. Sodium sulfide was identified by the EPA as a *best demonstrated available technology*. This additive is chemically combined with toxic and other metals and immobilized in the new waste form to trap such toxic materials and prevent them from escaping into the environment. Since the formation of the new waste form from sulfur-modified concrete minimizes waste-binder interactions, the additive

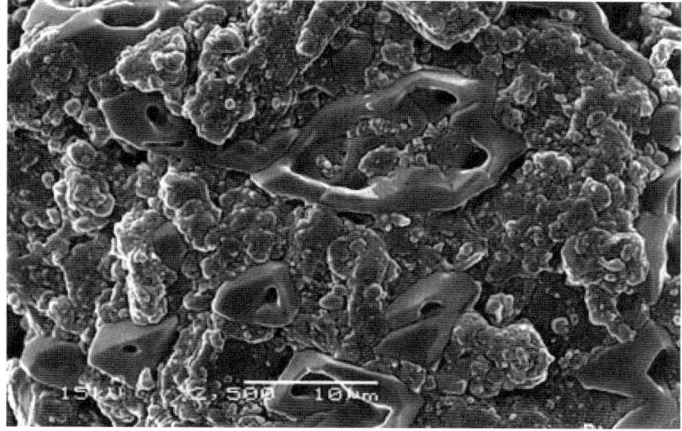

Figure 7.21. SEM photograph of CKD-based sulfur-modified concrete showing the formation of voids due to fast cooling (Mohamed and El Gamal, 2009).

remains chemically available to react with toxic metals after it is immobilized in the waste matrix. Consequently, the additive (i.e., the anti-leaching agent) is immobilized in the waste form and became available to react with toxic and other metals. As long as no moisture permeates the waste form, the anti-leaching additives and available metals in the waste form remain unreacted. When moisture permeates the waste form, the toxic and other metals in the waste form precipitates. Thus, when metals contact the immobilized and untreated anhydrous anti-leaching additives, as a result of moisture seeping into the waste form, it will react with the additives in the aqueous medium and reduce their leachability potential.

Additions of anhydrous sodium sulfide ranging from 1 to 10 wt% was tested; the optimal quantity of anhydrous sodium sulfide required depends on the concentration of toxic metals present in the waste, the preferred range was 4 to 6 wt%. The addition of anhydrous sodium sulfide as an anti-leaching agent resulted in the reduction of leached heavy metals without a reduction of the amount of cations and anions.

Failure of CKD-based Modified-Sulfur Concrete in Aquatic Mediums

The premature failure of sulfur-modified concrete in moist conditions is due to the chemical structure of CKD. The CKD tested by Mohamed and El Gamal (2009) was characterized by its high alkali and sulfate contents. The alkalis exist as alkali sulfates, such as arcanite (K_2SO_4), sodium sulfate (Na_2SO_4), and sylvite (KCl), as well as an assemblage of oxidized material such as lime that is unstable or highly soluble at earth surface conditions. When CKD contacts water, these

Table 7.9. Leached metal ions and anions (ppm) from CKD-based sulfur-modified concrete (CKD_SMC) specimens (Mohamed and El Gamal, 2009)

Test type	Time (hr)	Water/ (CKD–SMC)	Alkali metals			Heavy metals		Anions	
			Ca	K	Na	Sr	Cr	SO_4	Cl
Short term	6	2 L/kg	1418	2093	226	7.5	17	1737	1286
Long term	18	8 L/kg	1399	701	78.8	11.8	11	576	376

phases will dissolve completely and more stable and less soluble secondary phases will precipitate. Thus, the concentration of some constituent elements in CKD leachate will be controlled by the solubility of the secondary precipitates whereas the concentration of others will be controlled by their availability to the leachate solutions and by their diffusive flux into the solution from the leaching of primary phases over time (Eary et al., 1990).

To differentiate between these two classes of elemental behavior, Reardon et al. (1995) recommended conducting leaching tests on a particular waste at a minimum of two different solid/waste ratios. Leaching tests, reported by Mohamed and El Gamal (2009), were carried out in accordance with the British Standard BS EN12457: 2002 that examined the short- and long-term leaching behavior of CKD-based sulfur-modified concrete, for a liquid-to-solid ratio of 2 L/kg for 6 hr. It was then filtered and the residues were leached further at a liquid-to-solid ratio of 8 L/kg for 18 hr. The concentrations of leached Ca, Na, and K were high due to the high solubility of the bearing minerals, such as halite and sylvite. The release of other elements, such as Sr and Cr, was found in lesser amounts. Table 7.9 (Mohamed and El Gamal, 2009) shows the short- and long-term leaching behavior of CKD-based sulfur-modified concrete in distilled water.

7.8.2 Water Absorption

Penetrating water acts as a lubricant in sulfur concrete and tends to disintegrate the composite. In combination with cycling temperature, water poses a severe threat to the durability of sulfur concrete. It is, therefore, of great importance to limit the absorption. Modified sulfur does not pose any difficulty in this respect, but the aggregates must be subjected to absorption limits. This applies also to the mineral fillers and sulfur cement that includes the sulfur component and the possibility that the binder might fill the voids and the mineral filler the grain surfaces.

Mohamed and El Gamal (2009) reported that CKD is not recommended as a filler for use in sulfur concrete because of its affinity to water absorption and potential leachability of toxic metals into the environment. Also, Hellers and Makenya (1996) indicated that fly ash type F is not recommended for use in sulfur concrete because of its porosity and capillary effect. However, fly ash helps on the workability of the sulfur concrete. Therefore, the issue should be looked at from the viewpoint of whether or not the amount of sulfur in the mix was enough to coat the surfaces of the fly ash. By doing so, the exposed surfaces will be hydrophobic (i.e., water repellent) and the capillary issue will be resolved and the water absorption issue will be eliminated by the use of dense-graded mineral aggregates and proper techniques of mixture design and compaction.

In addition, the existence of swelling clays in the aggregates is a main concern and this is why the ACI guide for mixing and placing sulfur concrete mandates that swelling clays should not exist in sulfur concrete composition. Reported results indicated that problems associated with swelling clays can be observed after the second and fourth day of casting. Generally, swelling clays have an absorptive capacity that can be satisfied by molecular water permeating through the interlayer structure. As the clay absorbs water, expansion occurs (Yong et al., 1992; Mohamed and Antia, 1998). Soderberg (1983) has shown that linear expansion increases steadily in the initial stage (2 to 7 days) after which it remains constant. Weight gain is soon followed by spalling, hairline cracks, and an additional weight gain of 3 to 5%. If large amounts of swelling clays are present, extensive cracking can occur (ACI, 1988). This is the main reason why crushed aggregates are sometimes recommended to be washed to remove dust prior to use in sulfur concrete.

The desirable limiting values of water absorption (Soderberg, 1983) are 0.05 wt% after one day and 0.5 wt% after 14 days, using the test method specified by ASTM C 642. The first value is so small that it challenges the testing method and must be applied with the utmost care to be effective. The latter value for 14 days reveals the presence of swelling clays.

The preceding highlights that one of the disadvantages of sulfur cement mortars and concretes is that the presence of even small amounts of water-expandable clay (e.g., 1% by weight or more) in the aggregate causes the solidified sulfur cement mortars and concretes to disintegrate when exposed to water. This problem is particularly serious because transportation costs, therefore economic necessity, usually requires the use of aggregate sources close to the casting or job site regardless of the presence of expansive clay. The expansive clay can be removed from the aggregate by washing procedures but such procedures are also generally inconvenient and uneconomical. Thus, if the local sources of aggregate contain expansive clay, the use of sulfur cement mortars concretes is severely restricted.

U.S. Patent No. 4,188,230 reported that this problem may be overcome by the incorporation of petroleum or polyol additives. Such procedures have not, in fact, proved entirely satisfactory (Nimer et al., 1981). The problem of water-expansive clays is also considered by Shrive et al. (1977). Albom (1981) discloses that the water stability of sulfur cement-aggregate products containing up to 5% weight expandable clay can be substantially improved by treating it with a salt solution prior to admixture with the sulfur cement.

Nimer et al. (1981) have reported that the water stability of sulfur cement-aggregate compositions containing up to 5% expansive clay based on the weight of the aggregate can be substantially improved by the simple incorporation of an effective amount of an organosilane having at least one reactive functional group. This is important to the commercialization of sulfur cement mortars and concretes; most aggregates contain < 5% by weight of expansive clay and generally contain < 3% by weight.

Nimer et al. (1981) have reported a process for preparing a sulfur cement-aggregation composition containing an aggregate having up to 5% by weight, based on the aggregate, of expansive clay that comprises the improvement of admixing with an organosilane that has at least one reactive functional group to substantially reduce the water expandability of expansive clay.

The composition can be prepared by simply adding the appropriate organosilane to the ingredients of the composition. Because of the small amount of organosilane used, it is typically mixed with the sulfur cement, although it could also be added directly to the aggregate or the molten mixture of the sulfur cement and aggregate. The molten mixture is mixed to distribute the aggregate and organosilane throughout the sulfur cement. The organosilane can be added as a solid or liquid and generally will be applied in its usual commercial form. Since the particular organosilane is supplied commercially as a solid, it can be applied as a liquid (e.g., by melt liquefying or dissolution or suspension in a volatile solvent or carrier).

As is conventional, it is preferred to heat the aggregate prior to admixture with the molten-sulfur cement to avoid random cold spots, to remove entrained moisture, and to improve bonding of the sulfur cement to the aggregate. With the exception that it is preferred to add the organosilane to the sulfur cement, the order of the addition of the various other ingredients is not significant. Although, plasticized sulfur is used as the sulfur cement, it is generally preferred to add the plasticizer to the sulfur before adding the aggregate. The sulfur cement, aggregate, and any other ingredients are mixed together at temperatures above the melting point of the sulfur cement (i.e., sulfur or plasticized sulfur) and below the decomposition or boiling point of the materials. Typically, mixing is effected at temperatures in the range of 110 to 180°C and preferably, from 125 to 160°C. The molten mixture is then cast into the desired shape or form, in the case of concrete, or applied, in the case of mortar. Upon cooling, the composition solidifies into a final product that has improved water stability.

Typically, from 0.0015 to 0.015 g-mol (preferably 0.0038 to 0.0075 g-mol) of organosilane is used per kilogram of expandable clay-contaminated aggregate (Nimer et al., 1981). In terms of a more convenient weight-to-weight basis, generally 0.25 to 3 parts by weight (preferably 0.65 to 1.5) of the organosilane are used per 1000 parts by weight of expandable clay-contaminated aggregate.

Appropriate organosilanes that can be used are those that have at least one molten-sulfur reactive group that reacts with sulfur or that converts in the presence of molten-sulfur to such a group and, of course, a boiling point and decomposition temperature above the melting point of the particular sulfur cement used. The term molten-sulfur reactive group, as used herein, means a group that reacts with sulfur or that converts in the presence of molten sulfur to a group that reacts with sulfur. Examples of such molten-sulfur reactive groups include, for example, amino, diamino, epoxy, carbonyl, methacryloxy, aryl, (e.g., phenyl) mercapto, double bonds, and alkyl having at least two carbon atoms.

Suitable organosilanes that can be used include those that have the formula:

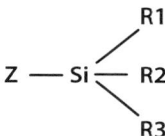

wherein R1, R2, and R3 are selected independently from the group consisting of lower alkoxy, aryloxy, ara-lower alkoxy, and halo; and Z is an organic radical attached to Si via a carbon atom and has at least one molten-sulfur reactive group such as amino, epoxy, double bond, triple bond, mercapto, cyano, hydroxy, aryl, substituted aryl, aralkyl, substituted aralkyl, carbonyl, alkyl having 2 through 20 carbon atoms (including the carbon atom of R attached to Si) cycloalkyl.

Typical Z groups include aminoalkyl, aminoalkylene–aminoalkyl, N-lower alkyl aminoalkyl, epoxyalkyl; epoxyalkoxyalkyl; alkenyl, alkynyl, mercaptoalkyl, alkylthioalkyl, cyanoalkyl, hydroxylalkyl, aryl, substituted aryl, arylalkyl, substituted arylalkyl, alkyl having at least 2 carbon atoms, cycloalkyl, cycloalkenyl, arylalkyl, aryloxyalkyl, preferably having 2 through 12 carbon atoms. In a case in which Z is hydrocarbyl, it is believed that its sulfur reactivity is due to a single bond that converts to a double bond in the presence of molten sulfur.

The preferred organosilanes are those wherein R has at least one molten-sulfur reactive group selected from the group of amino, epoxy (especially glycidoxy), and/or metha cryloyloxy mercapto. In terms of the R1, R2, and R3 groups the preferred organosilanes are those wherein R1, R2, and R3 are lower alkoxy or aryloxy and especially lower alkoxy, for example, methoxy and ethoxy. Preferably, organosilanes that have a combination of at least one preferred molten-sulfur reactive group and preferred R1, R2, and R3 groups are used.

The organosilanes are commercially available as coupling agents and adhesion promoters used with various polymeric materials such as epoxy resins, polyesters, polycarbonates, nylons, sulfur-cured elastomers, and mineral-filled compositions. Suitable organosilanes can thus be purchased or prepared via conventional procedures. Suitable organosilanes are, for example, vinyl triethoxy silane, gamma meth acryloxy propyl trimethoxy silane, beta (3, 4 epoxy cyclo hexyl) ethyl tri methoxy silane, gamma glycidoxy propyl trimethoxy silane, gamma amino propyl tri methoxy silane, gamma mercapto propyl trimethoxy silane, and compatible mixtures thereof. The preferred organosilanes are those wherein R is mercapto, epoxy, methacryl, amino or diamino, and/or R1, R2, and R3 are independently selected from the group of alkyl and aryl, especially methyl and ethyl.

Nimer et al. (1981) prepared sulfur concrete containing 25 wt% plasticized sulfur (95 wt% sulfur, 2.5 wt% dicyclopentadiene and 2.5 wt% cyclopentadiene oligomer); 3.0 wt% of the expansive clay, bentonite clay (4.0 wt% based on aggregate) and the remainder Kaiser top sand that have a U.S.A. Standard Testing sieves size range of 4 to 100 mesh and respectively containing 0.1 parts by wt of the organosilane, identified in Table 7.10, per 100 parts by wt of the sulfur cement-sand-clay composition.

The test compositions were prepared by mixing the organosilane with molten-sulfur cement and then mixing the organo-silane sulfur cement mixture with a preheated mixture of sand and clay. The molten mixture (125 to 135°C) was then cast into three 2 by 4 in cylinder molds and aged overnight at room temperature (approximately 20°C). A control composition was prepared and cast into three cylinders in the same manner but without the addition of the organosilane.

Representative cylinders for the control composition and each of the test compositions were selected and immersed in tap water at room temperature and visually inspected daily for fractures, cracks, and crumbling. At the first evidence of any of these, the cylinder was considered to have failed. The results of these trials are reported in Table 7.10.

As can be seen from Table 7.10, the compositions with organosilane had greatly superior water stabilities as compared to the corresponding control composition. The control composition containing 4% bentonite clay aggregate had a life of only about 4 hr whereas the test compositions containing 4% bentonite clay aggregate exhibited lives of 3 to 28 days.

Nimer et al. (1981) investigated the effect of organosilane loadings. Using the same experimental procedure and using the organosilane gamma-glycidoxy propyl trimethoxysilane sold under the Trademark DC Z6040 by Dow Corning. A number of test compositions were prepared containing various amounts of the organosilane. In each case the compositions contained 25% sulfur cement (95 wt% sulfur, 2.5 wt% dicyclopentadiene and 2.5 wt% cyclopentadiene oligomer); 75 wt% aggregate (72% Kaiser top sand plus 3% Bentonite clay) and variable amounts of organosilane (0–0.5 wt%). A control sample was prepared in the same manner but without the organosilane.

In each case, three cylinders were cast per composition. A representative cylinder was selected for each composition and immersed in tap water and then

Table 7.10. Performance of sulfur concrete with various organosilane (Nimer et al., 1981)

Organo silane	Parts by wt *organo- silane	Reactive functional group	Days to failure
Control	None		About 4 hr
N-(2-aminoethyl)(3-aminopropyl)trimethoxysilane	0.1	diamino	26
Gamma-glycidoxypropyltrimethoxysilane	0.1	epoxy	24
Gamma-glycidoxypropyltrimethoxysilane	0.1	epoxy	25
Gamma-methacryloxypropyltrimethoxysilane	0.1	methacrylic	28
Phenyltriethoxysilane	0.1	phenyl	13
Gamma-aminopropyltriethoxysilane	0.1	amino	10
Ethyltriethoxysilane	0.1	ethyl	3
Vinyltriethoxysilane	0.1	vinyl	5
Gamma-mercaptopropyltrimethoxysilane	0.1	mercapto	28

*Parts by weight organosilane per 100 parts of total composition (excluding organosilane)

observed for failure. The results of these tests are shown in Figure 7.22. It can be seen that the use of 0.02 to 0.5 parts of the organosilane per 100 parts of composition increased the life of the composition from approximately 6 hr to almost 7 days. The best results were obtained using 0.05 to 0.1 parts of organosilane, increasing the life of the composition to nearly 24 days.

7.8.3 Frost Resistance

Sulfur concrete resistance to frost action is of importance for structures that are subjected to such climatic conditions. Under moist conditions, it is important that the material endures frost action. McBee et al. (1983) conducted a study on freeze-thaw durability based on the C666-76 test specifications. The composition of the aggregates was dense-graded. Using crushed quartz aggregates of various compositions, the behavior of sulfur concrete undergoing 300 freeze-thaw cycles between −18 to 4.4°C was investigated. The relative dynamic modulus of elasticity from these tests was evaluated with respect to the initial water absorption (24 hr). The results indicated that when the moisture absorption of sulfur concrete exceeds the limit (0.05%), its resistance to freeze-thaw damage decreases dramatically. Therefore, good quality control for water absorption is essential in producing durable sulfur concrete. The key is a sufficient amount of binder and a restricted amount of filler to attain the needed material density with a dense-graded composition.

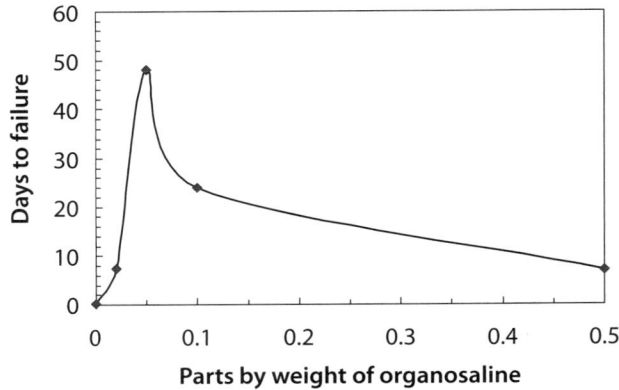

Figure 7.22. Performance of organosilane loadings in sulfur concrete (data from Nimer et al., 1981).

7.8.4 Service Temperature

As discussed in Chapters 5 and 6, sulfur concrete is a thermoplastic material that exposes a distinct softening with increasing temperature—from room temperature (20°C) to melting (120°C). With this limited span (100°C) between solid matter and full disintegration, it is obvious that the actual service temperature becomes a consideration of great importance. It should lie within the parameters of ambient climatic conditions with a margin to suit safe applications of sulfur concrete products. In addition to strength considerations, operating temperature can be a crucial factor in resistance to certain chemicals.

A practical limit for pre-stressed sulfur concrete would be 80% retention strength that is 50°C (Makenya, 1997). Different sources advise various limits for using unstressed sulfur concrete: Soderberg (1983) indicated 80°C (50% retention strength); Pickard (1983), 93°C; Chempruf (1988), 88°C; and Aarsleff (1989), 90°C. The reason for this variance is the duration of the temperature load and the absorption of heat into the material that depends on surface characteristics such as color. Sulfur concrete is darker than normal Portland cement concrete and obviously has a higher coefficient of absorption.

The maximum operating temperature of 88°C expressed by Chempruf (1988) is attributed to the poor thermal conductivity of sulfur concrete that allows short exposure to elevated temperatures. Available information on sulfur concrete conductivity is sketchy. Reported thermal conductivity values are:

    a. 0.4 to 2.0 W/m°C (Malhotra, 1983)
    b. 1.2 to 2.0 W/m°C (Soderberg, 1983)
    c. 1.5 W/m°C (Anderberg and Pettersson, 1992)
    d. 0.346 to 0.865 W/m°C (Chempruf, 1988)
    e. 0.372 W/m°C (Weber, 1993)

The thermal conductivity value (0.372 W/m°C) reported by Weber (1993) is close to one-fourth of the dry Portland cement concrete value (1.5 W/m°C). It is compatible with a reverse linear relation, adding 16% for aggregates (1.5 W/m°C) and 84% from modified beta-sulfur (0.16 W/m°C) to a sum of 0.374 W/m°C. This value means, in theory, that the rate of progression in a temperature shock situation is halved, or the time of cooling identical products is doubled. In practice, it is more than doubled because the cooling process of sulfur concrete comprises a temperature span of 100°C, whereas the cooling of Portland cement concrete rarely goes beyond 40 to 50°C with respect to strength loss and crack prevention.

Considering sulfur concrete is used for thin-walled structures, either as a genuine product or as a lining on a core of Portland cement concrete, it is recommended (Makenya, 1997) to restrict the use of pre-stressed sulfur concrete to a conservative level, 70°C, if the temperature load is permanent or semi-permanent (e.g., the sample is exposed to the sun). The retention strength would then be 65% or close to two-thirds.

7.8.5 Fire Load

As discussed in Chapter 3, elemental sulfur melts at about 119°C. Therefore, when sulfur concrete is subjected to such temperature for any length of time, it melts and loses its structural strength. However, the rate of temperature transfer within the sulfur concrete material is slow because of its low thermal conductivity. This, in turn, will allow time for defense actions. For temperatures higher than 260°C, sulfur concrete ignites and increases the fire load. It burns with rather cold and small, blue flames. Most of the released energy is consumed by the aggregates. The temperature remains so low that even a weak breeze or a naked hand can extinguish the fire (Makenya, 1997). In addition, because sulfur combustion is self-sustaining, once ignited it will continue to burn until extinguished. In the presence of air, sulfur will burn to sulfur dioxide, a toxic gas with a sticky smell, that is found at low concentrations. However, it is anticipated that sulfur concrete will not be used in structural elements whereby a risk of fire exists such as in residential and commercial buildings. Nevertheless, it can be used for water and subsurface structures.

Fire Retardant Additives

Simic (1982) utilized decabromodiphenyl oxide to provide a plasticized sulfur composition with both improved flame retardancy and improved crazing resistance. The sulfur composition was prepared by charging the indicated weight of sulfur to a stainless steel beaker. The sulfur was heated until molten at a temperature of approximately 145°C. Then, the indicated parts of plasticizer were added to the stirred, molten sulfur, and the resulting mixture was heated until homogeneous—10 to 15 min. The decabromodiphenyl oxide and then the filler were added, and the entire mixture was stirred and heated at 145°C until the filler was dispersed uniformly throughout the mixture—15 min to 1 hr.

The sulfur compositions were tested for flame retardancy by the *flammability of plastics using oxygen index method* as described in ASTM D-2863-70. This method determines the relative flammability by measuring the minimum concentration of oxygen (oxygen index) in an oxygen/nitrogen atmosphere that will barely support combustion. The sample to be tested, a rod 6.35 mm in diameter and 76.2 mm long, is mounted in a Pyrex chimney. Then a mixture of nitrogen and oxygen is introduced at the bottom of the chimney and allowed to rise slowly past the sample at a constant rate of 40 ± 10 mm/sec. The sample is then ignited at the top and allowed to burn in a candle-like fashion. This procedure was repeated with varying concentrations of oxygen until the concentration that just supports combustion is found (i.e., the oxygen index).

The results of applying this test to sulfur compositions with decabromodiphenyl oxide as well as to other sulfur bromine-containing flame retardant mixtures are shown in Table 7.11. The results are tabulated by wt% of the flame retardant. The percent bromine in each composition is also given for comparison. It is worth noting that the sulfur oxygen index is 15. The data shown in Table 7.11 from No. 1 to 4 show the superiority of decabromodiphenyl oxide as a specific flame retardant for sulfur compositions as compared to analogous flame retardant compositions on a percent bromine basis.

Simic (1982) also utilized a combination of decabromodiphenyl oxide and dicyclopentadiene or cyclooctadiene to provide a plasticized sulfur composition with both improved flame retardancy and crazing resistance. At low concentrations, the incorporation of these materials into sulfur compositions gives a significant increase in the oxygen index of a sulfur-decabromodiphenyl oxide composition as shown in Table 7.12.

7.8.6 Crazing Resistance

Simic (1982, U.S. Patent 4348233) tested three sulfur compositions for crazing resistance when used as coating materials. To test crazing resistance, 152.4 by 101.6 mm concrete bricks were covered to a 6.35 mm depth with a molten layer of a sulfur composition. Upon hardening, the bricks were tested on a 24 hr temperature cycle according to the following procedures:

a) 8 hr at 70°C
b) 15 hr in a freezer at 0°C. The freezing cycle required 2 hr to reach 0°C; the temperature was held at 0°C for 8 hr and then allowed to

Table 7.11. Flame retardation of sulfur composites using organic bromide additives (Simic, 1982; US Patent No. 4348233)

No.	Additive	Additive %				
		0.5	1.0	2.0	4.0	6.0
Sulfur composite I consisted of: 79% sulfur, 2% phenol-sulfur adduct, 1% thiokol LP-3 and 18% mica						
1	Decabromodiphenyl oxide Br (oxygen index)	0.41 (22)	0.82 (24)	1.64 (28)	3.28 (32)	
2	Octabromodiphenyl oxide Br (xygen index)	0.39 (17)	0.79 (19)	1.48 (22)	3.16 (28)	4.74 (31)
3	Pentabromodiphenyl oxide Br (oxygen index)	0.35 (16)	0.71 (18)	1.42 (22)	2.84 (25)	4.26 (26)
4	Hexabromobenzene Br (oxygen index)	0.43 (17)	0.87 (18)	1.74 (20)	3.48 (28)	5.22 (33)
5	Brominated phenoxy alkane Br (oxygen index)	0.35 (16)	0.70 (17)	1.40 (20)	2.80 (26)	4.20 (27)
6	2,4,6-Tribromophenol Br (oxygen index)	0.36 (15)	0.72 (16)	1.43 (20)	2.86 (24)	4.35 (25)
7	Tetrabromobisphenol A Br (oxygen index)	0.29 (16)	0.58 (17)	1.16 (19)	2.32 (24)	3.48 (27)
Sulfur composite II composition: 79% sulfur, 2% phenol-sulfur adduct, 1% triphenyl phosphate and 18% mica						
8	Decabromodiphenyl oxide Br (oxygen index)	0.41 (26)	0.82 (27)	1.64 (31)	3.28 (35)	

warm up to 10 to 15°C over a period of 5 hr. After each cycle, the bricks were examined for crazing, cracking, or de-lamination under a 7x magnifier.

The results are shown in Table 7.13. Testing was stopped after 100 cycles.

7.8.7 Creep

Creep is a measure of the ability of a material to sustain a permanent load with integrity and limited deformation. Reported data in the literature are conflicting. For example, Malhotra (1983) reported that sulfur concrete exhibits more creep after one year than Portland cement concrete at the same stress level. However, Soderberg (1983) reported that after 13 weeks, sulfur concrete has a lower creep than Portland cement concrete at room temperature and a relative humidity of 60%. Weber et al. (1990) reported that Chempruf-modified sulfur concrete creeps in equal amount or less than Portland cement concrete. In addition, it was

Table 7.12. Flame retardation with decabromodiphenyl oxide synergistic effect with dicyclopentadiene (Simic, 1982; US Patent No. 4348233)

No.	Sulfur Composite I (%)	Deca-bromo-diphenyl oxide (%)	Di-cyclo-pentadiene (%)	Oxygen index (ASTM D 2863)
1	100.00			15
2	99.50	0.5		22
3	99.00	1.0		24
4	98.00	2.0		28
5	96.00	4.0		32
6	99.40	0.5	0.10	30
7	98.90	1.0	0.10	32
8	97.90	2.0	0.10	36
9	95.90	4.0	0.10	37
10	99.00	1.0		24
11	98.95	1.0	0.05	29
12	98.90	1.0	0.10	32
13	98.80	1.0	0.20	33
14	98.50	1.0	0.50	35
15	98.00	1.0	1.00	37

indicated that the addition of fiber glass into sulfur concrete increases creep over unreinforced sulfur concrete; however, it remained less than Portland cement concrete. Hellers and Ekblad (1996) reported that sulfur concrete creeps in the same amount as Portland cement concrete when the two concretes have a similar compressive strength of approximately 60 MPa.

7.8.8 Fatigue Strength

Pickard (1981) reported that sulfur concrete fatigue behavior is about one- to threefold the fatigue resistance of Portland cement concrete. Shrive et al. (1977) reported that elemental sulfur concrete has fatigue lives of 1.7 million cycles at a stress level of 95% of the modulus of rupture. However, for sulfur concrete, the fatigue life reduces to 0.2 million cycles that is still significantly above that of Portland cement concrete for the same 95% level. A similar conclusion was reported by Breitfuss (1981) for sulfur concrete made from Vroom SRX modifier. Vroom (1996) claims that there is an endurance limit for sulfur concrete at 85 to 90% of modulus of rupture compared to 50 to 55% for conventional concrete. He further pointed out that sulfur concrete is highly applicable to railway applications due to its corrosion resistance, low electrical conductivity, and a good bond to reinforcing steel, high strength, and lack of fatigue.

Table 7.13. Crazing resistance of sulfur composites formulated with decabromodiphenyl oxide (Simic, 1982; US Patent No. 4348233)

No.	Composition (%)			No. of cycles	
	Sulfur composite	Decabromodiphenyl oxide	Dicyclopentadiene	Failure	No failure
Composition A: 82.5% sulfur, 1.0% phenol-sulfur adduct, 0.5% thiokol LP-3 and 16% mica					
1	100.00			3	
2	99.50	0.50			100
Composition B: 79.0% sulfur, 2.0% phenol-sulfur adduct, 1.0% thiokol LP-3 and 18% mica					
3	100.00			6	
4	99.50	0.50			100
5	99.40	0.50	0.10		100
6	99.15	0.75	0.10		100
7	96.00	4.00			100
Composition C: 79.0% sulfur, 2.0% phenol-sulfur adduct, 1.0% triphenyl po and 18% mica					
8	100.00			6	
9	99.50	0.50			100

7.8.9 Reinforcement

The purpose of reinforcement is to compensate for a lack of bending strength (Makenya, 1997). However, sulfur concrete has a high capacity to resist bending stresses because of its high density and low potential of crack formation upon solidification. Therefore, the use of reinforcement in sulfur concrete is not mandatory. Soderberg (1983) reported that reinforcement is necessary in many precast products and in poured-in-place products to arrest cracking and increase load carrying capacity. Malhotra (1983) has disqualified the use of steel reinforcement in sulfur concrete whenever the environmental conditions are not completely dry. Swamy and Jurjees (1986) reported that sulfur concrete is sensitive towards moisture and frost. This could be explained based on the high content of mineral fillers and low content of binder in the tested sulfur concrete.

Makenya (1997) concluded that unstressed steel reinforcement is acceptable with a dense—well compacted—surrounding sulfur concrete used for nonstructural purposes (no bending) such as in slab on soil, foundations, and low container walls. Epoxy-coated or even stainless reinforcing steel may be used for additional protection.

Glass Fiber Reinforcement

The ability to reinforce a constructional material with glass fiber is of great practical importance to enhance the tension characteristics of the produced products. The use of glass fiber to improve the structural integrity of sulfur-modified concrete in the construction industry was reported by Jong et al. (1985). A small quantity (0.5 wt%) of glass fibers, manufactured by Owens Coming, N.Y., measuring 12.7 mm in length was added to the preferred mixture for making sulfur-modified concrete. Accordingly, glass fibers were used to eliminate material cracking under saturated conditions (U.S. patent No. 5678234, Colombo et al., 1997), and U.S. patent No. 4414385, Swanson 1983).

Glass fiber reinforced composite materials consist of high strength glass fiber embedded in a cementitious matrix. In this form, both fibers and matrix retain their physical and chemical identities, yet they produce a combination of properties that cannot be achieved with either of the components acting alone. In general, fibers are the principal load-carrying members whereas the surrounding matrix has the following properties: (a) it keeps the fibers in the desired locations and orientation, (b) it acts as a load-transfer medium between the fibers, and (c) it protects the fibers from environmental damage. In fact, the fibers provide reinforcement for the matrix and other useful functions in fiber-reinforced composite materials. Glass fibers can be incorporated into a matrix either in continuous lengths or in discontinuous (chopped) lengths.

Glass fiber reinforcement in the amount of 1 to 3% is used regularly to control crack formation. The use of glass fibers in sulfur concrete has been effective as an enhancement in field applications for controlling shrinkage cracks and improving ductility and impact resistance (ACI, 1988). Glass fibers of a length between 13 to 38 mm can be added to sulfur concrete in loadings from 9 to 12 kg/m^3, which is around 2.5% of the binder weight, to attain this effect. Pickard (1981) reported that glass fiber loadings of 5 to 7% seem to be optimum. It is not necessary to use the alkali-resistant glass fiber commonly used in hydraulic concretes because there is no alkali present to cause the corrosion of glass. For the same reason, alkali/aggregate problems are unknown with sulfur-modified concrete.

Figure 7.23 (Mohamed and El Gamal, 2009) shows the variations of compressive strength with the amount of glass fiber in sulfur-modified concrete. The results indicated that the addition of glass fibers to the mix causes an increase in the compressive strength of sulfur-modified concrete up to a certain fiber glass content and then it reduced. The most preferred percentage of 0.35 wt% provides suitable structural integrity without adversely affecting mix ability. Larger quantities tended to clump, reducing the mixture workability and lowering the strength as shown in Figure 7.23. In this form, both fibers and matrix retain their physical

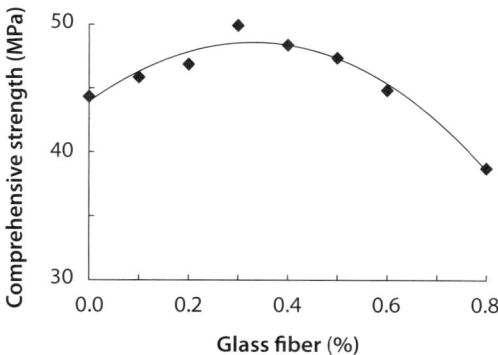

Figure 7.23. Effect of glass fiber content on developed compressive strength of modified-sulfur concrete (Mohamed and El Gamal, 2009).

and chemical identities, yet they produce a combination of properties that could not be achieved with either of the components acting alone.

Also, polypropylene fibers could be used for modifying the mixing process and keeping the hot plasticized material together. This is a benefit especially for the use of sulfur concrete underwater. It should be noted that at 140°C polypropylene is beyond the softening point, hence, an increase of temperature beyond this point must be avoided.

7.8.10 Abrasion Resistance

In the chemical- and food-processing industries, surfaces are subjected to a combinations of loads coming from chemical and mechanical sources and are sometimes aggravated by climatic factors such as moist and temperature. Wet floors must endure severe wheel pressures from trucks in an environment of chemical stains and moisture from high-pressure washing. In such cases, abrasion resistance is important. Portland cement concrete floors are sensitive to chemicals, and high-pressure washing tends to remove the binding matrix close to the surface between remaining singular stones that afterward form a rough surface. The abrasion resistance of sulfur concrete is twice that of Portland cement concrete (Cric and Whitmore, 1996). Abrasion by high-pressure washing after a submersion in organic acidic baths with pH values of 2.5, 3.5 and 4.5 was studied by Thylén (1991). The results indicated that sulfur concrete did not show any deterioration whereas Portland cement concrete deteriorated rapidly especially at low pH values. The use of sulfur concrete in federally inspected meat and poultry plants was approved in 1987 by the U.S. Department of Agriculture.

7.8.11 Chemical Resistance

Bacon and Davis (1921) reported the development of acid-resistant mortars containing 40% sulfur and 60% sand for use in the chemical industry. However, it was found that moisture and ambient temperature variations were destructive to the composite of sulfur and sand. An improvement was introduced with the addition of a modifier, Thiokol (olefin polysulfide), to the mortar and it stabilized the grout. This, in turn, has led to a wider use of sulfur as a binder of minerals in chemical plants.

In 1921, Kobbe mixed Portland cement with elemental sulfur for a binder of concrete with mineral aggregates. The composition was favorable to the compressive strength but was sensitive to moisture absorption, hence, poor durability.

However, the produced sulfur concrete was unstable due to the transition of sulfur at 95.5°C from monoclinic to orthorhombic and crystalline form as discussed in Chapter 3. The later form is dense and occupies less volume and that leads to contraction stresses and fatigue under thermal cycling. The development of microcracks due to contraction opens paths to the intrusion of moisture that will decrease the contact stresses and lead to material destruction.

Therefore, as discussed in Chapter 4, it was critical to develop a modifier for sulfur that would prevent and/or delay the development of internal stresses and disintegration of the sulfur concrete matrix. The behavior of various modifiers is discussed in Chapter 5.

In general, sulfur concrete is resistant to many acids, saline environments, and environments embossed by other chemicals. This property forms the basis of the sulfur concrete industrial applications. However, we have to pay close attention to the durability of the material. Table 7.14 shows the behavior of Chempruf, which was developed by McBee et al. at the U.S. Bureau of Mines, to a wide range of chemicals (Makenya 2001). Table 7.15 shows the general behavior sulfur concrete to a wide range of chemicals (Crick and Whitmore, 1998).

7.8.12 Corrosion Potential

Data on steel reinforcement preservation in sulfur concrete are not numerous and their nature is contradictory. Thus, according to the data (Vroom 1976), signs of reinforcement corrosion have not been found in products made of sulfur concrete that has been kept for 2 years in the open, which can be explained by low concrete air and water permeability. According to Swamy and Jurjees (1986), considerable corrosion of reinforcement has been observed while storing beams with steel reinforcement. When analogous beams—made of modified-sulfur concrete—were stored in water, the passivating effect of concrete with regard to reinforcement has been noted.

Table 7.14a. Resistance of Chempruf to municipal wastes (data from Makenya, 2001)

Chemical environment	References
Animal wastes	Pickard (1981); Soderberg (1984)
Deicing salts	Travnicek (1987)
Food wastes	Pickard (1981); Soderberg (1984)
Sewage water	Ecker and Steidl (1981); Weber (1993)

Table 7.14b. Resistance of Chempruf to organic acids (data from Makenya, 2001)

Chemical environment	Temperature	References
Acetic acid		
100%	27	Soderberg (1984)
50%	60	Chempruf Manual (1988)
10%	93	Pickard (1985); Ecker and Steidl (1986)
Acidic and biochemical	27	Weber et al. (1990)
Benzoic acid		Ecker and Steidl (1986)
Butyric acid	27	Pickard (1981); Soderberg (1984)
Carbonic acid		Ecker and Steidl (1986)
Citric acid (100%)	93	Pickard (1985); Ecker and Steidl (1986)
Formic acid (65%)	27	Chempruf Manual (1988)
Lactic acid	27	Pickard (1981); Soderberg (1984)
Oxalic acid (100%)	93	Chempruf Manual (1988)
Silage acid	27	Pickard (1981); Soderberg (1984)
Tannic acid	27	Pickard (1981); Soderberg (1984)
Wine acid		Ecker and Steidl (1986)

Table 7.14c. Resistance of Chempruf to saturated mixes with sulfuric acid (data from Makenya, 2001)

Chemical environment	References
Copper sulfate-sulfuric acid	McBee et al. (1992)
Manganese oxide-sulfuric acid	Ward et al. (1992)
Uranium sulfate-sulfuric acid	McBee et al. (1992)
Vanadium sulfate-sulfuric acid	McBee et al. (1992)
Zinc sulfate-sulfuric acid	McBee et al. (1992)

Table 7.14d. Resistance of Chempruf to inorganic acids (data from Makenya, 2001)

Chemical environment	Temperature	References
Hydrofluoric acid (35%)	93	Soderberg (1984)
Hydrofluosilicic acid		Weber (1993)
Manganic acid		Ecker and Steidl (1986)
Nitric acid		
50%	21	Soderberg (1984); Chempruf Manual (1988)
25%	25	Chempruf Manual (1988)
Nitric-citric acids (mixed)		Design and Construction (1988)
Phosphoric acid (85%)	93	Soderberg (1984); Chempruf Manual (1988)
Phosphorous acid (20%)		Pickard (1985); Ecker and Steidl (1986)
Sulfuric acid		
93%	49	Soderberg (1984); Chempruf Manual (1988)
60%	93	Soderberg (1984); Chempruf Manual (1988)

Table 7.14e. Resistance of Chempruf to other chemicals (data from Makenya, 2001)

Chemical environment	Temperature	References
Acetone		Ecker and Steidl (1986)
Vinegar		Ecker and Steidl (1986)
Ethanol (50%)	27	Chempruf Manual (1988)
Glycol (60%)	38	Chempruf Manual (1988)
Glycol (50%)	60	Ecker and Steidl (1986)
Formaldehyde		Ecker and Steidl (1986)
Glycerol		Ecker and Steidl (1986)
Lime water	20	Pickard (1985)
Linseed oil		Ecker and Steidl (1986)
Methanol (75%)	27	Chempruf Manual (1988)
Paraffine		Ecker and Steidl (1986)
Phenol (32%)	27	Chempruf Manual (1988)
Sea water		Pickard (1981)
Starch		Ecker and Steidl (1986)
Sugar solution		Ecker and Steidl (1986)
Transmission oil		Ecker and Steidl (1986)
Urea		Pickard (1985)
Vaseline		Ecker and Steidl (1986)
Water glass		Ecker and Steidl (1986)

Table 7.14f. Resistance of Chempruf to salts (data from Makenya, 2001)

Chemical environment	References
Aluminum (acetate, bromide, chlorate, chloride, fluoride, hydroxide, iodide, nitrate, and sulfate)	Ecker and Steidl (1986)
Ammonium (nitrate and sulfate)	Pickard (1985)
Barium (acetate, carbonate, chloride, chromate, cyanide, fluoride, hydroxide, nitrate, oxide, sulfate and sulfide)	Ecker and Steidl (1986)
Calcium (acetate, arsenate, bicarbonate, bromide, carbonate, chloride, and hydroxide)	Ecker and Steidl (1986)
Copper (chloride, cyanide, fluoride, nitrate, and sulfate)	Ecker and Steidl (1986)
Ferric (acetate, chloride, and nitrate)	Ecker and Steidl (1986)
Ferrous sulfate	Ecker and Steidl (1986)
Lead (chloride and nitrate)	Pickard (1985)
Magnesium (acetate, bicarbonate, carbonate, chloride, fluoride, hydroxide, nitrate, oxide, phosphate, sulfate, thiosulfate, and sulfate)	Ecker and Steidl (1986)
Potassium (acetate, bicarbonate, biphosphate, carbonate, chlorate, chloride, chrome sulfate, cyanide, dichromate, fluoride, iodine, nitrate, phosphate, and sulfate)	Ecker and Steidl (1986)
Silver (cyanide and nitrate)	Ecker and Steidl (1986)
Sodium (acetate, aluminate, benzoate, bicarbonate, biphosphate, bisulfate, bisulfite, carbonate, chlorate, chloride, chlorite, chromate, cyanide, fluoride, nitrate, nitrite, oxalate, perborate, phosphate, silicate, sulfate, sulfide, sulfite and thiosulfate)	Ecker and Steidl (1986)
Tin (chloride and sulfate)	Pickard (1985)
Zinc (acetate, chloride, nitrate, and sulfate)	Ecker and Steidl (1986)

Analysis of the results obtained by Swamy and Jurjees (1986) allows the assumption that the reason for reinforcement corrosion is its composition with ash dust as a filler (fly ash from heating-and-power plants) rather than poor passivating properties of the sulfur concrete. Fly ash is known to contain various sulfurous compounds, greatly increasing the danger of corrosion for reinforcement under a continuous effect of water. Substitution of silica flour by fly ash in cement concretes is also known to reduce the pH of porous liquid and to increase porousness and permeability of concrete structure (Yakub and Alekseev, 1971).

Mohamed and El Gamal (2008) conducted corrosion experiments on two cylindrical reinforced concrete specimens, one made of normal Portland cement concrete and another with bitumen-modified sulfur concrete. The specimens for the corrosion resistance consisted of concrete cylinders (100 by 200 mm) in which a steel reinforcing bar (10 mm in diameter and approximately 250 mm in length) was embedded (the specimen is usually referred to as a lollipop specimen). The steel bar was embedded into the concrete cylinder such that its end is

Table 7.15. Resistance of sulfur concrete to other chemicals (data from Crick and Whitmore, 1998).

Chemical Environment	Concentration (%)	Temperature
Acids		
Boric acid	100	Room temperature
Hydrochloric acid	32	Room temperature
Nitric acid	50	Room temperature
Phosphoric acid	85	Room temperature
Sulfuric acid	93	Room temperature
Salts		
Ammonium sulfate	100	Room temperature
Calcium chloride	100	Room temperature
Copper sulfate	100	Room temperature
Ferric chloride	100	Room temperature
Magnesium chloride	100	Room temperature
Potassium chloride	100	Room temperature
Nickel chloride	100	Room temperature
Nickel sulfate	100	Room temperature
Sodium chloride	100	Room temperature
Sodium sulfate	100	Room temperature
Zinc chloride	100	Room temperature
Zinc sulfate	100	Room temperature

at least 50 mm from the bottom of the cylinder. The reinforcing steel was sealed with an anticorrosion epoxy coating except on the area where corrosion was monitored. Figure 7.24 shows the typical reinforced concrete cylinders used.

Accelerated Corrosion Cell

An accelerated corrosion cell was used to compare the rate of corrosion of both the conventional Portland cement concrete and the bitumen-modified sulfur concrete. In this cell, the specimen was immersed to its half height into a 5% sodium chloride (NaCl) solution at room temperature and connected to a constant 12 volt DC power supply such that the steel bar acts as the anode. A steel plate electrode was used as the cathode. The steel plate was cleaned periodically to prevent the deposition of calcium on the surface. The specimens for the corrosion resistance consisted of concrete cylinders measuring 100 by 200 mm, whereby a steel reinforcing bar measuring 10 mm in diameter and 250 mm in length was

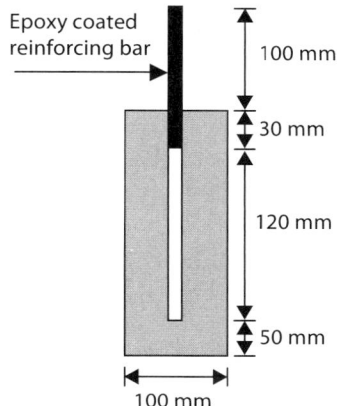

Figure 7.24. Schematic diagram of the reinforced concrete specimen (Mohamed and El Gamal, 2008).

embedded into the concrete cylinder so that its end is at least 50 mm from the bottom of the cylinder. Figure 7.25 shows a schematic diagram of the corrosion cell.

Linear Polarization Resistance Technique

The test setup used for the linear polarization resistance (LPR) techniques included a Gill 8, an electrochemical measurement potentiostat with a windows-based corrosion software version 4.0 for data acquisition, and a printer. Figure 7.26 shows a schematic diagram of the test arrangement for LPR. The exposed surface area, the equivalent weight of steel and the density of the reinforcing steel (d), were 18.84 cm², 27.93 g and 7.8 g/cm³, respectively.

Polarization studies were carried out in a commercially available electrochemical cell. For the present work, the cell was a simple device consisting of a 1000 ml glass beaker fitted with a perplex lid with openings for the working electrode, the counter/auxiliary electrode, and the reference electrode. Electrochemical experiments were performed in a conventional three-electrode assembly with concrete specimens as the working electrode, platinum as the counter electrode, and saturated calomel electrode as the reference electrode to which all potentials are referred.

In the LPR method, a controlled potential scan (100mV/min), E_{oc} (open circuit potential) or E_{corr} from −250 mV to +250 mV, was applied to the reinforcing steel in the specimen, and the resulting current was plotted against the potential. For each test specimen, the polarization resistance was obtained following each conditioning cycle. From the polarization diagrams, the corrosion rate was calculated by using the Tafel Extrapolation Method. The specimens were submerged

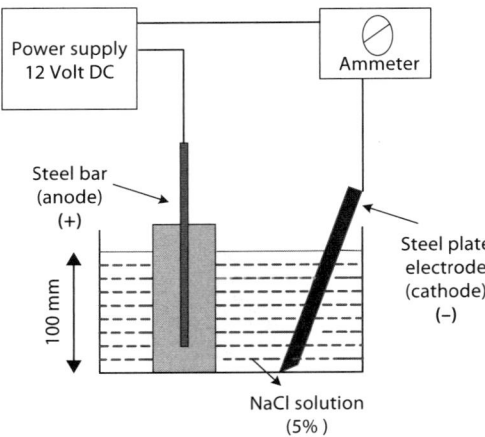

Figure 7.25. Schematic diagram of the accelerated corrosion cell (Mohamed and El Gamal, 2008).

in the chloride solution and the level of the solution was adjusted so that 80 mm of the specimen's height was in the solution during the test.

Accelerated Corrosion Cell Results

The acceleration corrosion cell proved to be a good and simple test to assess the durability of concretes with respect to chloride ion penetration and steel reinforced protection against corrosion. Figure 7.27a shows the curve of corrosion current versus time for Portland cement concrete. A fast longitudinal crack was observed after 17 days as indicated in Figure 7.27b. The following equations describe the electrochemical process involved in steel corrosion (Mehta, 1991).

Anode reaction: $\quad Fe \rightarrow 2e^- + Fe^{2+}$

Cathode reaction: $\quad \frac{1}{2}O_2 + H_2O + 2e^- \rightarrow 2(OH)^-$

Adding reactions: $\quad Fe + \frac{1}{2}O_2 + H_2O + Fe^{2+} + 2(OH)^- \rightarrow$ iron oxide

For sulfur-modified concrete, the experimental results indicated that no current was observed for 30 months, indicating high electric resistively. It is worth noting that the corrosion of reinforcing steel occurs in concrete as an electrolytic cell is formed between the reinforcing steel and the surrounding concrete. It is widely recognized that the permeability of hardened cement paste greatly affects the durability of concrete and, in turn, is strongly affected by the pore structure. The pores in sulfur concrete are not connected, providing low hydraulic conductivity characteristics of 1.4×10^{-13} m/s as reported by Mohamed and El Gamal (2009).

Technological Aspects of Sulfur Concrete Production **321**

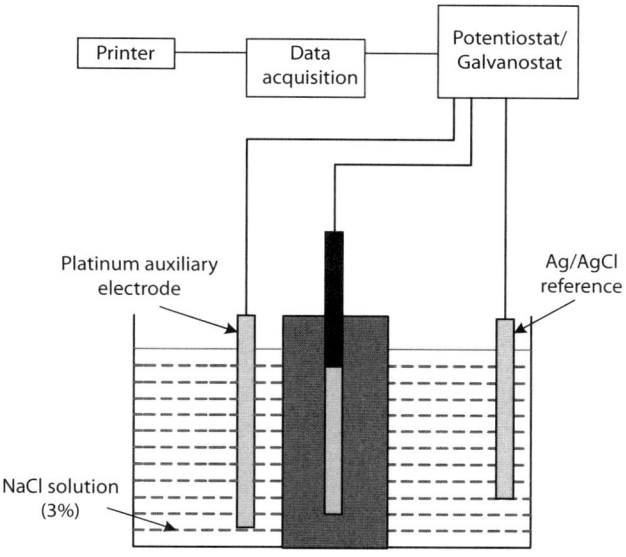

Figure 7.26. Schematic diagram of test setup for corrosion rate measurement using LPR and TP techniques (Mohamed and El Gamal, 2008).

Linear Polarization Resistance Technique Results

Corrosion rates by the LPR technique were determined by applying a current to produce a polarization curve (the degree of potential change as a function of the amount of current applied) for the metal surface whose corrosion rate is being determined. When the potential of the metal surface is polarized by the application of the current in a positive direction, it is said to be anodically polarized; a negative direction signifies that it is cathodically polarized. The degree of polarization is a measure of how the rates of the anodic and the cathodic reactions are retarded by various environmental (concentration of metal ions, dissolved oxygen, etc. in solution) and/or surface processes.

Test results based on the LPR technique for Portland cement concrete and sulfur concrete after 6 months indicates that after 1 month there was no change in the corrosion current with respect to a potential for sulfur concrete specimens, but there was some observed corrosion for ordinary Portland cement concrete. After 6 months of immersion, it was observed that the corrosion rate of sulfur concrete is low with respect to normal concrete as indicated in Figures 7.28 and 7.29, which means that corrosion resistance of sulfur concrete is much higher than that of normal concrete. The results are summarized in Table 7.16.

The results indicated that polarization resistance, corrosion current and the corrosion rate are small in comparison with Portland cement concrete specimens.

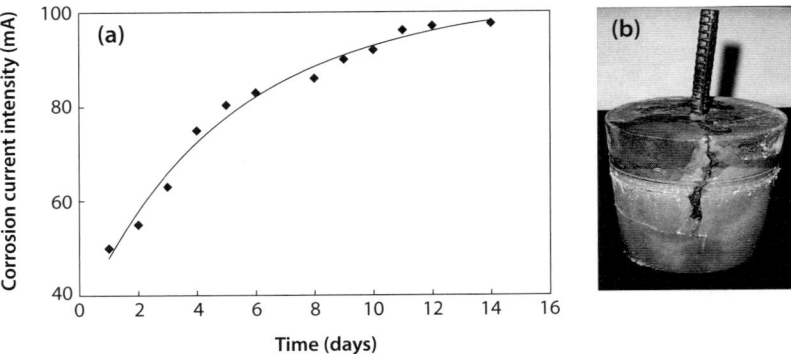

Figure 7.27. (a) Typical curve of corrosion current with time for Portland cement concrete; (b) longitudinal crack in Portland cement concrete specimen after 17 days of testing (Mohamed and El Gamal, 2008).

In quantitative terms, they are seven orders of magnitudes less than that of Portland cement concrete.

The role of the modified-sulfur cement binder in corrosion-resistant sulfur concrete technology can be described by the following three mechanisms: (a) it bonds the aggregate particles together, (b) it fills the voids in the mineral aggregate to minimize moisture absorption, and (c) it provides sufficient fluidity in the mix to give a workable sulfur concrete mixture.

It should be noted that to achieve corrosion-resistant sulfur concrete with good strength and durability, the mix should be designed for moisture absorption less than 0.1%. Such criteria was used to determine the best mix design based on adjusting the binder level to provide the most effective balance between maximum compressive strength, maximum specific gravity, minimum absorption, and a workable mixture.

Proceeding from modern views, the state of reinforcement in concrete can be influenced by ions H_2S^-, SO_4^{-2}, S^{-2}, and atomic hydrogen. Investigation of hydrogen sulfide corrosion reveals multiple accelerations of corrosion in a moist medium in the presence of hydrogen sulfide. It results not only in general corrosion acceleration, but in a distinctively revealed tendency of reinforced steel to become brittle; that results from the fact that atomic hydrogen reacting with metal penetrates into steel structure, recombines into molecular hydrogen, and creates high pressures (Orlowski et al., 2004).

To improve the protective properties of sulfur concrete regarding steel reinforcement, it is necessary to develop compounds with optimum density and strictly observing technological regulations in the manufacturing process such as in the case of cement concrete. Concrete structures must be faultless. Adequate

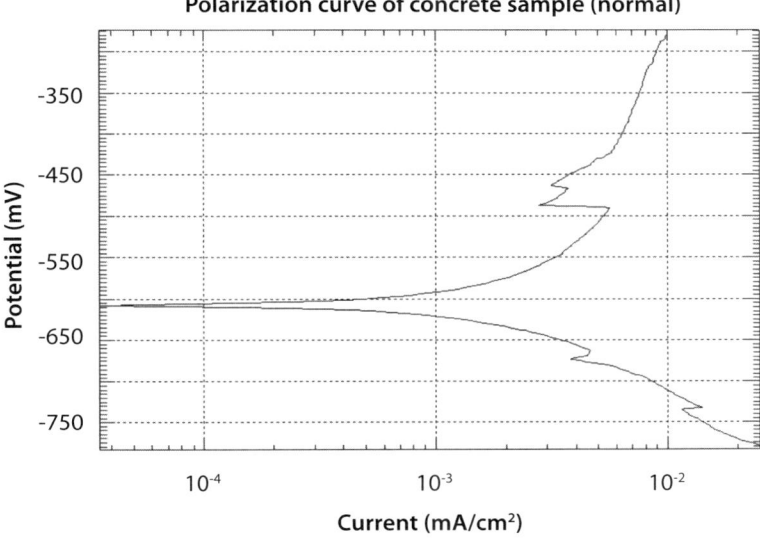

Figure 7.28. Polarization curve for normal concrete after 6 months of immersion (Mohamed and El Gamal, 2008).

Table 7.16. Corrosion results for Portland cement concrete and bitumen-modified sulfur concrete specimens after 6 months (Mohamed and El Gamal, 2008).

Specimens	Polarization Resistance, R_p, ohm.cm²	Corrosion Potential (mV)	Corrosion current (mA/cm²)	Corrosion rate (mils/yr)
PCC	2.023E+04	-528.49	1.290E-03	5.885E-01
BMSC	4.253E+11	-615.26	6.133E-11	2.799E-08

reinforcement adhesion to concrete and density of their contact zone are also of primary importance.

7.9 SERVICE LIFE

Malhotra (1983) reported an increase of the life expectancy of sulfur concrete over Portland cement concrete by a factor of 2 to 3. Also, Weber (1993) reported that sulfur concrete demonstrated superior performance after a period of 9 years under conditions that destroyed Portland cement concrete in less than 3 years. It should be emphasized that the resistance of sulfur concrete to any chemicals is

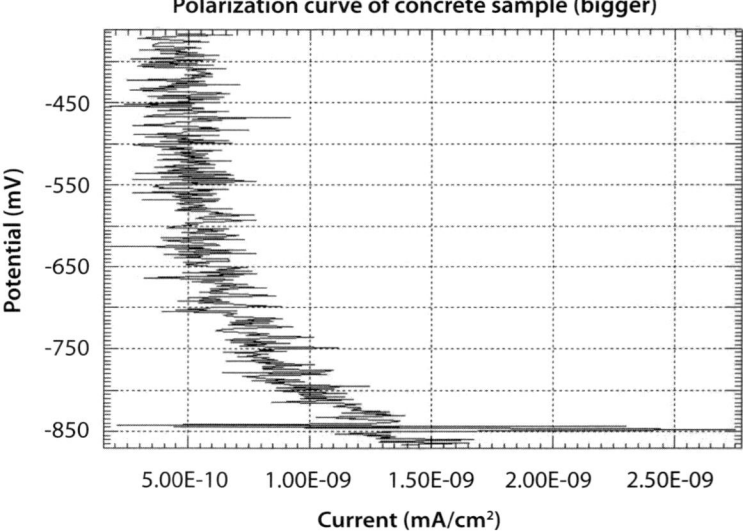

Figure 7.29. Polarization curve for sulfur polymer concrete after 6 months of immersion (Mohamed and El Gamal, 2008).

highly dependent on material composition, concentration, and service temperature.

7.9.1 ISO Approach

The ISO 15686-1 for 2000 gives definitions for specific terms used in the service life, performance, and degradation:
1. *Service Life*: Period of time after installation during which a building or its parts meets or exceeds the performance requirements
2. *Reference Service Life*: Service life that a building or the parts of a building would attain in a certain set of in-use conditions
3. *Estimated Service Life*: Service life that a building would be expected to have in a set of specific in-use conditions calculated by adjusting the reference in-use conditions in terms of material, design, environment, use, and maintenance
4. *Design Life*: Intended service life or expected service life; or service life intended by the designer.
5. *Predicted Service Life*: Service life predicted from the recorded performance over time
6. *Forecast Service Life*: Service life based on either the predicted service life or the estimated service life

7. *Degradation or Deterioration*: Changes in the composition, microstructure, and properties of a component or material over time that reduces its performance
8. *Durability*: Capability of a building or its parts to perform its required function over a specified period of time under the influence of the agents anticipated in service

The ISO definitions are relevant in the process of estimating the service life of sulfur concrete. According to ISO 15686-1, forecasting is described as the objective of service life planning that assures, as far as possible, that the estimated service life of the building or component will be at least as long as its design life. Since the length of the service life can't be known precisely in advance, the objective then becomes to make an appropriately reliable forecast of the service life using available data.

A forecast service life is referred to as the predicted service life if it is based on exposure and performance tests. If the predicted service life or another source of information is used to provide a reference service life, it can be adjusted to reflect local, project-specific factors using the factor method (ISO 15681-1[2000]). It is then referred to as the estimated service life. Buildings, materials, and components that are innovative pose a particular problem because forecasting is invariably based on the interpretation of the known performance of dissimilar applications, or on accelerated exposure alone (Makenya, 2001).

It is important to realize that innovative solutions may, at times, offer superior performance. However, it should be possible to indicate a minimum period during which it is anticipated that a building or component can remain in service on the basis of stated assumptions (ISO 15681-1[2000]).

7.9.2 Estimation of Sulfur Concrete Service Life

The estimation of service life of sulfur concrete is crucial because it is a basic factor as far as the material's industrial application and the markets are concerned. As outlined in the ISO 15681-1(2000), the factor method allows an estimate of the service life to be made for a particular component (or assembly) under specific conditions.

The factor method is based on a reference service life (RSL) and a series of modifying factors that relate to the specific conditions of a given case. RSL is a documented period, in years, that a component (or assembly) can be expected to last in a reference case under a certain well-defined set of in-use conditions. It is important to take into consideration that the reliability of the RSL figure is critical because it may affect the estimate proportionality. The factor method is a way of bringing together the consideration of each of the variables that is likely

Table 7.17. ISO 15681-1(2000) modifying factors for service life estimation

Factor	Indication
A	Quality of components
B	Design level
C	Work execution level
D	Indoor environment
E	Outdoor environment
F	In-use conditions
G	Maintenance level

to affect service life. The modifying factors as specified in ISO 15681-1(2000) are shown in Table 7.17.

For the sulfur concrete factor, G is omitted because there is no relevant maintenance information regarding the material. The value of factors varies from 0.8 (poor) to 1.2 (good). Five years are chosen as the actual design life for sulfur concrete (DL_{SC}) that is anticipated to meet the industrial demand for economical construction purposes in hostile environments (Makenya, 2001).

The estimated service life for sulfur concrete can be calculated as follows:

$$ESL_{SC} = RSL_{SC} \times A \times B \times C \times D \times E \times F$$

where RSL_{SC} is the reference service life for sulfur concrete based on previous service life performance observations in years, and factors A, B, C, D, E, and F are modifying factors matching sulfur concrete.

The RSL_{SC} is decided for 10 years (Chempruf Manual, 1988) and 6 years (Weber, 1993). Past experience has shown that the service life period ranges from 3 to 10 years. A logical approach is to take the average value for the reference service life, which is 6 years (i.e., RSL_{SC} = 6 years). Therefore:

$$ESL_{SC} = 6 \times 1 \times 1 \times 1 \times 1 \times 0.8 \times 1.2 = 5.7 \text{ years} \approx 6 \text{ years}$$

It follows then, that the estimated RSL_{SC} is approximately equal to or greater than the design life for sulfur concrete (DL_{SC}), or:

$$ESL_{SC} \geq DL_{SC} = 6 \text{ years}$$

7.10 SULFUR CONCRETE ASSESSMENT PROTOCOL

As discussed previously, sulfur concrete is resistant to many acidic, saline, and mix chemicals. This property forms the basis of a general industrial application of sulfur concrete whenever attention must be paid to durability and maintenance. It should be emphasized, however, that sulfur concrete is influenced by chemical concentration, service temperature, aggregate type, and mix proportions. Because moisture intrusion into sulfur concrete is undesirable, it should be given the highest priority when assessing the performance of sulfur concrete.

A systematic approach to assessing sulfur concrete should start with its composition, mixing and placing technology, and environmental loads. Figure 7.30 presents an assessment protocol for sulfur concrete. The protocol starts with an evaluation of the sulfur concrete composition and the mixing and placing technology to resist moisture penetration, temperature fatigue, and integration. The material must be properly mixed and placed according to the specific conditions and appropriate technology required for sulfur concrete. The material must be allowed to solidify before the application of any environmental load.

The second step is to evaluate the hydro-thermal properties of sulfur concrete. In this regard, two tests should be conducted: (1) moisture absorption and (2) permeability. These tests should be evaluated with reference to service temperature. If the tests fail, the material must be rejected. Hence, the designer must go back to review and quantify sulfur concrete composition.

The third step is to evaluate the thermo-mechanical behavior of sulfur concrete. This will be accomplished by performing the strength tests as a function of service temperature and to identify the temperature limits.

The fourth step is to evaluate the sulfur concrete performance with respect to thermo-chemical environments that are acidic, basic, salt, and/or mixed chemicals. The alternative of using sulfur concrete in this connection appears to be of value because Portland cement concrete is sensitive to organic acids. In case this test fails, there is a possibility to lower the concentration or temperature values. In the case of inorganic acids, if the test fails, material composition must be checked.

Sulfur concrete is used in many cases for the storage of salts in dry or saturated forms. A combination of chemical and mechanical loads aggravates the performance of sulfur concrete. Past experience suggested that glass fiber and stainless steel reinforcement will lead to a required structural coherence.

For composite chemicals, we suggest sulfur concrete be subjected to the procedures of toxicity characteristics and leachability procedures used by the U.S. Environmental Agency for testing solidified or stabilized wastes. In this test, the material is subjected to a sequence of harsh chemical environments. If the test fails, material composition should be reviewed.

When all the specified tests have been passed, a quality product of sulfur concrete has been achieved and is ready for use for industrial applications.

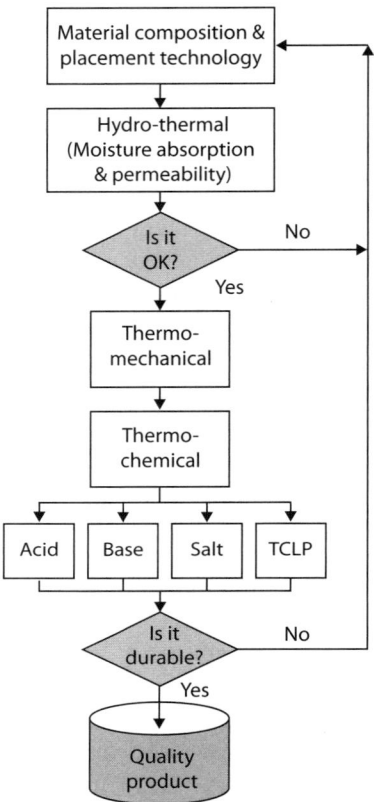

Figure 7.30. Assessment protocol for sulfur concrete.

7.11 SUMMARY AND CONCLUDING REMARKS

Sulfur concrete is a building material that offers exceptional chemical-resistant properties for structures, making it ideal for use in construction, in fertilizer or metal refining, and other industries in which Portland cement concrete is subject to attack. Specialized equipment has been developed for the manufacture and placement of sulfur concrete, and continued developments are serving to make commercial-scale manufacture and installation of sulfur concrete more efficient and practical.

To evaluate the economic feasibility of using sulfur concrete effectively for a particular project, the price of sulfur concrete should not be compared directly to that of Portland cement concrete. Rather, the finished in-place cost of sulfur concrete should be equated to the finished in-place cost of Portland cement concrete with a protection system. Even when the special requirements of working with sulfur concrete are taken into account, sulfur concrete can be installed for costs comparable to Portland cement with a protective lining system.

More research is required to develop sulfur concrete engineering and design further to: (a) take advantage of sulfur concrete's improved mechanical properties, (b) allow for more use in structural elements, and (c) help further a reduction in installation costs. In aggressive acid and chemical environments, sulfur concrete can provide an economical, high-performing, long-term alternative to conventional Portland cement concrete with protective linings.

8

SULFUR-MODIFIED ASPHALT

8.1 INTRODUCTION

As discussed in previous chapters, sulfur construction materials offer improvements over more traditional materials, especially in specific applications. Sulfur construction materials include sulfur cement, sulfur concrete, and sulfur-extended asphalt pavements. Asphalt is a heavy organic byproduct of petroleum refining. It is an attractive product to the road construction industry because of its high production volume and low cost. However, as demand for high-quality petroleum products such as gasoline grew, pressure mounted for the petroleum industry to improve the refining technology to produce more high-quality products and less asphalt as well as other low-value byproducts and wastes. These events, coupled with the increasing availability and decreasing costs of recovered sulfur, have made sulfur-extended asphalt a more attractive alternative to the traditional asphalt paving materials. The structure of sulfur-extended asphalt depends on its formula and heating temperature. Road pavements, based on sulfur-extended asphalt, have shown higher durability as compared to that of conventional binders.

In this chapter, basic concepts of sulfur-modified asphalt pavements are discussed.

8.2 ASPHALT

8.2.1 Asphalt Composition

Asphalt is a complex mixture of organic molecules that vary widely in composition from non-polar saturated hydrocarbons to highly polar, highly condensed ring systems. Elemental analyses of several representative petroleum asphalts are shown in Table 8.1. Although asphalt molecules are composed of carbon and hydrogen predominantly, most molecules contain one or more of the so-called *hetro-atoms* (nitrogen, sulfur, and oxygen) together with trace amounts of metals, principally vanadium and nickel. These atoms are called hetro-atoms (hetro meaning different) because they can replace carbon atoms in the asphalt molecular structure. However, their chemistry is different from carbon, and because hetro-atoms are responsible for most of the hydrogen bonding, they contribute to many of asphalt's unique chemical and physical properties by forming an association between molecules. This bonding occurs because the presence of hetro-atoms in a molecule can make the molecule polar and, thus, more likely to react with other molecules. Both the kind and amount of hetro-atoms present in a given asphalt are a function of both the crude oil(s) from which the asphalt was produced and the state of aging of asphalt.

The hetro-atoms, especially sulfur, also play an important role in the aging of asphalt because they are more chemically reactive than hydrogen or carbon, and they can oxidize, or incorporate oxygen, more easily than the hydrocarbons. It is worth noting that the terms *aging* and *oxidation* do not mean the same thing. Aging is the overall process that occurs when asphalt is exposed to the environment. It can include loss of volatiles (low boiling point materials remaining in the asphalt after refining) through evaporation or other degradation, attack by water and light, as well as the process of oxidation. Oxidation is the incorporation of oxygen into the molecular structure of the asphalt. When asphalts oxidize, two main products are formed: (a) sulfur atoms (S) are oxidized to sulfoxide (S=O); and (b) carbon (C) is oxidized to carbonyl (C=O).

As shown in Table 8.1, the hetro-atoms, although a minor component compared to the hydrocarbon moiety, can vary in concentration over a wide range, depending on the source of asphalt. In as much as hetro-atoms often impart functionality and polarity to the molecules, their presence may take a disproportionately large contribution to the differences in chemical and physical properties among asphalt obtained from various sources. Elemental analyses are average values and reveal minimal information about how the atoms are incorporated into the molecules or what type of molecular structures are present. Molecular type and structure information are necessary for a fundamental understanding of how composition affects physical properties and chemical reactivity.

Table 8.1. Elemental analysis of representative petroleum asphalts (Plancher et al., 1976)

Elements	Asphalt			
	B-2952 (Mexican)	B-3036 (Ark. La.)	B-3051 (Boscan)	B-3602 (Californian)
Carbon (%)	83.77	85.78	82.90	86.77
Hydrogen (%)	9.91	10.19	10.45	10.94
Nitrogen (%)	0.28	0.26	0.78	1.10
Sulfur (%)	5.25	3.41	5.43	0.99
Oxygen (%)	0.77	0.36	0.29	0.20
Vanadium (ppm)	180.00	7.00	1380.00	4.00
Nickel (ppm)	22.00	0.40	109.00	6.00

The heavy metals that asphalt contains are atoms such as vanadium, nickel, and iron (V, Ni, and Fe). They are present in small quantities in the asphalt, usually less than 1%. They may play an important role in the aging process, and they also serve as a *fingerprint* for the asphalt, because the amount of metals present in the asphalt is usually indicative of the crude source from which the asphalt was refined.

The hydrocarbons, hetro-atoms, and metals are all combined in the asphalt in a wide range of different molecules. When describing these molecules, we can group them into three broad groups: (a) aliphatics, (b) cyclics, and (c) aromatics.

The term aliphatic (literally, oily) can best be described as a linear or chain-like molecule in which the carbon atoms are linked end to end. As an example, the aliphatic molecule hexane (C_6H_{14}) contains six carbon atoms and fourteen hydrogen atoms.

The cyclic molecule cyclo-hexane (C_6H_{12}) contains the same six carbons but gives up two hydrogen atoms to form a cycle, or ring. The loss of two hydrogen atoms has had a significant effect on the chemical properties of the molecule. This is due to both the change in shape of the molecule as well as the interaction of the individual atoms with each other in the molecule.

The aromatic molecules are so named for their strong odor. As an example, benzene (C_6H_6) contains six carbon molecules and six hydrogen atoms in its structure. The electrons used to bond to hydrogen in hexane and cyclo-hexane have formed a *ring* of aromaticity, or shared electrons, generally represented by a circle inside the structure. The unique property that aromatics bring to asphalt is their flat shape. Cyclics and aliphatic molecules are three-dimensional and form shapes that keep the molecules apart. On the other hand, aromatics are flat and can stack closely on top of one another.

Concentrations of aromatic carbon range from 25 to 30% for petroleum asphalt (Ramsey, 1967). The aromatic carbon is incorporated in a condensed aromatic ring system that contains 1 to 10 rings per aromatic moiety (Ramsey, 1967). These ring systems may be associated with saturated naphthenic (cyclo-alkyl) ring systems (Petersen, 2000). Both the aromatic and naphthenic ring system may have attachments composed of a variety of normal or branched hydrocarbon side chains. Concentrations of carbon associated with naphthenic ring systems in asphalts are typically in the range of 15 to 30% (Ramsey, 1967). Normal- and branched-chain hydrocarbons are present either in individual molecules or as moieties associated with naphthenic or aromatic rings. The non-aromatic and non-naphthenic carbon content of asphalt would typically range from 35 to 60%.

The hydrocarbon molecular structures are complicated by the hetro-atoms, sulfur, nitrogen and oxygen that are often present in sufficient compound amounts so that, on the average, one or more hetro-atoms per molecule may be present. These may be incorporated within the ring or non-ring components or in more discrete chemical functional groups attached to these compounds. The hetro-atoms, particularly nitrogen and oxygen, and the aromatic ring systems contribute considerable polarity or polarizability to the molecules that produce the major association forces affecting physical properties.

8.2.2 Asphalt Fractionation

A variety of procedures have been employed in attempts to fractionate asphalt into less complex and more homogeneous fractions. Some procedures are simple whereas others are more complex. Many are specialized for a given research endeavor in which they were used to prepare fractions for further characterizations. Several, however, have found a more general use to characterize and classify asphalts. Generally, asphalt can be separated into four main fractional groups; saturates, aromatics, resins, and asphaltenes (Whiteoak, 1990; Isacsson and Lu, 1995).

Classical Asphalt Structure Model

Researchers used the micellar model to describe the structure of asphalt. In this model, the aromatic asphaltenes existed as a discrete phase in the asphalt and were surrounded (and solubilized) by the resins. The asphaltenes were made up of highly aromatic groups of molecules that were quite large and insoluble in the remainder of the asphalt. The resins acted as intermediates in the asphalt, serving to homogenize the otherwise insoluble asphaltenes. The resins and asphaltenes existed as islands floating in the final asphalt component.

Concentrations of aromatic carbon range from 25 to 30% for petroleum asphalt (Ramsey, 1967). The aromatic carbon is incorporated in a condensed aromatic ring system that contains 1 to 10 rings per aromatic moiety (Ramsey, 1967). These ring systems may be associated with saturated naphthenic (cyclo-alkyl) ring systems (Petersen, 2000). Both the aromatic and naphthenic ring system may have attachments composed of a variety of normal or branched hydrocarbon side chains. Concentrations of carbon associated with naphthenic ring systems in asphalts are typically in the range of 15 to 30% (Ramsey, 1967). Normal- and branched-chain hydrocarbons are present either in individual molecules or as moieties associated with naphthenic or aromatic rings. The non-aromatic and non-naphthenic carbon content of asphalt would typically range from 35 to 60%.

The hydrocarbon molecular structures are complicated by the hetro-atoms, sulfur, nitrogen and oxygen that are often present in sufficient compound amounts so that, on the average, one or more hetro-atoms per molecule may be present. These may be incorporated within the ring or non-ring components or in more discrete chemical functional groups attached to these compounds. The hetro-atoms, particularly nitrogen and oxygen, and the aromatic ring systems contribute considerable polarity or polarizability to the molecules that produce the major association forces affecting physical properties.

8.2.2 Asphalt Fractionation

A variety of procedures have been employed in attempts to fractionate asphalt into less complex and more homogeneous fractions. Some procedures are simple whereas others are more complex. Many are specialized for a given research endeavor in which they were used to prepare fractions for further characterizations. Several, however, have found a more general use to characterize and classify asphalts. Generally, asphalt can be separated into four main fractional groups; saturates, aromatics, resins, and asphaltenes (Whiteoak, 1990; Isacsson and Lu, 1995).

Classical Asphalt Structure Model

Researchers used the micellar model to describe the structure of asphalt. In this model, the aromatic asphaltenes existed as a discrete phase in the asphalt and were surrounded (and solubilized) by the resins. The asphaltenes were made up of highly aromatic groups of molecules that were quite large and insoluble in the remainder of the asphalt. The resins acted as intermediates in the asphalt, serving to homogenize the otherwise insoluble asphaltenes. The resins and asphaltenes existed as islands floating in the final asphalt component.

It has been recognized that asphalt exhibits properties that deviate from those of a true solution. The colloidal nature of asphalt was first recognized by Nellenstyn (1924 and 1928) who considered asphalt a dispersion of micelles in an oily medium. The aromatic asphaltene fraction was associated with the dispersed or micelle phase early on (Mack, 1932). It was also recognized that the inability of the resinous components to keep these highly associated asphaltene components dispersed in the oily phase largely determined the gel or non-Newtonian flow characteristics of asphalt (Pfeiffer and Saal 1940; and Saal et al., 1946).

Asphalt components are separated primarily by the solubility of the various molecules in different solvents. A typical separation method is Corbett (ASTM D-4124). These components are then used to explain the asphalt's performance properties. For example, Corbett (1970) described the effects of the four fractions separated on the physical properties as:

a. Asphaltenes function as solution thickeners; fluidity is imparted by the saturate and naphthalene aromatic fractions that plasticize the solid polar aromatic and asphaltene fractions
b. Polar aromatic fraction imparts ductility to the asphalt
c. Saturates and naphthene-aromatic in combination with asphaltenes produce complex flow properties in the asphalt

It is further concluded by Corbett (1970) that each fraction or combination of fractions perform separate functions in respect to physical properties, and it is logical to assume that the overall physical properties of one asphalt are thus dependent upon the combined effect of these fractions and the properties in which they are present.

8.2.3 Asphalt Component Interaction

Atoms combine to form molecules through the formation of strong covalent bonds. These molecules can then interact with one another through the formation of other, much weaker types of bonding (*pi-pi* and *hydrogen* bonding and *Van der Waals* forces). These types of interactions are important to chemists and asphalt technologists because they are responsible for determining many of the asphalt's physical properties. Weak bonds require relatively little energy to break, beause they are susceptible to both heat and mechanical forces.

Aromatic molecules form a stack of molecules due to their flat shape, and the electrons in the aromatic rings interact with one another to form pi-pi bonds. The pi-pi interaction is unique to aromatic molecules and leads to some interesting chemical properties. Graphite is important commercially because its aromatic rings form flat sheets of molecules that can easily slide around on top of one another. That provides the slippery property that makes it such an excellent dry lubricant.

The molecules containing hetro-atoms are generally polar (have unevenly distributed electrical charges) and play an important role in affecting asphalt's physical properties by interacting with other molecules through the formation of hydrogen bonds. This occurs when a hetro-atom on one molecule interacts with a hydrogen atom next to a hetro-atom on a different molecule. This means that two hetro-atoms are required for the formation of hydrogen bonds. Hydrogen bonding is the most important of the forms of weak molecular interactions in asphalt.

Van der Waals forces are the final form of weak interactions between molecules that are important in asphalt chemistry. In this type of weak bonding, long chains of aliphatic hydrocarbons (such as hexane) intertwine and are held together weakly. This type of bonding is dependent on the amount and type of aliphatic molecules in a given asphalt; just as pi-pi bonding depends on the amount and type of aromatic molecules and hydrogen bonding depends on the number of hetro-atoms.

As mentioned above, all of these types of bonds are weak and are easily broken by heat or stress. They will form again when the molecules cool or the stress is removed. This property of formation and breaking of weak bonds between the asphalt molecules is the key to understanding the physical properties of asphalt. It is also important to remember that the bonds between atoms that form the molecules are called *covalent bonds* and are not broken or reformed during the processes that normally take place in asphalt. The covalent bonds between atoms are 10 to 100 times stronger than the weak bonds between molecules.

All of the hydrocarbons, hetro-atoms, and metals are combined into the millions of different molecules that make up asphalt. Each asphalt has a unique assortment of the molecules, and the molecules change as asphalt ages. The hetro-atoms are incorporated into the aliphatic, cyclic, and aromatic hydrocarbon molecules. They affect the chemical behavior of the individual molecules as well as the associations with the molecules' form.

The asphalt chemical composition is described as follow (Jones, 1992):

a. Asphalt consists of two functional families of molecules that are polar and nonpolar
b. Polar molecules differ according to: (a) strength and number of polar groups, (b) molecular weight, and (c) degree of aromaticity
c. Nonpolar molecules differ according to: (a) molecular weight and (b) degree of aromaticity
d. The *compatibility* of the polar and non-polar fractions, or the degree to which they can dissolve in each other, is controlled by the relative

aromaticity (a measure of the amount of aromatic versus aliphatic and cyclic molecules) of the two fractions

Polar vs. Non-polar Molecules

All of the molecules in asphalt behave in one of two ways. If they are polar at service temperatures, they participate in the formation of a network through hydrogen and pi-pi bonding that gives the asphalt its elastic properties. The non-polar materials form the body of the material in which the network is formed and contribute to the viscous properties of the asphalt.

Polar Molecules

Polar molecules participate in the formation of the network of associated molecules and comprise a wide range of molecular types and sizes. They contribute to the performance of the asphalt through the network. The most important attribute of the polar molecules is the relative strength and the number of polar sites per molecule, because this directly affects network formation. The amount of polar materials present in asphalt are determined by an analytical technique known as *non-aqueous-acid-base titration*. In this method, the material is dissolved in a solvent other than water (nonaqueous) and the amount of acids and/or bases present are determined by titration of either whole asphalts or selected fractions. To determine the amount of acid in a sample by titration, small amounts of base are added and allowed to react with the acids in the sample. Excess base in the solution signals when all the acids have reacted. The amount of base added is then proportional to the acids present in the sample originally. The process is simply reversed to determine the amount of base present.

The second important parameter affecting the polar molecules is their degree of aromaticity. This can be determined by another measurement technique known as nuclear magnetic resonance (NMR) analysis. In an NMR analysis, the sample is subjected to an intense magnetic field and the response of the material is measured. Because of their pi-pi bonds, aromatic molecules have a distinct signature in NMR analysis. The molecules that are not aromatic are aliphatic or cyclic. This distinction allows characterization of the polars as containing some percentage of aromaticity, the balance being aliphatic and cyclic.

The molecular weight of the polar molecules is probably not as important a parameter in controlling the performance of asphalt as the other two factors of strength of polarity and aromaticity. It will play a minor role in the performance of the asphalt cement.

Non-polar Molecules

An important feature of the non-polar molecules is their molecular weight, because this directly affects the low-temperature cracking properties of asphalt. A preponderance of high-molecular weight, non-polar molecules will lead to asphalts that stiffen at low service temperatures and perform poorly due to brittleness.

The relative amount of aromatic character of the non-polar materials is also important. If the nonpolars are waxy (a type of aliphatic molecule) in nature, they may precipitate or crystallize at low temperatures, contributing to poor performance. If the nonpolars are cyclic or aromatic in nature, they may resist the effects of low temperature more effectively. There may well be an interaction with the molecular weight of the polar molecules as well, however, evidence to date indicates the major effects are due to the molecular weight of the nonpolars.

Compatibility

The non-polar molecules are also important to asphalt's performance because they are the substance in which the polar molecules must interact. As such, they combine with the polar network, and their compatibility with the polar molecules must be considered. If the molecules are similar in chemistry, they will mix easily and be compatible. On the other hand, if they are distinctly different, they will be incompatible, and the molecules may not want to stay in the solution. The determination of aromaticity will be important because the compatibility of the two types of molecules will have an effect on the performance of the asphalt. A compatible asphalt will have both polar and non-polar molecules that are similar.

8.2.4 Asphalt Aging

Asphalt composition changes over time when the asphalt is exposed to atmospheric oxygen in the pavement. Asphalt reacts with atmospheric oxygen that stiffens or hardens the asphalt. Atmospheric oxidation is the major factor responsible for the irreversible hardening of asphalts (Van Oort, 1956) and is the reason why pavement void content correlates so strongly with asphalt pavement hardening (Goode and Owings, 1961; Vallegra et al., 1970). Hardening from the loss of volatile components, the physical factor that might affect the correlation of hardening and void content, is not considered a significant factor when asphalts that meet current specifications are used (Petersen, 2000). The potentially volatile components would be part of the saturate fraction. Thus, in dealing with asphalt durability, a major factor that must be addressed is the change that takes place in asphalt composition from oxidative aging.

Table 8.2. Chemical functional groups formed in asphalts during oxidation aging (Plancher et al., 1976)

Asphalt	Concentration (mol/l)				Average hardening index[a]
	Ketone	Anhydride	Carboxylic acid	Sulfoxide	
B-2959	0.50	0.014	0.008	0.30	38.0
B-3036	0.55	0.015	0.005	0.29	27.0
B-3051	0.58	0.020	0.009	0.29	132.0
B-3602	0.77	0.043	0.005	0.18	30.0

[a] Ratio of viscosity after oxidative aging to viscosity before oxidative aging

Chemical studies (Plancher et al., 1976; Petersen et al., 1974; Dorrence et al., 1974; Petersen et al., 1975; Petersen and Plancher, 1981; Petersen, 1981) have yielded considerable information on the specific chemical changes that take place in asphalt on oxidative aging by the reaction with atmospheric oxygen. The major oxygen-containing functional group formed during aging are listed in Table 8.2 for four asphalts of different crude sources and aged under identical conditions (air, 130°C, 24 hr, 24 mm thin film).

The data (Plancher et al., 1976) represent averages for the four asphalts aged on four different aggregates and the same asphalt shown in Table 8.1. The level of oxidation has been judged to be equivalent to that typically found in asphalts after five years or more of pavement service (Davis and Petersen, 1967). The chemical functionality developed during laboratory oxidation at 130°C is similar to that developed during usual pavement aging at normal temperatures. The results shown in Table 8.2 indicate that ketones and sulfoxides are the major oxidation products formed during oxidation aging; anhydrides and carboxylic acids are formed in smaller amounts. The mechanism of asphalt oxidation has been highly elusive because of the molecular complexity of asphalts, and the details of any proposed mechanism are disputed frequently among researchers.

8.2.5 Lime-Modified Asphalt

The results in Table 8.3 show the effect of the lime treatment in reducing the hardening rate of asphalts when subjected to the laboratory gas chromatographic column oxidation procedures (Plancher et al., 1976), during which the asphalts were supported as thin films on the surface of four different aggregates. It should be noted that except for asphalt B-3602, lime treatment reduced the hardening index of the asphalts by more than 50% (Plancher et al., 1976). Functional group analyses, however, showed that the oxidation reaction, as measured by the formation of ketones, was reduced by approximately 10% by lime treatment. It was believed that lime removed carboxylic acids and other highly polar functionality

Table 8.3. Reduction of hardening rate of asphalt by treatment with hydrated lime (Plancher et al., 1976)

Sample	Hardening index (viscosity after oxidative aging to viscosity before oxidative aging)		
	Untreated	Lime treated	Reduction (%)
Asphalt			
B-2959	37	17	54
B-3036	27	10	63
B-3051	132	35	73
B-3602	29	18	39
Aggregate			
Quartzite	57	22	61
Hol limestone	58	22	61
Riverton limestone	36	13	63
Granite	75	22	70

that would otherwise have interacted with oxidation products to increase asphalt viscosity. Separate studies showed that the introduction of carboxylic acid functional groups into asphalt molecules greatly increases asphalt viscosity (Petersen, 2000).

The flow properties and hardening rate of asphalt might be significantly altered also by the manipulation of molecular interactions through chemical modification or the addition of modifiers that can interact with polar chemical functionality in the asphalt and alter its activity as suggested by the data on lime addition. The ability of surface active materials such as anti-stripping agents (often amines) to alter asphalt viscosity, frequently to an extent not expected from simple additive effects, is widely known. Because anti-stripping agents have polar chemical functionality, they might be expected to affect the dispersibility of asphaltene-like compounds in asphalt by associating with polar functionality, thus altering the association and micelle structure within the asphalt.

8.3 SULFUR-MODIFIED ASPHALT

Asphalt was previously available at a reasonable cost because asphalt is a residue in petroleum refining and certain petroleum-refining residues could only be utilized economically for the production of asphalt. However, currently higher percentages of petroleum are utilized for the reduction of other more profitable forms of petroleum products. As this trend continues, the price of asphalt is expected to increase even though under constant demand. This expectation is supported by the evolution of the average price of asphalt over the past 34 years. During this period, the price has risen from approximately $23/ton in 1968 to approximately $152/ton in 2000, an increase of 561%. It is generally recognized that there is no economical paving binder that can be substituted currently for asphalt, and that there is no low-priced asphalt paving binder that can replace high priced asphalt paving binders effectively.

8.3.1 Beneficial Use of Sulfur-Modified Asphalt

Sulfur asphalt binders for pavements emerge as a significant development because of the potential for application and the associated economic implication in highway construction and maintenance. In the last few decades, the availability of sulfur has considerably grown in many countries. This is mainly due to the trends in the current environmental restrictions regarding petroleum and gas refining processes that limit the maximum quantity of sulfur present in fuels. The process involves the extension of asphalts by dispersion of liquid elemental sulfur into hot asphalt-yielding sulfur asphalt binder.

Additionally, roads produced with asphalt binders are subjected to many harsh environmental conditions such as traffic loading, ingress of water, chemical attack, and widely fluctuating temperatures. One of the limitations to the use of asphalt as a binder for paving materials is that it softens and flows within a wide range of ambient temperatures. This limitation makes transporting this type of conventional asphalt-based material difficult and can also give rise to serious environmental problems. Conventional asphalt oftentimes cannot provide the desired resistance to these conditions therefore the modification of the asphalt properties becomes necessary. In addition to the asphalt and aggregate, additives may be added to improve the properties of the final product.

8.3.2 Sulfur Asphalt History

Asphalt and paving technologists have used sulfur to modify the properties of paving asphalt composition since the early 1900s (Abraham 1961; Bacon and Bencowitz, 1939). In early trials, it was observed that sulfur (up to 50 wt% in asphalt) did improve the physical and mechanical properties of the resulting mixture. The renewed interest in this subject was triggered in the 1970s and 1980s by the escalating cost of oil as well as the increasing awareness of the energy crisis. These factors were coupled to an over-supply of sulfur, the price of which continues to be low. (The sulfur asphalt binder concept was proposed by Bencowitz in 1938). Sulfur as a replacement for aggregate in preparing sand sulfur asphalt paving has been reported by Burgess and Deme (1975), Sullivan et al. (1975), and

Saylak (1975). A number of projects using sulfur asphalt as a binder were carried out in Canada and the United States (Anonymous, 1973, 1975; Kennepohl et al., 1975, 1978; McBee et al., 1980; Miller and Grawford, 1981; McBee and Sullivan, 1982; Lartaut, 1985).

In 2008, Shell Thiopave was selected to resurface three test sections of the Sea to Sky highway during the months of October and November, 2008. The locations were carefully selected to evaluate the ability of the material to be used under adverse weather conditions—the average daytime winter temperature is just 3°C. Replacing a proportion of the bitumen in the asphalt mix with Shell Thiopave increases the overall mix strength and results in an improved load-bearing capacity of the finished road. Adding Shell Thiopave enables the asphalt to achieve compaction at a considerably lower temperature when compared to traditional bitumen-based roads. Due to the special additives, Shell Thiopave also enables lower temperature production, delivering meaningful reductions in greenhouse gas emissions. This allows the compound to cool and harden quicker and helps reduce the chance of thermal cracking—even in adverse weather conditions.

The pilot for Shell Thiopave involved laying 3,300 tons on three test sections of the Sea to Sky highway. Over a period of two weeks, the resurfacing team conducted the trial, working diligently to achieve the desired performance and results.

Despite being Shell Thiopave's first major pilot, the field data continues to bear out the excellent performance characteristics of Shell Thiopave. The Sea-to-Sky consortium has indicated interest in considering more extensive use of Shell's sulfur-enhanced asphalt for the rest of the project.

Patented Shell Thiopave technologies have been developed to help road construction companies and road owners to improve road performance through the innovative use of sulfur in the asphalt mix. The potential of sulfur to improve asphalt mix performance has been recognized for some time. These technologies offer a combination of benefits in the road sector: (a) reduced rutting, (b) reduced thermal cracking, (c) extended road life for thinner pavements, (d) increased load-bearing capacity, (e) reduced CO_2 emissions, (f) reduced VOC emissions, (g) reduced bitumen usage, and (h) energy savings at the hot mix plant.

These technological innovations mean the latest workplace health and safety standards can now be met, and the benefits of sulfur in the asphalt mix can now be accessed around the world on a significant scale.

8.3.3 Sulfur Behavior in Liquid State

From the viewpoint of modification of asphalt with sulfur, the most interesting problem is the behavior of sulfur in a liquid state. At temperatures < 95°C, sulfur exists as a cyclooctasulfane crown with an S-S bond length of 0.206 nm and an S-S-S bond angle of 108 degree; at temperatures < 119°C, sulfur crystallizes. At 119°C (the melting point of sulfur), liquid sulfur is dispersed thoroughly in bitumen, forming an emulsion, and cyclooctasulfane partly turns into polymeric zigzag chains (bond length 0.204 nm) (Voronkov et al., 1979). The crystallization features are affected by such factors as the chemical reaction of sulfur with bitumen components and its dissolution or dispersion. At heating temperatures < 140°C, elementary sulfur forms polysulfide that initiates the formation of a network. Such structures differ considerably from unmodified sulfur in the chemical and thermal stability. At 119 to 159°C, molten sulfur exists essentially as cyclooctasulfane (λ-S). Above 159°C, eight-member rings rapidly break down into bi-radicals. In their turn, bi-radicals recombine to form polymeric chains with the maximal length of 106 sulfur atoms:

$$S_8 \leftrightarrow S^0 - S_6 - S^0 \tag{8.1}$$

$$S^0 - S_6 - S^0 + S_8 \leftrightarrow S^0 - S_6 - S - S^0 \text{ etc.} \tag{8.2}$$

At 159 to 180°C, less stable (than S_8) cyclic structures having less or more sulfur atoms can occur in the melt. Linear sulfur bi-radicals are in equilibrium with large cyclic structures. The equilibrium sulfur species detected by ESR spectroscopy are as follows:

$$nS_8 \text{ (ring)} \leftrightarrow nS_8 - \text{(chain)} \leftrightarrow (S_8)_n \text{(chain)} \tag{8.3}$$

At temperatures > 200°C, sulfur is mahogany in color and its melt contains S_3, S_4, and, possibly, S_2 fragments.

Elemental sulfur was demonstrated to be the most reactive in the temperature range from 220 to 260°C. (A remarkable exothermic effect in the DTA curve is shown in Figure 8.1.)

8.3.4 Sulfur Asphalt Interaction

Sulfur in sulfur-asphalt blends has been found to occur in three different forms (Gawel, 2000): (a) chemically bonded, (b) dissolved in asphalt, and (c) crystalline sulfur that generally exists in the form of discrete tiny particles dispersed in asphalt.

Chemically-bonded Sulfur

Part of the added sulfur reacts chemically with asphalt. Substantial variances have been found with regard to the reported percentages of reacted sulfur in the literature. Petrossi et al. (1972), Qarles and Vlugter (1965), and Gawel (1989) reported

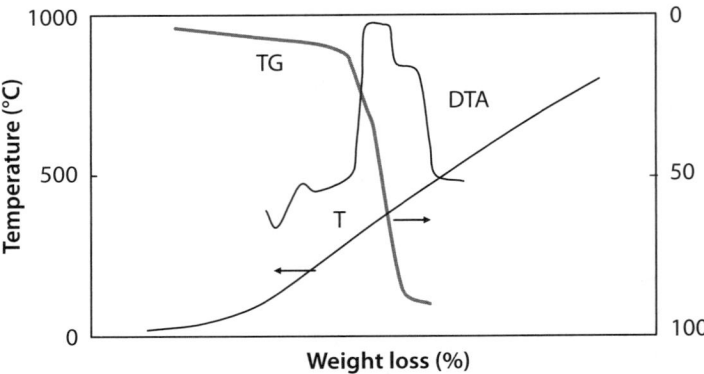

Figure 8.1. Thermal analysis of elemental sulfur (adapted from Syroezhko et al., 2003).

high values of bonded sulfur up to 85% at a temperature below 150°C, whereas Kennepohl and Miller (1978) and Lee (1975) calculated that only 20% of sulfur reacted with asphalt and entered into its molecule. The differences that have been observed are most likely caused by variations in the asphalt composition as well as by the reaction of temperature and time.

It has been confirmed by electron paramagnetic resonance (Kennepohl and Miller, 1978; Petrossi et al., 1972) that sulfur reacts with hydrocarbons competitively; by addition to give carbon-sulfur bonds or by abstraction of hydrogen and subsequent hydrogen-sulfide formation. On the basis of the results of complex investigations into the products of sulfur reactions with model hydrocarbons, the most likely mechanism of the reaction between sulfur and asphalt has been reported as (Rahimian, 1976; Korn et al., 1966):

$$S_8 \rightarrow 4\,^0SS^0 \tag{8.4}$$

$$R - CH_3 + {}^0SS^0 \rightarrow R - CH_2^0 + HSS^0 \tag{8.5}$$

$$HSS^0 \rightarrow HS^0 + S \tag{8.6}$$

$$R - CH_3 + HS^0 \rightarrow R - CH_2^0 + H_2S \tag{8.7}$$

$$R - CH_2^0 + {}^0SS^0 \rightarrow R - CH_2 - SS^0 \tag{8.8}$$

$$R - CH_2 - SS^0 + R - CH_2^0 \rightarrow R - CH_2 - S_2 - CH_2 - R_{disulphide} \tag{8.9}$$

$$2R - CH_2 - SS^0 \rightarrow R - CH_2 - S_4 - CH_2 - R_{polysulphide} \quad (8.10)$$

It has been suggested that the most reactive part of asphalt is the naphthene-aromatic fraction. At low sulfur-to-asphalt ratios, the prevalent reaction depends on temperature. At a low temperature range (120 to 150°C), sulfur introduced into asphalt enters the molecules. Naphthene-aromatics are partially transformed into polar-aromatics, mainly polysulfides. This can be assumed by considering the increase in the sulfur content of the polar-aromatic fraction corresponding to 4 to 5 sulfur atoms per molecule (Petrossi et al., 1972). It has been observed that at lower temperatures sulfur reacts not only with hydrocarbons but also with organic compounds, for example, with indole to simultaneously form polysulfides (Qarles and Vlugter, 1965).

Polysulfides undergo partial decomposition to form a polysulfide with a lower degree of polymerization and free sulfur. Cleavages are not affecting the newly formed bonds between sulfur and carbon atoms. Investigations with the use of sulfur isotopes have shown (Qarles and Vlugter, 1965) that in $R-S-S_x-S-R$ compounds only atoms designated as S_x can take part in decomposition reactions with the separation of elementary sulfur, whereas the atoms of sulfur connected directly with carbon do not take part in the reaction. The increase in the content of asphaltenes at temperatures of up to 150°C is negligible. Asphaltenes isolated from asphalt treated with sulfur at 130°C contain less sulfur than maltenes (Qarles and Vlugter, 1965).

It was also demonstrated by Stepan'yan (1988) and Perov et al. (2001) that alkanes C_7-C_{12} actively react with sulfur at lower temperatures (110 to 190°C) with the formation of mercaptans, sulfides, and di-sulfides.

At temperatures above 180°C, dehydrogenation with the emission of hydrogen sulfide takes place and only a negligible part of the introduced sulfur remains in the asphalt (Petrossi et al., 1972). It has been found (Qarles and Vlugter, 1965) that raising the temperature by approximately 20°C causes a fourfold increase in the dehydrogenation reaction rate.

At temperatures of above 220°C, the action of sulfur is analogous to that of oxygen in the air-blowing process. There are, however, some differences; the reaction sets in at a lower temperature than in the case with oxygen, and more sulfur is combined as compared with oxygen (Siegmann, 1950).

At elevated temperatures, polysulfide linkages that might have been formed first at lower temperatures are broken to produce cyclic sulfides as well as complex cross-linking between separate molecules. The infrared analyses data of the various fractions of the sulfurized asphalts indicated the presence of sulfur in the cyclic tiophenic-type form (Tucker and Schweyer, 1965). With increasing temperatures, naphthene-aromatics are transformed into asphaltenes through dehydrogenation and cyclization processes. For different asphalt samples treated with sulfur at a temperature of 240°C, a varied content of asphatlenes ranging from several to 40% is reported (Oyekunle and Onyehanere, 1991; Tucker and Schweyer, 1965). The amount of asphaltenes formed is dependent on the composition of the original asphalt and on the quantity of sulfur being added. The increase of the sulfur content leads to an increase in the quantity of asphaltenes. It has been stated (Oyekunle and Onyehanere, 1991) that the lower the asphaltene content of unreacted asphalt, the higher the growth of the asphaltene content of the product at the same reaction temperature.

The change of the chemical structure of asphalt caused by introducing sulfur involves a change of its colloidal structure. The increase in the reaction temperature results in the increase in the asphaltenes/resins ratio that causes their structure to be changed into the gelstructure. A comparison of the transmission electron photomicrographs of asphaltenes isolated from the asphalt treated by sulfur and from the initial asphalt has shown a more pronounced degree of association of the sulfurized asphalt (Papirer et al., 1980).

Dissolved Sulfur

Sulfur has a 20% or greater solubility in most asphalts in the temperature range of 130 to 150°C (Kennepohl et al., 1975; McBee and Sullivan, 1978). The dissolved sulfur plays the role of the binder. Polysulfides that are formed as a result of the reactions at temperatures above 120°C partly dissolve unreacted sulfur (Love, 1979). The solubility of sulfur depends upon the type of asphalt as well as its origin. A slowly developing re-crystallization of dissolved sulfur can be observed when a sulfur-asphalt mixture is being cooled to ambient temperatures. However, it is possible to prevent or to reduce re-crystallization by increasing the mixing time (Qarles and Vlugter, 1965). The relative amounts of crystalline and dissolved sulfur components are shown in Figure 8.2 (Kennepohl and Miller, 1976).

An investigation into the aging characteristics and the stability of the non-crystalline portion of the sulfur asphalt binder with three grades of paving asphalts; 40/50, 85/100, and 300/400 pen, at four levels of sulfur concentrations; 20, 40, and 50 wt% indicates that the amount of dissolved sulfur is independent of the asphalt grade and sulfur content and that no crystallization of the dissolved sulfur occurs at ambient temperatures over time.

Earlier studies showed that 2 to 14% of the added elemental sulfur reacted chemically with the asphalt. However, a recent study by Kennepohl and Miller (1976) reported a higher percent of bonded sulfur in sulfur asphalt composition. The differences most likely are caused by variation in the asphalt composition with regard to polar aromatic and naphthene compounds as well as by the reaction temperature and the contact time. It is worth noting that free sulfur is,

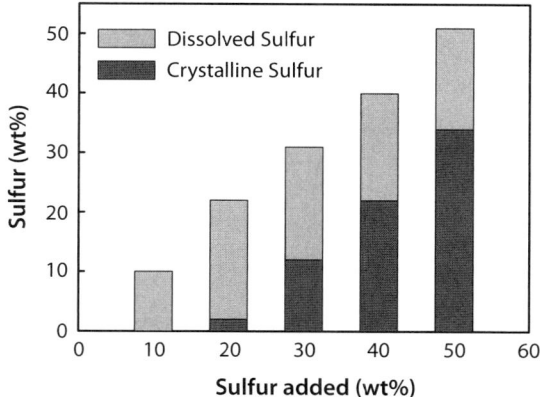

Figure 8.2. Relative amount of dissolved and crystalline sulfur in sulfur asphalt blends (adapted from Kennepohl and Miller, 1976).

generally, determined by the extraction with sodium sulfite followed by titration with iodine.

Comparison of the DTA curves of DTA pen 60/90 and the sulfur-extended composite (5% S) shows that at 240°C paving asphalt activity reacts with sulfur. In the DTA curve shown in Figure 8.3, the exothermic effect of the composite is slightly weaker as compared to the initial asphalt. The maximal thermal effect due to the reaction of asphalt components with sulfur was observed at 320 to 340°C that was accompanied by an active liberation of hydrogen sulfide.

At 390 to 425°C, active thermal decomposition of relatively thermal-unstable paraffin-naphthlene compounds takes place that is reflected in clearly pronounced endothermic effect at 425°C. In this case, primary, radical sulfur species react with unsaturated fragments of resins and alkenes (=C–S–S–C= or C–S_{n+1}–C=). Consequently, the major part of sulfur in an asphalt concrete mix serves as filler.

Crystalline Sulfur

The excess sulfur that is not dissolved in asphalt exists in the crystalline state (Syroezhko et al., 2003). The crystalline sulfur finely dispersed in asphalt acts as a filler or a structuring agent, playing the role of the aggregate. Elemental sulfur disperses readily in asphalt, although the surface tensions of the two substances are quite different (Papirer et al., 1980).

Microscopic studies of sulfur-asphalt mixtures have shown the apparent differences in the degree of the dispersion of crystalline sulfur, depending on sulfur concentration (Courval and Akili, 1982). In a blend of 10% sulfur/90% asphalt

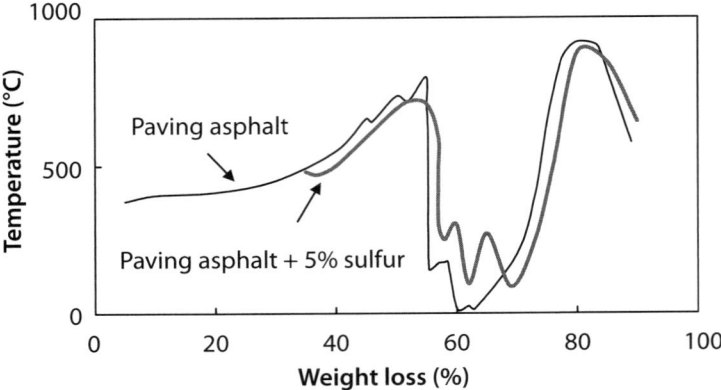

Figure 8.3. Thermal analysis of BND 60/90 paving asphalt after and before modification with sulfur (adapted from Syroezhko et al., 2003).

(wt%), sulfur particles with the diameters of approximately 1 to 2 mm are homogeneously distributed throughout the asphalt. When the concentration of sulfur in the sulfur-asphalt blend increases to 25%, some heterogeneity in its structure can be observed. Sulfur occurs in the form of discrete particles as well as groups of many particles with an average size of 6 mm. In the 50% sulfur/50% asphalt (wt%), blend, the heterogeneity of dispersion is quite evident (Courval and Akili, 1982).

8.3.5 Sulfur Asphalt Mix Concepts

A sulfur asphalt mix can be prepared in two ways: (1) the sulfur and asphalt can be mixed directly with the aggregate and (2) the sulfur-asphalt binder can be prepared before being mixed with the aggregates. In the first method, sulfur may be introduced into the asphalt in the pulverized or liquid form; however, the latter is preferred. Preparation of sulfur-asphalt binder consists of emulsifying sulfur into the asphalt. The hot emulsion is then mixed with the aggregates. Both methods have proved successful and produce a fine dispersion of sulfur throughout the asphalt.

The sulfur-asphalt emulsion is prepared by mixing liquid asphalt with molten sulfur to attain a fine sulfur dispersion of particles of several micrometers in diameter. Since the density of sulfur is approximately twice that of asphalt, a fine droplet size of sulfur must be obtained. High shear during mixing is essential to obtain good sulfur dispersion and a stable binder. Sulfur dispersion in asphalt can be obtained through various mechanical methods. The best results are achieved by mixing the components in a colloid mill. The freshly prepared sulfur-asphalt binder exhibits a similarity to most finely dispersed aqueous-asphalt emulsions. After a period of time, dependent on the sulfur-to-asphalt ratio as well as on the mixing time, the coagulation becomes noticeable. The solubility of dispersion can be improved by prolonging the mixing time as well as by using additives. The storage of the liquid binder leads to the settlement of sulfur particles and, therefore, the binder must be used immediately after preparation. Another disadvantage of this method is the formation of hydrogen sulfide in the storage tank.

The direct addition of sulfur and asphalt to the aggregates has appeared to be more practical in the preparation of paving mixes than the use of emulsification methods. No special blending apparatus is required and the sulfur-asphalt-aggregate mixes can be prepared at the same mixing plant as that used for conventional blends.

The total volume of the sulfur-asphalt binder in the mix is the same as that of the conventional asphaltic binder. Since the density of sulfur is twice that of asphalt, approximately double the weight of sulfur is required to replace an equal volume of asphalt.

The sulfur-to-asphalt ratio varies from approximately 0.1 to 4.0. The optimum sulfur concentration is dependent on its function in the mixture and the type of sulfur-asphalt-aggregate mixture. For a particular asphalt or mixing temperature, there is an optimum sulfur content.

The optimum sulfur content in sulfur-asphalt binders is found somewhere between 20 and 30 wt% (Kennepohl et al., 1975; Celard, 1978). Below 20 wt%, no hardening effect is obtained and above 30 wt%, the improvement of mix workability is reduced.

Sulfur has also been used in substantial concentrations in sulfur-asphalt-aggregate mixes to directly improve their mechanical properties. The sulfur-to-asphalt ratios from 1:1 to 4:1 was recommended in several patents (Canadian 755,999, 1967; German 1,295,463, 1970; U.S. 3,738,853, 1973). To obtain the best results, a sufficient amount of sulfur should be added to completely fill the void spaces between the aggregates.

The mixing temperature should be maintained in the range of 127 to 150°C. The former temperature represents the sulfur melting point plus the tolerance to avoid its structuring effect whereas the latter one is the temperature above which sulfur undergoes an abrupt and appreciable increase in viscosity resulting in the deterioration of the workability of the paving mix. The temperature of 150°C should not be exceeded because of the formation of hydrogen sulfide and sulfur dioxide. It has been found that to attain the best workability and strength of mixtures containing a sulfur-asphalt binder with the typical 30% sulfur content, the mixing temperature should be maintained in the 130 to 140°C range (Love, 1979). When a mixture with a higher sulfur-to-asphalt ratio is prepared, an approximate 150°C mixing temperature is preferred (U.S. Patent 3,738,853, 1973).

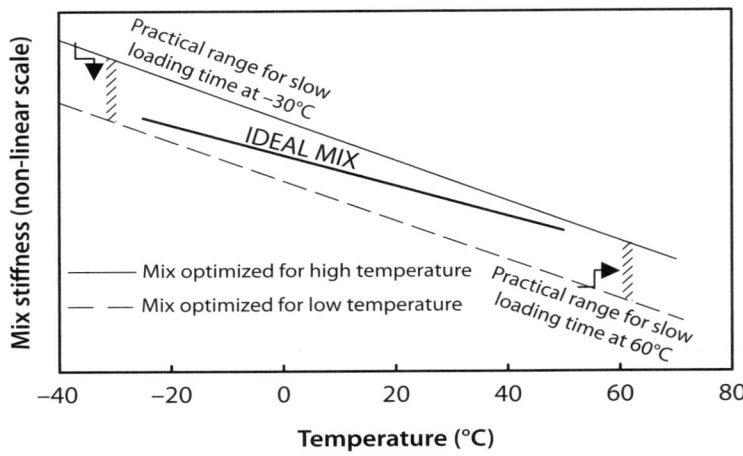

Figure 8.4. Schematic of consistency versus temperature requirements for asphalt paving mixtures (adapted from Kennepohl and Miller, 1976).

The mixing times range from a dozen seconds to a few hours, depending on the method used. McBee and Sullivan (1978) have found that a 2 min mixing time is sufficient to obtain beneficial results. The optimum mixing time to attain fine sulfur dispersion in asphalt, with a diameter satisfying the requirements for particle size of emulsions, was 15 to 30 sec after the sulfur addition (Deme, 1978).

The use of sulfur in the sulfur-asphalt design mix provides the designer with an extra degree of flexibility. The grade of asphalt and the sulfur-asphalt ratio can be revised to gain desired mix characteristics at low and high service temperatures. The design implications can be demonstrated best by the concept of an ideal mix reported by Higmell et al. (1972) and is illustrated in Figure 8.4. The bottom dashed line gives the characteristics of a mix optimized for low temperature application with stiffness low enough to resist low temperature cracking but with an insufficient stability at a high service temperature. The top solid line shows a mix optimized for high temperature application whereas the middle line represents the characteristics of a mix that is optimum for both low and high temperature conditions.

Sulfur-asphalt mixes tend to approach the ideal mix. Figure 8.5 shows results from mix properties of conventional asphalt and sulfur-asphalt mixtures. The results indicate that at a low temperature the 300/400 sulfur-asphalt mix has properties similar to the conventional 300/400 mix, and at high temperatures the 300/400 sulfur-asphalt mix has a stiffness in the range of that for conventional 40/50 mix. Thus, the 300/400 sulfur-asphalt mix will crack less than a conven-

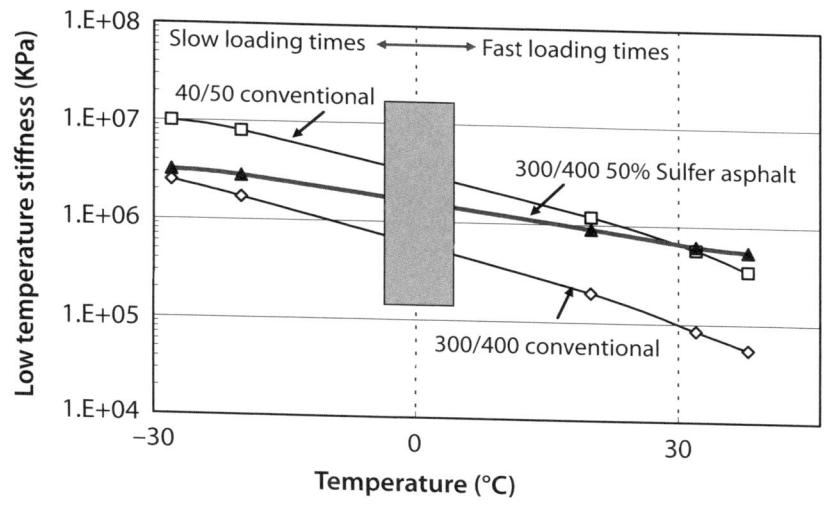

Figure 8.5. Effect of sulfur on the modification of temperature susceptibility of the asphalt mixtures (adapted from Kennepohl and Miller, 1976).

tional 40/50 mix at a low temperature and will have high strength and stability at high temperatures.

8.3.6 Rheology of Sulfur Asphalt Binder

Conventional test methods such as softening point, viscosity, penetration, and ductility have been used to characterize the rheology of sulfur-asphalt binder (Kennepohl el al., 1974). The physical structure of sulfur-asphalt binders is complicated. Asphalt itself consists of a complex colloidal dispersion of resins and asphaltenes in oils. The introduction of liquid elemental sulfur that on cooling congeals into finely dispersed crystalline sulfur particle and, in part, reacts with the asphalt, necessarily complicates the rheology of such sulfur-asphalt binder.

Density

The density of sulfur-asphalt binder increases with the increasing amount of sulfur due to a higher density of elemental sulfur. This growth, however, is less than expected with the sulfur content up to 10 wt%, and it increases normally with the higher sulfur content (Garrigues and Vincent, 1975).

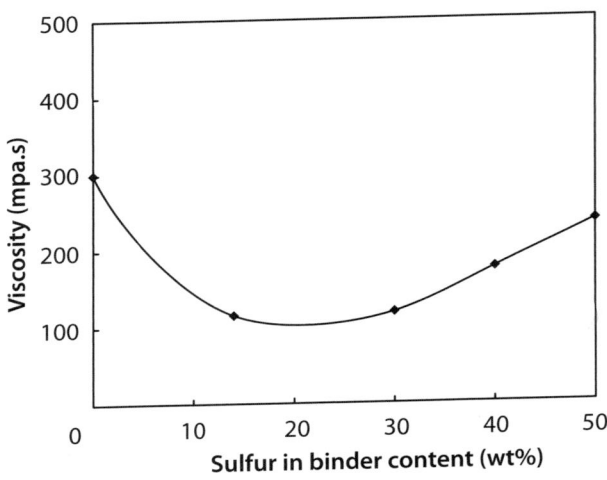

Figure 8.6. Effect of sulfur content on binder viscosity at 135°C (Kennepohl et al., 1975).

Viscosity

Differences and changes with sulfur-asphalt binder preparation, curing time, and temperature must be expected and may be demonstrated by viscosity characteristics. Viscosity studies of asphalt and sulfur asphalt as a function of shear rate at 90°C and 25°C indicate that the asphalts and sulfur asphalt exhibit some non-Newtonian behavior. The degree of non-Newtonian behavior does not appear to be affected by sulfur content. The viscosity depends on temperature, sulfur content, and asphalt grade. Figure 8.6 shows that viscosity decreases with an increase in the sulfur in binder content hitting a minimum with approximately 15 wt% of sulfur content (Kennepohl el al., 1975). Further addition of sulfur, however, increases the viscosity, surpassing the viscosity of the original asphalt at the sulfur level of approximately 56 to 60 wt%. These results are not in full agreement with those reported by Garrigues and Vincent (1975) who found, for the same penetration grade asphalt at 130°C, that the phenomenon of viscosity reduction is inverted as soon as the sulfur content exceeds 25 to 30 wt%.

At higher temperatures, the viscosity for all binder containing dispersed sulfur drops below that of the same asphalt. The extent and temperature at which this phenomenon occurs depends upon sulfur content. The viscosity of the sulfur-asphalt binder at temperatures > 100°C is lower by a factor of 1.6 than that of the initial paving asphalt, allowing a reduction in power consumption by virtue of decreasing temperature of binder heating and asphalt concrete mix preparation by 25 to 30°C. Therefore, the addition of 5 to 10% sulfur to asphalt allows up to 30% saving of paving asphalt without deterioration in the charac-

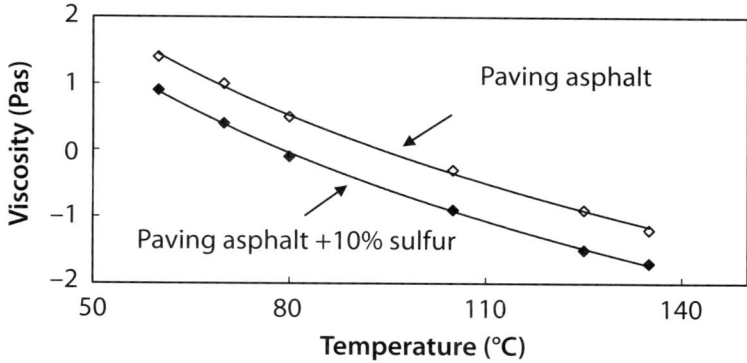

Figure 8.7. Effect of sulfur modification on the viscosity of the binder as a function of the temperature (adapted from Syroezhko et al., 2003).

teristics of the resulting asphalt concrete. At a lower temperature, sulfur-asphalt binders containing more than 20% sulfur have viscosities above that exhibited by the asphalt. This increased viscosity appears to be maximized in the temperature range 25 to 60°C. At 4°C, the viscosity differential between the original asphalt and sulfur asphalt has narrowed. It appeared that sulfur additions improve not only the binder ductility, but also its viscosity temperature characteristics as clearly illustrated in Figure 8.7 (Syroezhko et al., 2003).

Compressive Strength

It was demonstrated from the studies of the strength of the sulfur-modified asphalt concrete mixture over time that slow cooling of sulfur asphalt results in the formation of rigid mechanical contacts by virtue of sulfur crystallization. The structure of these contacts can be changed in rapid cooling or mechanical failure in compaction. With time, monoclinic sulfur transforms into orthorhombic sulfur that is accompanied by an increase in its strength. Therefore, structure formation in sulfur containing organic binder is an important problem in view of development of durable sulfur organo-mineral materials (Syroezhko et al., 2003).

Based on pen 60/90 paving asphalt, prepared by mixing asphalt with sulfur at 140°C followed by adding granite heated to the same temperature into the stirred apparatus, the strength of the resulting asphalt concrete passes through a minimum at a sulfur content of approximately 5% and then increases as shown in Figure 8.8. The strength is higher compared to conventional asphalt concrete.

354 Sulfur Concrete for the Construction Industry

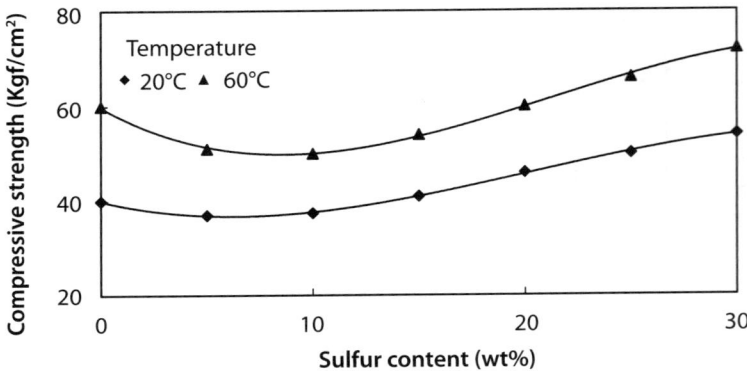

Figure 8.8. Effect of sulfur content on the strength of the binder at various temperatures (adapted from Syroezhko et al., 2003).

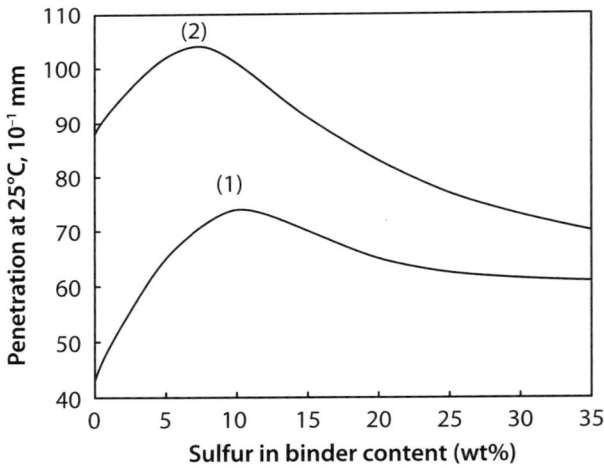

Figure 8.9. Effect of sulfur content on binder penetration. Curve 1 = 40-50 pen grade asphalt sulfur (Rubbio, 1977); Curve 2 = 85-100 pen grade asphalt with sulfur (Garrigues and Vincent, 1975).

Penetration

The penetration of sulfur-asphalt blends increases up to 10% of sulfur in the binder then decreases in a nearly linear fashion as shown in Figure 8.9 (Garrigues and Vincent, 1975). However, reported results by Kennepohl el al. (1974) indicated that the penetration increases up to 40% of sulfur content and, after reaching this value, it decreases steadily. Therefore, it seems that there are dis-

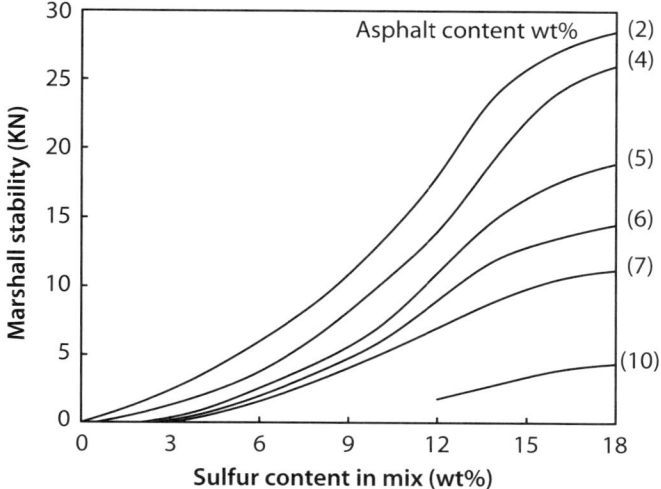

Figure 8.10. Effect of sulfur content on mix stability for materials consisted of 150-180 pen asphalt and medium-coarse sand (adapted from Burgess and Deme, 1975).

crepancies between the results obtained by various researchers in regard to the value at which the curve of the penetration dependence on the sulfur content in the binder attains a maximum.

Marshall Stability

Sulfur has a favorable effect on the mechanical properties of sulfur-asphalt aggregates mixes in terms of the improvement in Marshall stability, increase of the resilient modulus, and decrease in the rutting tendency when compared with conventional asphalt concrete. The mixtures exhibit a high water resistance due to the presence of sulfur.

It has been shown by microscopic studies that when the mix cools, the sulfur solidifies in the void spaces between the binder-coated aggregate particles, conforming to the configuration of the voids (Burgess and Deme, 1975). By conforming to the shape of the voids, the sulfur causes a mechanical interaction between the particles that results in a higher Marshall stability of the mixture (Papirer et al., 1980).

The Marshall stabilities of the sulfur-asphalt binder-based mixes are considerably higher than those made with asphalt only. The stability increases with an increase in the sulfur content for all sulfur contents as shown in Figure 8.10 (Burgess and Deme, 1975). The amounts of sulfur substitution, because they are larger than 50 wt% of the binder, yield high stabilities of the mixes but these would have to be worked at higher compaction temperatures.

8.4 SULFUR ASPHALT PROCESSING TECHNOLOGY

8.4.1 Manufacturing Evaluation

The use of elemental sulfur in paving mixes provides new engineering properties and also gives an interesting evaluation to identify and quantify the significant factors required to develop a design technology for sulfur-asphalt-based paving mixes and their application. Various types of manufacturing evaluation, both field and laboratory, are required to establish the engineering properties of the mixes made with the sulfur-asphalt binder in order to permit an assessment of structural response (i.e., fatigue, permanent deformation, and shrinkage fracture) for making comparisons with standard conventional paving mixes and, eventually, to conduct fundamental structural analyses.

8.4.2 Preparation Apparatus for Sulfur Asphalt Binders

An apparatus was disclosed by Kennepohl et al. (1979) for continuous preparation of a blend of liquid sulfur in fluid asphalt in specified proportions to provide a dispersion of fine sulfur droplets in the asphalt, particularly suitable for coating onto aggregate to form hot mix for an asphalt concrete of improved strength. It consisted of a pumping means to supply a continuous metered stream of both molten sulfur and fluid asphalt as well as blending the stream continuously in proportions of 25 to 60 parts of sulfur per 75 to 40 parts of asphalt. They are thoroughly mixed to emulsify liquid sulfur that does not combine homogeneously with the asphalt in the blended stream as droplets from 1 to 50 microns. Additionally, the temperature is regulated to maintain the stream of molten sulfur from 121 to 154° C, the stream of fluid asphalt from 121 to 177°C, and the blended stream from 121–154°C.

8.4.3 Premixing

The development and engineering of the sulfur-asphalt processing technology is based on extensive laboratory and field experience with sulfur-asphalt binder and sulfur-asphalt mixes. The key aspects and requirements for the incorporation of sulfur into paving mixes for industrial scale mix production depend on the sulfur-asphalt ratio and the coating of the aggregate by the binder, mixing temperature, and mixing time. Premixing is the most economical route for commercial application as a function of application requirements as shown in Figure 8.11.

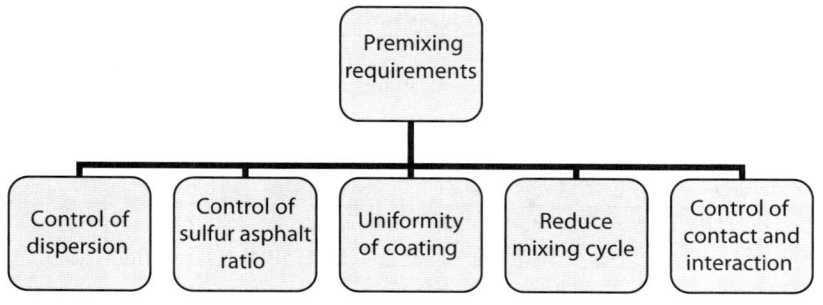

Figure 8.11. Schematic showing premixing requirements of sulfur asphalt mixes.

8.4.4 Sulfur Asphalt Module

An integral part of sulfur-asphalt binder hot mix production is the sulfur-asphalt mixing module (SAM). In conventional mixing stations, asphalt is metered into the pug mill where it is mixed with dried aggregate. The aggregate and asphalt are fed to the pug mill with an appropriate feed control. SAM permits the addition of sulfur. The asphalt feed system can be bypassed and replaced by SAM that can deliver the desired sulfur-asphalt binder or asphalt only. SAM is designed to produce sulfur-asphalt binder at commercial rates up to 80 T per hour, enough binder to produce 1000 T per hr hot mix, and can be adapted to continuous and batch-operated mixing plants.

8.4.5 Mix Production

The feed control system of the mix plant is integrated with the automatic controls on SAM. SAM automatically controls the sulfur-asphalt ratio and matches the production of the sulfur-asphalt binder to the aggregate feed rate. It is mixed with the aggregate between 116 and 160°C. This reduces the levels of emissions and produces mixes that are well-coated and appear virtually indistinguishable from conventional asphalt mixes.

8.4.6 Construction Procedure

Paving with a sulfur-asphalt hot mix is exactly the same as paving with asphalt. Conventional dump trucks are used to haul the hot mix from the asphalt plant to the road. There the mix is spread with standard pavement equipment and compacted with conventional rollers. Actual measurements on the completed pavements indicated that sulfur-asphalt mixes have good workability the same as conventional asphalt.

Table 8.4. Comparison of emissions between asphalt and sulfur asphalt binder

Pollutants	Method of analysis	Typical measured values	
		Asphalt	Sulfur asphalt binder
Hydrogen sulfide	Methylene blue	0.2	1.8
Sulfur dioxide	West Gaeke	0.1	0.2
Carbon disulfide, carbon sulfide, and mercaptans	Potentiometric titration with silver nitrate	<0.1	<0.1
Elemental sulfur	Particulate filter total sulfur	–	0.5
Total hydrocarbons	Hydrocarbon Beckman M 400	6	6

8.4.7 Safety and the Environment

In the normal temperature range for handling sulfur-asphalt materials (127 to 149°C) there is little or no evolution of sulfur-containing gases. If the temperature of the sulfur-asphalt binder is allowed to rise above 154°C, evolution of hydrogen sulfide and sulfur dioxide may occur.

8.4.8 Emissions

To evaluate the effects of sulfur asphalt during and after construction, an environmental assessment has been reported by Filippis et al. (1998). Ambient air samples were taken around the pavers and analyzed for hydrogen sulfide (H_2S), sulfur dioxide (SO_2), carbon di-sulfide (CS_2), carbonyl sulfide (COS), mercaptans (RSH), and total hydrocarbons. The results indicated that no problems exist in terms of current health standards during or following construction. Emission monitoring at mixing plant and construction sites during the construction of several sulfur-extended asphalt pavements indicates that emissions are controllable within acceptable limits. Typical test results obtained at the pug mill for asphalt and sulfur-asphalt binder are given in Table 8.4.

H_2S evolution was measured as the difference between the weight losses of pure asphalt and the sulfur-asphalt mixture. When heated, sulfur-extended asphalt starts to react at approximately 130°C. The main reaction is the dehydrogenation that causes H_2S production and the hardening of the asphalt (Filippis et al., 1998).

The emission of hydrogen sulfide and sulfur dioxide can be reduced with the use of appropriate additives as well as by installing constructional protection. Cupric oxide, sodium carbonate, aluminum oxide, and calcium chloride were found effective in reducing hydrogen sulfide emission (Lee, 1975). The addition of 1% of these substances caused approximately a 50% reduction in hydrogen sulfide evolution. The hydrogen sulfide formation in the course of mixing sulfur with asphalt may be inhibited by the addition of a redox catalyst (German Patent 2,459,386, 1975). It has also been found that the addition of lime flour (often used as a filler in paving mixes) to sulfur-asphalt blends in the mixing temperature range reduces the hydrogen sulfide evolution (Alama et al., 1981). The addition of Fe(II) and Fe(III) compounds to the sulfur-asphalt blend at a temperature of 140 to 150°C caused a decrease in the hydrogen sulfide emission by 50% and in the emission of sulfur dioxide by 75% (Gawel, 1994).

For a full environmental assessment of the effects of using sulfur in pavement, soil and water quality measurements were carried out before and after road construction (Kennepohl and Miller, 1978). The results indicated no change in the sulfur content or pH.

On account of the potential environmental and safety risks during the production of sulfur asphalt and their utilization in pavement construction, no problems should appear unless the temperature of 140°C is exceeded. Below this temperature, the formation of hydrogen sulfide and sulfur dioxide is negligible.

8.5 IMPROVED PERFORMANCE

8.5.1 Low Temperature Stiffness

Adding sulfur (sulfur-to-asphalt ratio), changing the binder content, or varying the viscosity did not appear to affect the stiffness. Therefore, the nature of the asphalt in the sulfur-asphalt binder controls the potential for low-temperature cracking of the binder. Therefore, the addition of sulfur does not affect the low temperature characteristics of the asphalt concrete adversely.

8.5.2 Tensile Strength

The addition of sulfur in excess of 20% increased the strength significantly. The increase in strength produced by adding 50% sulfur was approximately equal to the increase associated with using a 40/50 penetration asphalt cement rather than an 85/100 penetration.

8.5.3 Resilient Modulus of Elasticity

The addition of sulfur produced an increase in stiffness in terms of resilient modulus of elasticity as shown in Figure 8.12. In some cases, 20% sulfur produced a substantial increase; generally this increase required the addition of 50% sulfur.

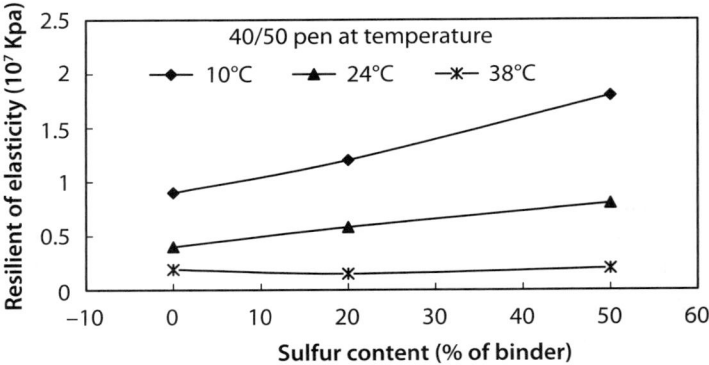

Figure 8.12. Relationships between resilient modulus and sulfur content (adapted from Kennepohl and Miller, 1976).

8.5.4 Fatigue Life

The addition of sulfur in excess of 20% produced a significant increase in fatigue life. The increase in fatigue life produced by adding 50% sulfur was approximately equal to the increase associated with using a 40/50 pen asphalt rather than an 85/100 penetration.

8.5.5 Resistance to Stripping

One major problem encountered with asphalt concrete pavements is the loss of the adhesive bond between the aggregate surface and the asphalt cement. This breaking of the adhesive bond is known as *stripping*. The stripping of asphalt binder from aggregate surfaces results in shorter pavement life. The tendencies of stripping by water can be reduced greatly by incorporating sulfur in the paving mix.

8.6 SULFUR-TO-ASPHALT RATIOS AND PROPERTIES

8.6.1 Sulfur-to-Asphalt Ratios

Binders have been used with sulfur-to-asphalt weight ratios as high as 50/50. Most field trials have used binders with 30/70 and 40/60 sulfur-to-asphalt weight ratios; since the specific gravity of sulfur is approximately twice that of asphalt, the sulfur in a 50/50 sulfur-extended asphalt binder has replaced about 33% of the asphalt by volume. The function of sulfur in the asphalt paving mixture depends on the sulfur concentration and the sulfur-asphalt ratio. At a low sulfur-to-asphalt ratio, the sulfur modifies the chemical and rheological properties of asphalt through chemical reactions whereas at high sulfur-to-asphalt ratio (larger than 1), the sulfur acts as filler and as a *structuring agent*, improving the workability of the sulfur-asphalt aggregate mixture at processing temperatures of 130 to 160°C and the mechanical strength of the mixture at service temperatures. The most successful formulation was that of a sulfur-to-asphalt ratio of 2.25 with a uniform-sized sand as aggregate (Hammond et al., 1971; Burgess and Deme, 1975; Deme, 1974; Gallaway and Saylak, 1974).

At low sulfur-to-asphalt ratios (up to 20% sulfur by weight of asphalt), the predominant reactions depend on the reaction temperature (Petrossi et al., 1972; Bocca et al., 1973) whereas at higher temperatures (above 240°C), the asphalt is dehydrogenated similar to air blowing (oxidized asphalt) (Abraham, 1961; Tucker and Schweyer, 1965). At lower temperatures (below 140°C), sulfur combines with asphalt. In either case, the rheological properties of asphalt (penetration, temperature susceptibility, and ductility) are modified mainly due to the change in the asphaltene-to-resin ratio.

For the paving composition of this example (Kennepohl et al., 1979), the sulfur and asphalt were blended in proportions of 50:50. By conventional Marshall mix design tests, using the foregoing aggregate and the foregoing 85-100 pen asphalt as the only binder, it was determined in the laboratory that 5.8% was the optimum proportion of this binder with this aggregate for the most effective paving composition properties.

In as much as sulfur has a considerably higher specific gravity than asphalt, and blends of sulfur and asphalt have a higher specific gravity than asphalt, it is natural that the weight proportion of sulfur-asphalt mix, which is optimum as a binder for the paving composition, should be higher than that for asphalt alone as a binder; equal volumes of binder are required to completely coat equal weights of aggregate with equal thicknesses of coating.

Thus, the optimum weight proportion of the 50:50 sulfur-asphalt binders for the foregoing aggregate, as determined by Marshall Tests, was in the range 7.0 to 8.0%, and a proportion of 7.7% was used in the paving composition of this example. To verify that the sulfur-asphalt binder contained the dispersed droplets of liquid sulfur of below 10 micron size only, random samples of the binder were taken as it was being added to the aggregate and examined visually under a microscope. Of particular significance in the Marshall test results was the fact that the stability determinations on the test samples showed that the optimum asphalt-aggregate proportions provided a Marshall stability of slightly less than 2000 whereas the optimum sulfur-asphalt-aggregate proportions provided a Marshall stability of over 3000, an increase of over 50%.

Furthermore, despite the greater Marshall stability, indicative of greater strength in the finished surface coat after rolling and hardening of the paving

Table 8.5. Effects of suppressants on H_2S emission from 50 Kg batches of a mixture of sulfur/asphalt/sand (composition 18/6/76wt%) (Gaw, 1976)

Suppressants (%wt between brackets)	H_2S	SO_2	Mixing temperature (°C)
None	40	0	126
Tetra-methyl-thiuram-disulfide (0.05)	0	0	123
Hydroquinone (0.1)	20	0	124
Zinc diethyl-dithio-carbonate (0.2)	0	0	126
Zinc diethyl-dithio-carbonate (0.6)	4	0	126
Iodine (0.01) / KI (0.01)/ethylene glycol (0.08)	0	0	129
Iodine (0.002) / KI (0.002)/ethylene glycol (0.016)	25	1	131
$CuCO_3$ (0.1)	4	0	120
$CuCO_3$ (0.01)	20	0	121
Fe_2S_3 (0.1)	0	0	127
FeS	12	5	127
$FeCl_3 \cdot 6H_2O$ (0.1)	0	0	130
$FeCl_3 \cdot 6H_2O$ (0.02)	0	0	128

composition, the sulfur-asphalt paving composition was no harder to compact while the paving composition was still hot, because of the presence of the finely divided liquid sulfur droplets in the binder; also, the hot paving composition containing the sulfur-asphalt binder could cool to a lower temperature than can normal asphalt-binder mixes before becoming too hard to roll and compact effectively.

8.6.2 Sulfur-Modified Asphalt Characteristics

Interest in the use of sulfur in highway pavement construction has been stimulated by unpredictable increases in cost and by an uncertainty as to the future availability of asphalt cement. Studies and experimental field tests (McBee et al., 1980) have shown that sulfur can, under certain circumstances, replace as much as 35 volume percent of the asphalt cement used in asphaltic concrete mixes. The unique properties of sulfur, both alone and when blended with asphalt, have also shown a potential for use in recycling old bituminous pavements.

The addition of sulfur to asphalt at a working temperature of 149°C resulted in:

    a. An increase in asphalting content, percent bonded sulfur, and evolution of hydrogen sulfide; however, more than 75% of added sulfur remained unreacted in colloidal suspension.

b. For a particular asphalt and a particular set of mixing temperatures, there is an optimum sulfur content.
c. The addition of a small amount of sulfur-to-asphalt (5 to 20% by weight of asphalt) at 149°C resulted in an increase in penetration and a reduction in viscosity and ductility.
d. Addition of sulfur-to-asphalt-concentration mixtures significantly increased the stability, resistance to water action, and tensile strength of the compacted mixtures. For practical considerations, the post-mix procedure with the sulfur added at the end of the normal mixing cycle is recommended.
e. There are definite improvements in certain sulfur-treated aggregates in reducing asphalt and water absorption and in increasing mechanical stability of the mixes.
f. Natural weathering of sulfur-asphalt concrete up to 30 months increased the mechanical stability of the mixes with little change in flow value.

8.6.3 Hydrogen Sulfide Emission Control

Hydrogen sulfide emission during sulfur-asphalt mixing increased with the sulfur concentration and mixing temperature. Free radical inhibitors and redox (oxidation-reduction) catalysts were found effective in controlling H_2S emissions (Lee, 1975; Gaw, 1976; Bailey, 2005). Examples for free radical inhibitors are tetramethyl thiuram disulfide, hydroquinone, and zinc diethyldithiocarbonate whereas for redox catalysts they are iodine, copper salts and copper oxides, iron salts and iron oxides, and cobalt salts and cobalt oxides. Table 8.5 shows the effects of some of these suppressants on H_2S emissions during simulated conditions to transport and unload sulfur asphalt (Gaw, 1976).

8.7 SULFUR ASPHALT DEVELOPMENT

The presence of sulfur in asphalt mixtures provides improved compaction and strength benefits that improve the durability and performance of the composition as a paving material. Research into the use of sulfur in asphalt-paving materials has resulted in the development of two distinct technologies whose basic difference lies in the primary role sulfur plays in the mixture as illustrated in Figure 8.13.

The first development is sand-asphalt-sulfur (SAS) that involves the use of sulfur as a structuring agent with poorly graded sands to produce a quality asphaltic-paving material (i.e., the role of the sulfur is that of aggregate). These

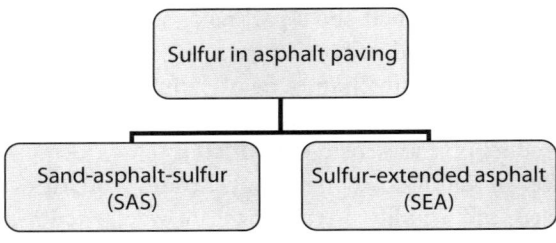

Figure 8.13. Sulfur asphalt technologies.

two distinct uses of sulfur in asphaltic mixtures are discussed at length in a review article by Rennie (1979).

8.7.1 Sand-Asphalt-Sulfur

SAS mix has been considered by many as an alternative to asphalt concrete mixes. The preparation of SAS materials involves a two-cycle process. First, all three ingredients are preheated to a temperature above the melting point of sulfur (116°C) and below (149°C). The upper limit is the temperature above which sulfur begins to undergo an abrupt and substantial increase in viscosity that could adversely affect mix workability.

During the first cycle of the preparation of SAS mixtures, sand and asphalt are mixed to coat the particles with asphalt. Liquid sulfur is then added and mixed with the asphalt and sand until the three ingredients are dispersed throughout the mix. Upon cooling, the sulfur that has not dissolved in the asphalt solidifies within the voids of the mixture thereby creating a mechanical interlock from which the material derives its strength. By acting as a conforming filler, the crystallized sulfur induces such a high degree of mechanical stability to the mix that high-quality paving materials can be achieved using poorly graded aggregates such as single-sized sands. Upon consideration of a number of criteria, it has been concluded that an optimum SAS system would have sulfur and asphalt contents between 12 and 14% and 5 and 7%, respectively. (Saylak and Conger 1982).

8.7.2 Sulfur-Extended Asphalt

The term sulfur-extended (SE), when applied to an asphalt cement, asphalt concrete mix, or pavement, denotes the replacement of a significant portion of the conventionally used asphalt with elemental sulfur. Typically, 20 to 40% of the weight of the asphalt is replaced by sulfur. The sulfur-extended asphalt (SEA) technology is a later development in which sulfur is utilized as an integral part of the binder to effect a partial replacement or extension of the asphalt cement in

Table 8.6. FHWA evaluated sulfur extended asphalt projects (FHWA-RD-90-110, 1990)

No.	Location	Age (years)	Blending method
1	California-Anaheim	4.3	Direct feed (liquid)
2	California-Baker	3.2	In-line blending (liquid)
3	Delaware	6.4	In-line blending (liquid) & direct feed (liquid)
4	Georgia	4.6	Direct feed (liquid)
5	Idaho	4.0	In-line blending (liquid)
6	Kansas	5.0	Direct feed (liquid)
7	Louisiana	6.0 for AC section & 7.2 for SEA section	In-line blending (liquid)
8	Maine-Benton	4.1	Direct feed (liquid)
9	Maine-Crystal	6.2	Direct feed (liquid)
10	Minnesota	7.0	Direct feed (liquid)
11	Mississippi	4.4	Direct feed (liquid)
12	North Dakota	5.2	In-line blending (liquid)
13	New Mexico	3.7	In-line blending (liquid)
14	Texas-College Station	7.4	Colloid mill pre-blending
15	Texas-Pecos	4.2	In-line blending (liquid)
16	Texas-Nocogdoches	5.2	Direct feed (liquid)
17	Wisconsin	3.6	In-line blending (liquid)
18	Wyoming	3.7	Direct feed (liquid)

conventional asphaltic pavement materials. Binders are formulated by replacing some of the asphalt with sulfur in conventional binders. The principle objective of using elemental sulfur as a binder for aggregate, without adversely affecting the flexibility of the in-place pavement, guided the basic research approach and led to the premixing of sulfur and asphalt.

Although commercial processes for modifying properties of asphalts with sulfur have been in existence for more than a century, the current developments in the use of SEA binders originated with the work of Bencowitz and Boe (1938). Using a wide variety of types and sources of asphalts, they produced stable blends containing 25% sulfur. Blends with as much as 40% sulfur were achieved with some asphalt. These early studies, together with more recent investigations, have established some conclusions regarding the effect of sulfur on the properties of SEA binders and the hot-mix concretes in which they are used.

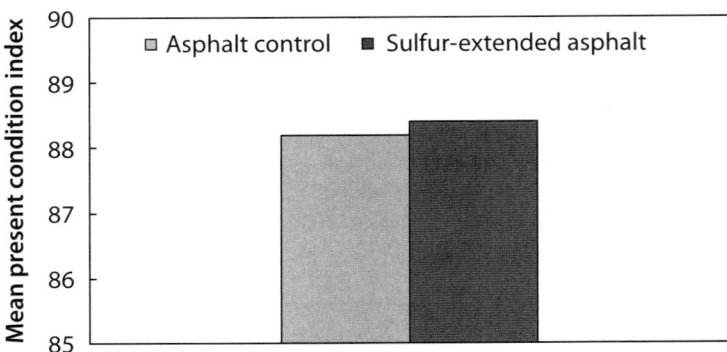

Figure 8.14. Comparison of the PCI for SEA and AC pavement for FHWA study (adapted from Beatty et al., 1987).

The first SEA test pavement on U.S. public roads was done in Texas in 1975 using a pre-blended SEA binder; a second test section, using a direct-mixing procedure was used in Nevada in 1977. Since that time, many other sections have been laid in many American states. The Federal Highway Administration (FHWA) has produced reports on SEA and is currently evaluating the performances of the previous SEA pavements (FHWA-IP-1980; FHWA-RD-90-110-1990).

In 1985, the FHWA organized a task force to conduct a comprehensive SEA field evaluation study (Beatty et al., 1987). The objectives were to compare the field performance of a representative group of SEA pavements with that of a control group of conventional asphalt concrete pavements and to determine what differences in performance and in durability existed between the two groups.

Twenty-six SEA projects in eighteen states were selected for evaluation after considering such criteria as geography, climate, sulfur form and content, age, and blending methods as shown in Table 8.6. A *present condition index* (PCI) was determined for each pavement. The PCI is determined through a comprehensive evaluation of visible pavement distress that includes cracking and rutting. The predominant distresses were longitudinal, transverse, and joint reflection cracking, and some rutting. No trends were observed in the occurrence of cracking and rutting with a variation in the sulfur-to-asphalt ratio from 20/80 to 40/60. Figures 8.14 and 8.15 suggest that the performance of the SEA pavements was comparable to the asphalt concrete control group, and that the performance may even be marginally better regardless of the sulfur content of the binder.

Beatty et al. (1987) then used statistical methods to test the significance of observed differences in PCI and to the deduct values between the SEA and the control groups. The analysis concentrated on the effect of sulfur on performance

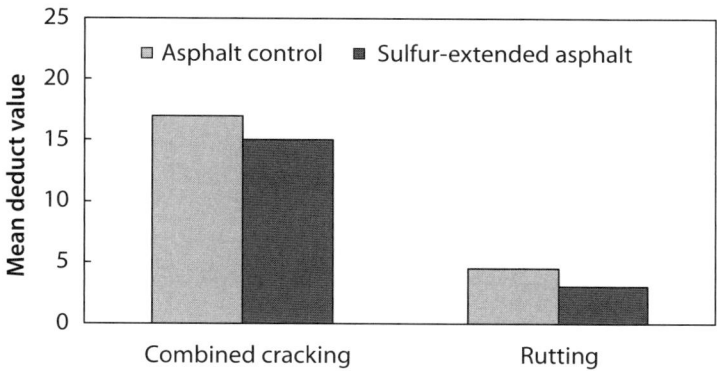

Figure 8.15. Comparison of mean distress deduct values for SEA and AC pavement sections for FHWA study (adapted from Beatty et al., 1987).

and on durability of the pavements. The test was employed to estimate the significance of the observed differences. The results indicated that no significant differences exist between the SEA and asphalt concrete control groups; the two types of pavements have performed comparably. Similarly, the performance of the SEA pavement was not significantly influenced by the sulfur content of the paving binder from 20/80 to 40/60. Finally, the researchers attempted to find correlations between the PCI and deduct values with the pavement age and freezing index. No evidence for causal relationships between the age and freezing index and the PCI and deduct values was found. The results of this study imply that, in most circumstances, the use of sulfur as an extender in asphalt paving mixtures is innocuous and that SEA pavements should perform in a satisfactory manner if they are constructed to proper design with adequate attention to detail.

The adhesion of the initial paving asphalt and sulfur-extended composites to mineral filler (granite) is shown in Figure 8.16 as a function of heating time. Adhesion/cohesion interaction in both the initial paving asphalt and sulfur-extended composites depends on the contact time of the binder with granite. All the curves demonstrate the tendency to saturation at a contact time of about 2 hr. It is important that, in the case of sulfur-extended composites with a sulfur content of 5-20%, the amount of asphalt retained on granite is higher as compared to the initial paving asphalt. With large sulfur additions (40%), the composite becomes loose and its adhesion to the mineral filler is minimal (Syroezhko et al., 2003).

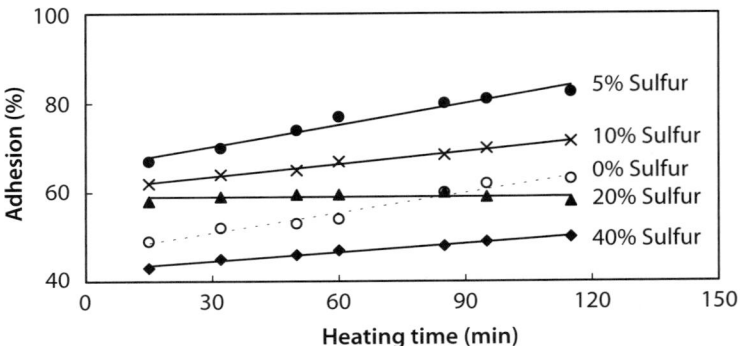

Figure 8.16. Effect of the heating time and sulfur percentage of paving asphalt composites on the adhesion of asphalt to mineral filler (adapted from Syroezhko et al., 2003).

8.8 SULFUR-EXTENDED ASPHALT AND TRADITIONAL ASPHALT MATERIALS

Tests showed that SEA was at least comparable to traditional asphalt concrete, with some properties being superior. SEA was stiffer, making it more resistant to rutting damage in warm climates. Many data on SEA paving materials have been studied.

The work described here is a part of a study on sulfur-modified asphaltic concretes using 3/8-in top size heavyweight and lightweight paving aggregates (Pronk et al., 1975). The resilient moduli are shown as a function of temperature in Figure 8.17. The sulfur-asphalt bound mix has a slightly higher stiffness than the normal asphalt concrete at low temperatures. As the temperature increases, the decrease in stiffness is not as great for the sulfur-asphalt material as for the ordinary asphalt concrete. Thus, it has a higher relative stiffness at the higher temperatures. Since for a given service strain criterion, the stiffer the material the thinner the necessary pavement layer thickness, it is clear that at low temperatures no significant saving is effected in the volume of material to be laid. However, at higher operating temperatures such a saving might be possible.

Vacuum saturating and freeze-thaw cycling has a much greater detrimental effect on the ordinary asphalt concrete than on the sulfur-asphalt concrete. Indeed, after soaking, the sulfur-asphalt concrete shows a slight increase in stiffness and the freeze-thaw cycling causes only a slight decrease in stiffness. It thus appears more durable than its ordinary asphalt counterpart.

The sulfur-asphalt bound material shows better durability under the imposed conditions. The retention of stiffness after freeze-thaw cycling by the sulfur asphalt bound concrete with 3/8-in aggregate is known in a second series of tests.

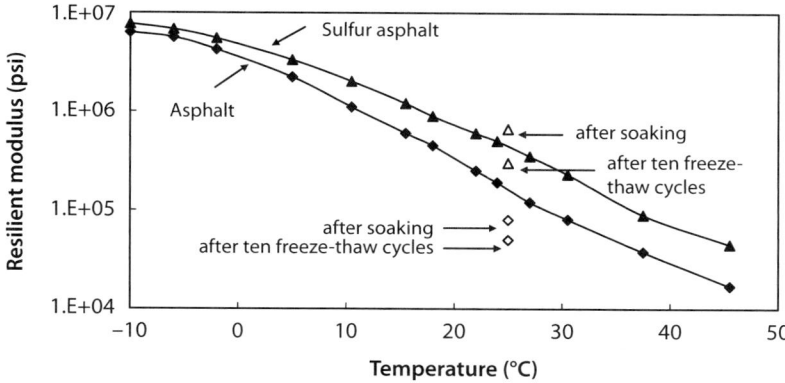

Figure 8.17. Variation of resilient modulus with temperature, soaking, and freeze-thaw cycling for a sulfur modified asphalt concrete and unmodified asphalt concrete (adapted from Shrive and Ward, 1977).

However, neither concrete with lightweight aggregate retained stiffness to the same degree. The sulfur-asphalt bound concrete was again stiffer, but it lost a greater proportion of its stiffness during the test. At the higher temperature, the sulfur asphalt-bound materials again show less change in general than the ordinary asphalt concrete.

Use of SEA mixes on airport pavements appears feasible. These mixes have a higher stability and lower air voids content than conventional mixes. This is a function of the sulfur in the material that lowers the viscosity of the binder in the hot-mix form thereby giving better compaction. On cooling to ambient temperature, the sulfur acts as a filler and gives a higher stability material. In addition, SEA mixes are reported to be more resistant to water stripping and its resistance to gasoline, diesel fuel, and other solvents is improved.

Transverse cracking was also a lesser problem than in more traditional asphalt paving. The addition of sulfur to the asphalt mix just prior to pouring the roadway reduces the viscosity of the hot paving material, making the pavement easier to install.

Equipment for preparing traditional asphalt requires little or no adaptation to handle SEA, an important factor in the acceptance of SEA on a large scale (Weber and McBee, 2000).

Even with the demonstrated advantages of SEA over more traditional paving materials, no large-scale highway projects have used SEA. A period of elevated sulfur prices in the late 1980s eliminated the financial incentives for the substitution (Weber and McBee, 2000). Recent increases in low-cost sulfur availability have reinvigorated interest, and large scale testing of SEA as a viable, economic alternative to more traditional asphalt formulations are underway.

8.9 PLASTICIZED SULFUR

8.9.1 Plasticization Concept

Sulfur is an additive that has been incorporated into the binder as a minority constituent. Mixing asphalt with sulfur, however, presents a number of problems. For the sulfur to effectively modify the asphalt, the sulfur must be chemically modified (plasticized or polymerized). This plasticization may occur when the sulfur is mixed with the hot asphalt. However, problems with the plasticization of the sulfur often result as the liquid sulfur, liquid asphalt, and aggregate are mixed. In certain mixtures, the sulfur and asphalt can separate due to the differences in their respective densities that tend to cause an uneven dispersion of the plasticized sulfur. As a result, the sulfur-depleted portions of the binder then retain the softening and flowing properties of asphalt. Not only does the presence of sulfur-depleted portions of binder diminish the overall effectiveness of the asphalt as a binder, but handling and transporting the binder remains difficult.

Chemicals such as dicyclopentadiene and heptane have been used as an attempt to keep the sulfur homogeneously dispersed in asphalt. Further, crushed limestone has been used for this purpose. However, the use of calcium-based materials leads to the formation of calcium sulfides and polysulfides that are detrimental to pavement longevity.

Additionally, when liquid sulfur, liquid asphalt, and aggregate are mixed simultaneously, even with other components, additional problems with the plasticization of the sulfur can occur. Specifically, when part of the liquid sulfur reacts with the aggregate before being completely plasticized by the asphalt, the non-plasticized sulfur bonds with the aggregate rather than completing its plasticization reaction. This non-plasticized sulfur works to weaken rather than strengthen the overall material strength.

Previous studies (Sereda and Beaudoin, 1980) indicated that with the best quality control for the binder, the hardness drops by a factor of approximately 2.5 because the two continuous phase systems are not attained as shown in Figure 8.18. Furthermore, even when the final sulfur-modified asphalt binder is successfully prepared, this process requires the handling of liquid sulfur on the site. The presence of liquid sulfur creates potential environmental and material handling concerns.

Therefore, there is a need to be able to provide a solid, pre-plasticized sulfur that can be readily mixed with the asphalt to effectively modify it. Such a

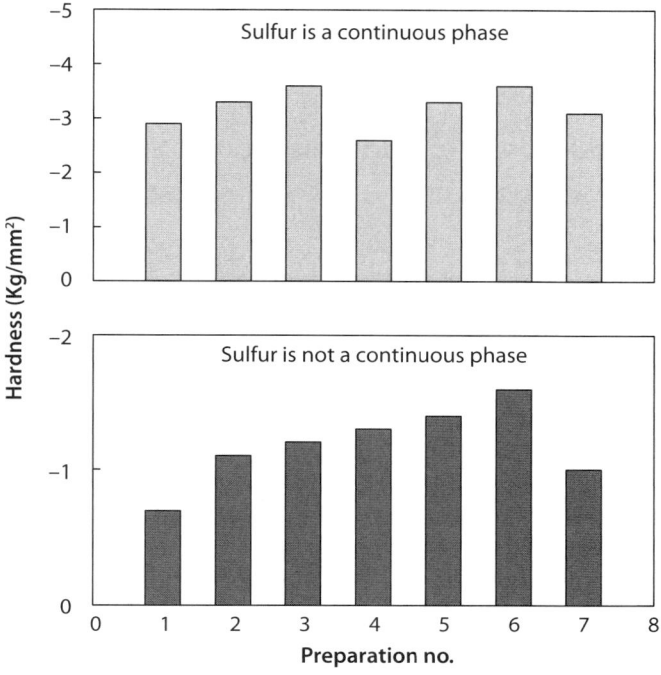

Figure 8.18. Hardness values for a number of preparations, where sulfur is a continuous and non-continuous phase (adapted from Sereda and Beaudoin, 1980).

pre-plasticized sulfur modifier can reduce the complexity of the binder preparation, eliminate the need for handling liquid sulfur in conjunction with liquid asphalt, and provide increased uniformity of binder without the concern that non-plasticized sulfur may weaken the paving material. Further, the ability to transport the solid, pre-plasticized sulfur-additive material that is in the form of a smaller-sized, non-sticky, non-flowing, and non-melting material facilitates the delivery of the pre-plasticized sulfur-additive materials to locations where it can be readily mixed with asphalt to create a material acceptable for the specific project specifications.

The plasticized sulfur material can be mixed with asphalt and a fine mineral constituent to create usable asphalt-based binder that will retain nonflow properties within a broad range of ambient temperatures. Plasticized sulfur usually has a lower melting point and a higher viscosity than elemental sulfur. Furthermore, plasticized sulfur requires a longer time to crystallize; the rate of crystallization of plasticized sulfur is slower than that of elemental sulfur. The plasticized sulfur material can be prepared independent of the asphalt into which it is added and can be shipped to locations where asphalt is available for preparation of modified asphalt-based paving binder. Additionally, it is ideal for remote locations where asphalt and, more particularly, quality asphalt is not readily available.

8.9.2 Chemicals Used for Plasticization of Sulfur

The plasticized sulfur composition is formed by mixing elemental sulfur and a plasticizing agent, such as:

a. Inorganic plasticizers include iron, arsenic, and phosphorus sulfides. Further, crushed limestone has been used for this purpose. However, the use of calcium-based materials leads to the formation of calcium sulfides and polysulfides that are detrimental to pavement longevity. This plasticized sulfur can be formed into solid particles and then added to asphalt to create a desired paving binder compositions.

b. Organic plasticizers include aliphatic polysulfides, aromatic polysulfides, styrene, alphamethylstyrene, cyclopentadiene, dicyclopentadiene, dioctylphthalate, acrylic acid, vinyl cyclohexene, heptane, epoxidized soybean oil, triglycerides, and tall, oil fatty acids. These chemicals have been used in an attempt to keep the sulfur homogeneously dispersed in asphalt.

Aliphatic polysulfides that will not form cross-linking are preferred as plasticizers. Thus, butadiene is not a preferred constituent to form the aliphatic polysulfide, because it may form cross-linking sulfur bonds. However, dicyclopentadiene is a preferred compound for forming the aplhatic polysulfide and is useful as a sulfur plasticizer. With molten sulfur, dicyclopentadiene forms an extremely satisfactory aliphatic polysulfide.

Aromatic polysulfides formed by reacting one mol of an aromatic carbocyclic or hetero-cyclic compound, substituted by at least one functional group of the class –OH or –NHR, in which R is H or lower alkyl with at least two moles of sulfur, are useful as sulfur plasticizers. Suitable aromatic compounds of this type include phenol, aniline, N–methyl aniline, 3-hydroxy thiophene, 4-hydroxy pyridine, p-aminophenol, hydroquinone, resorcinol, meta-cresol, thymol, 4,4'-dihydroxy biphenyl, 2,2-di(p-hydroxyphenol) propane, di(p-hydroxyphenyl) methane, p-phenylene diamine, and methylene dianiline. Phenol is an especially preferred compound to form the aromatic polysulfides.

The aromatic polysulfides are generally prepared by heating sulfur and the aromatic compound at a temperature in the range of 120 to 170°C for 1 to 2 hr, usually in the presence of a base catalyst such as sodium hydroxide. The polysulfide product made in this way has a mol ratio of an aromatic compound: sulfur of 1:2 to 1:10. Upon completion of the reaction, the caustic catalyst is neutralized with an acid such as phosphoric or sulfuric acid. Organic acids may also be used for this purpose.

Another type of aliphatic polysulfides useful as a plasticizer are the linear aliphatic polysulfides. Although they may be used alone as a sulfur plasticize, it is preferred to be used in combination with either: (a) dicyclopentadiene or (b) the aromatic polysulfides described earlier, especially with the phenol-sulfur adduct.

The aliphatic polysulfides may have branching indicated as:

$$C = C + S \rightarrow -S_x - C - C - S_x - \atop || \atop BB \qquad (8.11)$$

where x is an integer from 2 to 6 and B is H, alkyl, aryl, halogen, nitrile, ester or amide group. The chain with the sulfur is linear, but it can have a side group as indicated by B in Eq. 8.11. Also, this side group B may be aromatic. Thus, styrene can be used to form a phenyl-substituted linear aliphatic polysulfide. The preferred aliphatic polysulfides of this type are linear and nonbranched.

Unbranched linear aliphatic polysulfides include Thiokol LP-3 that contains an ether linkage and has the recurring unit:

$$-S_x CH_2 CH_2 OCH_2 CH_2 S_x \qquad (8.12)$$

where x has an average value of approximately 12. The ether constituent of this aliphatic polysulfide is relatively inert to reaction.

8.9.3 Plasticizing Agent Requirement

The paving binders should be readily made by incorporating additives into the asphalt. They would have the following characteristics:

a. Can be manufactured in forms that are nonsticky, nonmelting, and nonflowing within a wide range of ambient temperatures
b. Comes ready to use without the need of further reaction or modification
c. Should reduce the quality of asphalt in the final paving binders to reduce petroleum dependency and cost
d. Should not substantially incorporate constituents that, whether directly or when combined with other binder constituents, are known to affect quality and longevity of the pavement detrimentally
e. Desirable to manufacture finished, plasticized sulfur plus asphalt-based binder compositions in which the sulfur is a majority component and which can be readily shipped

8.9.4 Plasticization Perceptions

The quantity of plasticizer to be added varies with the nature of the plasticizer, but usually it is in the range 0.5 to 10 wt%, based on the final composition. Polyunsaturated olefins such as dicyclopentadiene are preferably used at the 1 to 5 wt% level. The upper limit on plasticizer concentration is determined by the viscosity of the resulting plasticized sulfur. Final viscosity must be such that the composition will flow at the desired temperature and will also be able to be mixed with the asphalt-coated aggregate.

The finished plasticized sulfur or final sulfur-rich paving binder composition products can subsequently be stored near the production site or at a remote site. It can be used alone or in combination with additional paving material at road sites, and it can be shipped to a hot-mix plant where the plasticized sulfur or final sulfur-rich paving binder composition is mixed with additional paving materials to manufacture asphalt pavements and surface-treating materials. Among the asphalt pavements, asphalt concrete is a high quality, thoroughly controlled, hot mixture of asphalt cement and is a well-graded, high quality aggregate that is thoroughly compacted into a uniform dense mass.

The plasticized sulfur can be mixed with various concentrations of an asphalt-based material during the mixing cycle with aggregate, sand, or other materials to create the desired product characteristics. Such a pre-plasticized sulfur modifier can reduce the complexity of binder preparation, eliminate the need for handling liquid sulfur in conjunction with liquid asphalt, and provide increased uniformity of the binder without the concern that non-plasticized sulfur may weaken the paving material. Further, the ability to transport the solid, pre-plasticized sulfur-additive material that is in the form of a smaller-sized, nonsticky, nonflowing, and nonmelting material facilitates the delivery of the pre-plasticized sulfur-additive materials to locations where it can be readily mixed with asphalt to produce a material acceptable for the specific project specifications.

8.9.5 Plasticized Mixing Conditions

To combine the plasticized sulfur with the asphalt material to create the desired binder or final product, both the plasticized sulfur and the asphalt must simply be liquefied and mixed with aggregate. This mixing and liquefying can be accomplished in almost any order of mixing. For example, it can be accomplished by combining the plasticized sulfur with the asphalt followed by the combination sulfur-asphalt binder with the aggregate. It can also be done by initially combining the asphalt and aggregate followed by the plasticized sulfur or a combination of all three elements at once.

The plasticized sulfur is added to the asphalt at not much more than a 50 wt% concentration. While it is generally recognized that a concentration of less than

20% plasticized sulfur provides little strength enhancement or modification to the asphalt, the plasticized sulfur can still be utilized at concentrations less than 20% to extend the amount of asphalt required. The strength of the plasticized sulfur or final sulfur-rich paving binder composition is already high upon solidification, reaching generally about 80% of the ultimate strength after a period of 24 hr after solidification.

8.9.6 Case Studies for Plasticization of Sulfur

Case 1

Ethyl cyclo-hexane dimmer-captan and a viscosity control agent such as acrylic acid and heating the mixture to between 149° and 160°C to form the plasticized sulfur were used (U.S. Patent 5304238; Marwil et al., 1994). The plasticized sulfur was then deodorized by distilling with the solvent-stripping agent to remove odor-causing materials such as unreacted mercaptans. The solid filler materials were then added to the deodorized plasticized sulfur, along with additional elemental sulfur. A thinning agent such as Santowax was also added to reduce the viscosity of the mixture. The mixture was then cooled to a temperature of 121 to 126°C and stirred vigorously to mix in the solid filler material.

Case 2

Sulfur was plasticized by the addition of carbon and further treated with amyl acetate to improve the handling and odor characteristics of the plasticized sulfur (U.S. Patent 6,863,724 B2; Bailey et al., 2005). This plasticized sulfur can be formed into solid particles and then added to asphalt to create the desired paving binder compositions. The solid plasticized sulfur has nonstick nonflow properties within a wide range of ambient temperatures.

Examples to this process, as reported by Bailey et al. are described below.

Example 1: Sulfur was heated and liquefied at a temperature of 140°C. The liquefied sulfur was treated with 0.25% amyl acetate and 0.5% carbon and within approximately 5 min the composition turned a shiny, dark gray color, indicating the completion of the plasticization reaction of the sulfur. The plasticized sulfur was then cast into slate approximately 0.63 cm thick. After cooling, the slate was broken up into pieces not bigger than the forms that would have their length and width approximately equal to their thickness. AC-20 asphalt cement, aggregate, and the plasticized sulfur were mixed with the overall composition of the asphalt cement containing approximately 2.7% AC-20 asphalt cement,

3.0% plasticized sulfur, and 94.3% aggregate, and the mixture was found to have stability of over 5400 lb and a flow of 12 at 50 blows.

Example 2: Plasticized sulfur was prepared as described in Example 1. AC-20 asphalt, aggregate, and the plasticized sulfur were mixed with the overall composition of the asphalt cement containing approximately 2.0% AC-20 asphalt, 2.0% plasticized sulfur, and 96% aggregate, and the mixture was found to have a stability of over 5800 lb and a flow of 12 at 50 blows.

Example 3: Plasticized sulfur was prepared as described in Example 1. Subsequently, 70% plasticized sulfur, 15% type F silica flour, and 15% AC-10 asphalt cement were mixed together for 3 min at approximately 140°C and then cast into slate approximately 0.63 cm thick. After cooling, the slate was broken up into pieces not bigger than the forms that would have their length and width approximately equal to their thickness. This sulfur-rich paving binder was then mixed with graded mineral aggregate in relative amounts of approximately 5% of sulfur-rich paving binder and 95% of aggregate, and the mixture was found to have a stability of over 5000 lb and a flow of 8 at 2 blows.

Example 4: Plasticized sulfur was prepared as described in Example 3. This sulfur-rich paving binder was mixed with graded mineral aggregate in relative amounts of 10% of sulfur-rich paving binder and 90% aggregate, and the mixture was found to have a stability of over 10,000 lb and a flow of 8 at 2 blows.

A detailed discussion regarding sulfur modification by various types of chemicals can be found in Chapter 5.

8.10 POTENTIAL ECOLOGICAL EFFECTS

The sulfur in sulfur-modified asphalt behaves like metals and organic compounds in the mixture. The Heritage Research Group (Asphalt Institute, 2003) used the toxicity characteristics and leaching procedures (TCLP) developed by the U.S. EPA to test fresh hot-mix asphalt for leachability of 40 metals and organic contaminants, including PAHs of the types found in asphalt as shown in Table 8.7. Of the 40 chemicals identified as hazardous by the EPA, only chrome was extracted from the asphalt by the TCLP. It had a concentration in the leachate of 0.1mg/l. Asphalt rarely contains elevated levels of chromium, therefore the chrome may have been leached from the slag aggregate in the asphalt pavement. Although the detection limit of PAHs was below 1 ppb, only naphthalene was detected in the leachate. It was present at a concentration of 0.25 ppb. Naphthalene is the least toxic of the PAHs in the petroleum products. None of the carcinogenic 4- through 6-ring PAHs were detected. Kriech et al. (2000) obtained similar results for leaching of PAHs from paving and roofing asphalt. Although the asphalt contained detectable concentrations of several PAHs (Table 8.7), only traces of low molecular weight PAHs were leached into acidified water. It is likely that sulfur-modified asphalt behaves like conventional hot-mix asphalt and that PAHs and the sulfur in it are immobile.

Lindgren (1996) reported that small amounts of several metals are released to the environment during the wearing of asphalt pavements. Most of the metals in the asphalt pavement were associated with aggregates. Several metals were released under acidic conditions of pH4. Thus, if the sulfur in sulfur-modified asphalt pavement is accessible to sulfur-oxidizing bacteria, some sulfuric acid could be generated mobilizing metals from the aggregates in the asphalt pavement or from adjacent soils. The sulfur in the asphalt pavement probably has a limited accessibility to bacteria and little metal is likely to be mobilized from it under normal environmental conditions.

As with Portland and sulfur cement, the main hazards to the environment and people come during the manufacture and application of sulfur-modified asphalt. Workers applying hot-mix asphalt for road construction are exposed to vapors that may contain some carcinogenic PAHs at levels above normal (Watts et al., 1998). Exposures to high molecular-weight vapor or particles adsorbed, respirable PAHs were quite low for workers applying conventional asphalt and crumb rubber asphalt.

There probably would be no difference in exposure to harmful fumes during the application of a sulfur-modified asphalt pavement. Worch (1979) reported that there was little or no evolution of sulfur containing gases when sulfur-asphalt materials were handled at the normal temperature of 127 to 150°C. However, if the temperature of the sulfur-asphalt binder is allowed to rise above 155°C, hydrogen sulfide and sulfur toxic gases (both considered highly toxic) may be emitted. Emission monitoring at mixing plant and construction sites during manufacture of several sulfur-extended asphalt pavements indicated that sulfur gas emissions can easily be controlled within acceptable limits (Worch, 1979).

Although road paving asphalt contains low concentrations of a few known carcinogenic PAHs, there is no evidence that asphalt bitumen alone is carcinogenic to mammals, including humans (IARC, 1987). Some extracts of bitumen are carcinogenic. Hot asphalt can cause burns and irritation to the skin and eyes. Continued exposure may cause dermatitis and acne-like lesions.

Table 8.7. Concentration ranges of PAH in six samples of paving asphalt and four samples of roofing asphalt (Kreich et al., 2002)

Compound	Roofing asphalt (mg/kg)	Paving asphalt (mg/kg)
Naphthalene	<1	0.68 – <1
Acenaphthylene	<1	<1
Acenaphthene	<1	<1
Fluorene	0.56–1.2	0.79–1.80
Phenanthrene	0.74–1.9	<1–9.9
Anthracene	<1	<1–1.5
Fluoranthene	<1	0.75-<1
Pyrene	0.86 – <1	0.64-9.1
Benz(a)anthracene	0.76 – <1	0.96-2.0
Chrysene	1.80–9.5	<1–12
5-Methylchrysene	0.72–3.5	0.75–9.2
1-Nitropyrene	<1	<1
Benzo(b)fluoranthene	0.62–1	0.74–1.5
Benzo(j)fluoranthene	<1	<1
Benzo(k)fluoranthene	<1	<1
7,12-Dimethylbenz(a)anthracene	<1	<1
Benzo(a)pyrene	<1	0.54–3.7
3-Methylcholanthrene	<1	<1–7.8
Dibenz(a,h)acridine	<1	<1
Dibenz(a,j)acridine	<1	<1
Dibenz(a,h)anthracene	0.59 – <1	<1
7H- Dibenz(c,g)carbazole	<1	<1
Benzo(g,h,i)perylene	0.56–1.4	<1–2.4
Dibenzo(a,j)pyrene	<1	<1
Dibenzo(a,e)pyrene	<1	<1
Dibenzo(a,h)pyrene	<1	<1
Dibenzo(a,e)fluoranthene	<1	<1
Benzo(r,s,t)pentaphene	<1	<1
Total detected PAHs	4.0–23	1.9–66

In summary, the toxicological properties of sulfur-modified asphalt and conventional hot mix asphalt are similar and are likely to exhibit similar environmental effects. Sulfur is not likely to increase the environmental or human health hazard of road-paving asphalt.

8.11 SUMMARY AND CONCLUDING REMARKS

The potential economic benefits of the utilization of sulfur-substituted asphalts in paving mixes consist of: (1) saving of materials and (2) a considerable improvement in the mechanical properties of paving mixes and, consequently, a substantially better performance of the pavement.

A saving of 20% of asphalt can be reached by incorporating sulfur into an asphalt paving mix. The partial replacement of asphalt with sulfur is economical only in the case in which the price of sulfur is, at the very least, half that of asphalt, because of the greater density of sulfur as well as the additional technological costs. There may also be savings in aggregates, because the sulfur-asphalt binder allows conventional aggregates to be replaced by inexpensive, low-quality aggregates such as windblown sand or aggregates with a poor gradation.

The use of a sulfur-asphalt binder instead of normal asphalt in paving mixes might bring significant economic benefits on account of the properties that are superior to those of conventional asphalt mixes. The improved performance characteristics of sulfur-extended asphalt materials offers a promise of more durable pavements and enable thinner road surfaces to be built, which would make it possible to conserve both asphalt and aggregates. It has been reported that the thickness of the layer can be reduced by as much as 20%.

From the viewpoint of the pavement performance many advantages result from the addition of sulfur:

a. Sulfur-substituted asphalt mixes exhibit significantly higher fatigue lives than comparable conventional mixes
b. The behavior of sulfur-containing pavements is more elastic compared with conventional pavements
c. The addition of sulfur makes it possible for softer asphalts to be used to reduce low-temperature cracking without the high–temperature deformation that occurs when virgin asphalt is used
d. The addition of sulfur with 30 to 40 wt% in paving mixes leads to the reduction in deformation approximately by half
e. Paving materials based on sulfur-asphalt and binder exhibit a better resistance to water compared with conventional mixes

The evaluation of the economic aspects of using sulfur-extended asphalts involves also the saving of energy. The addition of sulfur reduces the binder viscosity and this significantly improves the workability of the mix. Therefore, the hot-mix paving materials can be prepared and compacted at lower temperatures, which results in lower energy consumption (30% saving of fuels).

Currently, the world prices of asphalt and sulfur are such that the use of sulfur-extended asphalts in paving mixes is uneconomical. However, the increasing amount of sulfur as a byproduct may change these proportions. With a stricter control of sulfur pollution sources, sulfur is expected to become more plentiful.

Even though the use of pure sulfur in asphalt paving mixes is uneconomical, the utilization of industrial wastes with appropriately high sulfur content in sulfur asphalt mixes should be taken into consideration. It has been found that wastes from coking plants (obtained by desulfurizing coke oven gas), with approximately 30% sulfur, may be applied as an asphalt modifier-producing material that performs well in paving mixes. Another byproduct that proved to be useful in asphalt paving mixes was the waste material that remained after flotation of sulfur ore. It contains approximately 40% sulfur.

REFERENCES

Aarsleff, L. (1989) "ERMCO 89: Acid-proof, salt resistant sulphur concrete," 4K Beton A/S, Copenhagen.

Aarsleff, L. (1994) "Corrosion resistant sulphur concrete pipe," General Information Leif Aarsleff Consulting Co. DK-9230 Svenstrup.

Abdel-Jawad, Y. and Al-Qudah, M. (1994) "The combined effect of water and temperature on the strength of sulfur concrete," *Cement and Concrete Research*, Vol. 24, p. 165-175.

Abraham, H. (1961) *Asphalt and allied substances, Vol. II, 6th ed.* Van Nostrand, New York, p. 178.

ACI Committee 548. (1988) "Guide for mixing and placing sulfur concrete in construction," *ACI Materials Journal*, American Concrete Institute.

ACI Committee 548. (1993) *Guide for mixing and placing sulfur concrete in construction (ACI 548.2R-93),* American Concrete Institute, Farmington Hills, MI.

Adams, J.W. and Kalb, P.D. (2001) "Sulfur polymer stabilization/solidification (SPSS) treatability of Los Alamos National Laboratory mercury waste," *U.S. Dept. of Energy, Washington, D.C. Govt. Reports Announcements & Index, Issue 18.* Available from NTIS/de2002-23680.

Ahme, K. and Ringenson, O.F. (1994) *A Little Book on Market Investigations*, published by the Swedish Association for district heating (in Swedish).

Alama, K. and Wojdanowicz, G., (1981) *Właściwości lepiszcz siarkowo-asfaltowych*, Prace Instytutu Badawczego Dróg I Mostów (Warsaw), 3:42.

Allan, G.G., Neogi, A.N. (1970) *J. Appl. Polym. Sci.*, Vol. 14, p. 999-1005.

Al-Tayyib, A.J. and Khan, M.S. (1988) "Evaluation of corrosion resistance of reinforcing steel in sulfur concrete," *International Journal for Housing Science and Its Applications,* Vol. 12, Issue 4.

Anderberg Y. and Petterson, O. (1992) *Fire safety design of concrete structures Report T13: 1992, Part 1,* The Swedish Council for Building Research, Stockholm, SE (in Swedish).

Anderson, R.M., and Bukowski, J.R. (1997) "The utilization of torture tests as an addendum to the Superpave mix design system: a case study," Paper presented at the 1997 Transportation Research Board Annual Meeting, Washington, D.C. Available at http://www.asphaltinstitute.org/airesear/trb-mo.html.

Annual Report of the Sulphur Institute (1992) "Sulphur," Washington, D.C.

Anonymous (1973) *Sulphur Inst. J.,* Vol. 9, Nos. 3 & 4, p. 13.

Anonymous (1975) *Sulphur Inst. J.,* Vol. 11, Nos. 3 & 4, p. 2.

Asphalt Institute. (2003) *Asphalt use in water environments,* Asphalt Institute, Lexington, KY. Available at http://www.asphaltinstitute.org/environ-is-186.html.

Association of Swedish Chemical Industries. "Kemikontoret (Sveriges Kermiska Industri Kontor)" 1996/99 (www.chemind.se/).

ASTM C1159-98. (2003) Standard specification for sulfur polymer cement and sulfur modifier for use in chemical-resistant, rigid sulfur concrete.

ASTM C128-97. (1997) "Specific gravity and absorption of fine aggregate."

Bacon R. F. and Fanelli, R. (1943) "The viscosity of liquid sulfur," *J. Am. Chem. Soc.* Vol. 65, p. 639

Bacon, R.F. and Bencowitz, I. (1939), U.S. Patent No. 2,182,837.

Bacon, R.F. and Davis, H.S. (1921) "Recent advances in the American sulfur industry," *Chemical and Metallurgical Engineering,* Vol. 24, No. 2, p. 65-72.

Bacon, R.F. and Fanelli, J. (1943) "The viscosity of sulfur," *J. Am. Chem. Soc.,* 65, p. 639-648.

Bahrami, N., Mohtadi, M., and Mohammad, N. (2008) "Preparation of sulfur mortar from modified mulfur," *Iran J. Chem. Chem. Eng.* Vol. 27, No.1, p.123.

Bailey, W.R. and Pugh, N. D. (2005) "Sulfur additives for paving binders and manufacturing methods U.S. Patent No. 6,863,724 B2.

Baker, H.J. and Bro., Inc. (2001) experts in feed, fertilizer, and sulfur.

Ballone P. and Jones R.O. (2003) "Density functional and Monte Carlo studies of sulfur. II. Equilibrium polymerization of the liquid phase," *Journal of Chemical Physics*, 119, 8704.

Beall, P.W. and Neff, J.M. (2005) "Potential non-traditional uses of by-product E & P produced sulfur in Kazakhstan," SPE 94177. This paper was prepared for presentation at the 200 5 SPE/EPA/DOE Exploration and Production Environmental Conference held in Galveston, Texas, 7-9 March 2005.

Beatty, T.L., Dunn, K., Harrigan, E.T., Stuart K., and Weber H. (1987) "Field evaluation of sulfur-extended asphalt pavements," TRR 1115, TRB, National Research Council, Washington, D.C. 161-170.

Beaudoin, J.J. and Sereda, P.J. (1974) "The freeze-thaw durability of sulfur concrete," Building Research Note, Division of Building Research, National Research Council, Ottawa, 53.

Beaudoin, J.J. and Feldmant, R.F. (1984) "Durability of porous systems impregnated with dicyclopentadiene-modified sulfur," *The International Journal of Cement Composites and Lightweight Concrete*, Vol. 6, No. 1, p. 13-17.

Becker, Y., Médez, M.P., and Rodriquez, Y. (2001) "Polymer modified asphalt," *Vision Technologia*, Vol. 9, p. 39-50.

Bencowitz, I. and Boe. (1938) "Effect of sulphur upon some of the properties of asphalts," *Proceedings of the American Society of Testing Materials*, 39 (II), p. 539.

Berry, E.E., Soles J.A., and Malhotra, V.M. (1977) "Leaching of sulfur and calcium from sulfur infiltrated concrete by alkaline and neutral aqueous media," *Cement and Concrete Research*, Vol. 7, No. 2, p. 185-189.

Betong—handbok material, utgava 2. (1994) AB Svensk Byggtjanst och Cementa AB (in Swedish), *Concrete Manual, Materials*, 2nd ed.

Biermann, C.H.R., Winter, R., Benmore, C.H.R., and Egelstaff, P.A. (1998) *J. Non- Crystalline Solids*, Vols. 232-234, p. 309.

Blight, L., Currell, B.R., Nash, B.J., Scott, R.A.M., and Stillo, C. (1978) "Preparation and properties of modified sulfur systems," In New Uses of Sulfur–II, *Advances in Chemistry Series*, No. 165 (American Chemical Society), p.13-30.

Bocca, P.L., Petrossi, U., and Pinoni, V. (1973). Chim. Ind. (Paris), Vol. 55, p. 427.

Bodenlos, A.J. and Nelson, C.P. (1979) "Sulfur: economic geology," Vol. 74, No. 2, p. 459-461.

Boduszynski, M.M., McKay, J.F., and Latham, D.R. (1981) "Composition of heavy ends of a Russian petroleum, *J. Amer. Chem. Soc.* Vol. 26, p. 865-880.

Boegly, W.J., Francis, C.W., and Watson, J.S. (1986) "Characterization and disposal of by-product elemental sulfur," *U.S. Dept. of Energy, Washington, D.C. Govt. Reports Announcements & Index, Issue 22*. Available from NTIS/ de86012454.

Bomrungwad, C. (1995) "A study of the mixture of asphalt concrete and Sulfur," Thesis, University of Thailand, Bangkok.

Bordoloi, B.K. and Pearce, E.M. (1978) "Plastic sulfur stabilization by copolymerization of sulfur with di-cyclo-penta-diene," *Advances in Chemistry Series*, No. 165, American Chemical Society, Washington, D.C., p. 31-53.

Böttcher, M.E., Thamdrup, B., and Vennemann, T.W. (2000) "Stable isotope fractionation during anaerobic bacterial disproportionation of elemental sulfur: the influence of Fe(II), Fe(III), and Mn(IV) compounds," *J. Conference Abstracts*, Vol. 5, p. 231.

Bourne, D. J. (ed.) (1978) "New Uses of Sulfur-II," *Adv. Chem. Ser.* 165, ACS, Washington D.C..

Bradnovic, M. (1989) "The use of sulfur to improve aggregates: The results of research in North America," *Quarry Management*, Vol. 16, No. 1, p. 37.

Britton, T.H.S. (1952) *Hydrogen Ions, 4th ed.* Chapman and Hall, 313.

Brown T.L., Le May H.E., and Bursten B.E. (2000) *Chemistry the Central Science, 8th ed.*, Prentice Hall, Upper Saddle River, NJ.

Brown, T.L. and Lemay, H.E., Jr., (1981) *Chemistry: The central science*, Prentice Hall, Englewood Cliffs, NJ, p. 655.

Building Research and Information (1999) *The International Journal of Research, Development & Innovation, Special Issue (A global Agenda for sustainable Construction)*, Editor: Richard Lorch. Guest Editor: Christer Sjostrom, Volume 27, Number 6. Nov-Dec.

Burgess, R.A. and Deme, J. (1975) "Sulfur in asphalt-paving mixes," *New Uses of Sulphur. Adv. Chem. Ser.*, Vol. 140, p. 85.

Canadian Patent No. 755,999 (1967).

CBI. (1986) "Cement och Betong Institute" (The Swedish Cement and Concrete Research Institute), 1986 Report "Rapport Nr. 8628 G. klingstedt/S.S. Stockholm, SE, (in Swedish).

Celard, B. (1978) "Sulfur addition to asphalt paving mixes," *Eurobitume Seminar*, London, p. 318.

Celard, B. (1980) Schwefel-Asphalt, *Das Stationäre Mischwerk*, Vol. 3, p. 23.

Chempruf Concrete. (1988) *Chempruf manual*, National Chempruf Concrete Inc., Clarksville, TN.

Chempruf Concrete. (1993) Journal (Brochure).

Chempruf Sulphur Acid-Proof, Salt Resistant Concrete, 4K A/S. Copenhagen, available through The Sulphur Institute, Washington, D.C., <http://www.chemproofconcrete.com

CIB. (1999) "CIB Report Publication 237," Agenda 21 on Sustainable Construction, July 1999.

Colombo, P., Kalb, P.D., and Heiser, J.H. (1997) "Process for the encapsulation and stabilization of radioactive hazardous and mixed wastes," U.S. Patent No. 5,678,234, Oct. 14, 1997.

CONCAWE. (1985) "Health aspects of petroleum fuels: Potential hazards and precautions for individual classes of fuels," Report No. 85/81, The Hague, The Netherlands.

Concise Encyclopedia: *Chemistry.* 2nd ed.(1994) Translated by Mary Eagleson.

Corbett, L.C. (1970) "Relationship between composition and physical properties of asphalt," *Proc. Assoc. Asphalt Paving Technol.*, Vol. 39, p. 481-491.

Cotton, F.A.G. Wilkinson, C.A. Murillo, and M. Bochmann. (1999) *Advanced inorganic chemistry,* 6th ed. New York: Wiley.

Courval, G.J. and Akili, W. (1982) "Sulfur asphalt binder properties by the sliding plate rheometer," *Asphalt Paving Technol.*, Vol. 51, p. 222.

COWIconsult & 4K A/S. (1990) "Tegninnger for svovlbetontank, type 2 $FeSO_4$ - oplag (in Danish)." Drawings of a vessel for sulfuric Acid.

Crick, S.M. and Whitemore, D.W. (1996) "Sulfur concrete application in commercial scale project: Fields considerations," 1996 Fall Convention, American Concrete Institute, New Orleans, LA.

Crick, S.M. and Whitmore, D.W. (1998) "Using sulfur concrete on a commercial scale," *Concrete International,* Vol. 20, February, 1998, p. 83-86.

Crow, L.J. and Bates, R. C. (1970) "Strength of sulfur-basalt concretes," Bureau of Mines, Report No. RI 7349, U.S. Bureau of Mines, Washington, D.C., 21.

Cruzan, G., Lowe, L.K., Cox, G.E., Meeks, J.R., Mackerer, C.R., Craig, P.H., Singer, E.J., and Mehlman, M.A. (1986) "Systemic toxicity from sub-chronic dermal exposure, chemical characterization, and dermal penetration of catalytically cracked clarified slurry oil," *Toxicol. Indust. Health.*, Vol. 2, p. 429-444.

Cultural Heritage, No. 40. (2000) "Sustained care of the cultural heritage against pollution."

Cunningham, C. (2000) "Oil price underpins gulf brimstone output," *Sulphur*, No. 267, March-April, p. 13-18.

Currell B. R. Williams, A. J., Mooney, A. J., and Nash, B. (1974) "Plasticization of Sulfur," in J. R. West (ed.), Proceedings of symposium "New Uses of Sulfur," Los Angeles, April 1974, *Advances in Chemistry Series*, No. 140, Am. Chem. Soc., Washington, D.C. 1975, p. 1-17.

Currell, B.R. (1976) "The importance of using additives in the development of new applications for sulfur," *Symposium on new users, for sulfur and pyrites*, Madrid, p. 103-105.

Czaban, J. and Kobus, J. (2000) "Oxidation of elemental sulfur by bacteria and fungi in soil," *Acta Microbiol. Polonica.*, Vol. 49, p. 35-147.

Czarnecki, B. and Gillott, J.E. (1989) "Stress-strain behavior of sulphur concrete made with different aggregates and admixtures," *Quarterly Journal of Engineering Geology*, London, Vol. 22, p. 195-206.

Czarnecki, B. and Gillott, J.E. (1989) "Concrete, and aggregates," *CCAGDP*, Vol. 11, No. 2, p. 109-118.

Czarnecki, B. and Gillott, J.E. (1990) "Effect of different admixtures on the durability of sulfur concrete made with different aggregates," *Engineering Geology*, Vol. 28, p. 105-118.

Dadze, T.P. and Sorokin, V.I. (1993) "Experimental measurements of concentrations of H_2S, HSO_4-, SO_2 (sol), $H_2S_2O_3$, S0 (sol), and Stotal in the water phase of the system S—H2O at elevated temperatures, Geochem. Int., p. 30, 36.

Dale, J.M. and Ludwig, A.C. (1965) "Mechanical properties of sulfur," in Meyer, B., ed., Elemental sulfur, chemistry and physics, Interscience Publishers, New York, p. 161-178.

Dale, J.M. and Ludwig, A.C. (1966) "Feasibility study for using sulfur-aggregate mixtures as a structural material. Technical Report No. AFWL-TR-66-57, Southwest Research Institute, San Antonio, TX, p. 40.

Dale, J.M. and Ludwig, A.C. (1967) "Sulphur-aggregate concrete," *Civil Engineering*, Vol. 37, p. 66-68.

Dale, J.M. and Ludwig, A.C. (1968) "Advanced studies of sulfur aggregate mixtures as a structural material," Technical Report No. AFWL-TR-68-21, Southwest Research Institute, San Antonio, TX, p. 68.

Darnell, G.R. (1991) "Sulphur polymer cement, a New stabilization agent for mixed and low-level radioactive waste," In *Proceedings of the First International Symposium on Mixed Waste*, Baltimore, MD, August 26-29, 1991, A.A. Moghissi and G.A. Benda, eds., Univ. of Maryland, Baltimore.

Darnell, G.R., Aldrich, W.C., and Logan, J.A. (1992) "Full-scale tests of sulfur polymer cement and non-radioactive waste in heated and unheated prototypical containers," EGG-W-10109, Idaho Natl. Engineering Lab., Idaho Falls, ID.

Davani, B., Sanders, W., and Jungclaus, G. (1989) "Residual fuel oil as potential source of groundwater contamination," in *Proceedings U.S. EPA Symposium on Waste Testing and Quality Assurance*, July 24-27, 1989, Washington, D.C. American Chemical Society, Washington, DC., p. I-259-I-273.

Day, D.T. (1885) "Sulphur in mineral resources of the United States," *U.S. Geological Survey Professional Paper 820*, p. 864-905.

Deanna Dunlavy (1998) http://www.thecatalyst.org/experiments/Dunlavy/Dunlavy.html.

Debenedetti P.G., and Stanley, H.E. (2003) "Super-cooled and glassy water," *Physics Today* 56 (6), p. 40-46.

DeFilippis, P., Giavarini, C., and Santarelli, M.L. (1998) "Sulphur-extended asphalt: reaction kinetics of H2S evolution," *Fuel*, Vol. 77, No.5, p. 459-463.

Deme, I. (1974) Assoc. Asphalt Paving Technology, No. 43.

Deme, I. (1978) "Sulfur as asphalt diluent and a mix filler," In *New Uses of Sulfur II. Adv. Chem. Ser.*, No. 165, p. 172.

Design and Construction Manual (1986) "Corrosion-resistant sulphur concrete," Revised 1986, The Sulphur Institute (TSI), Washington, D.C.

Dharmananda, S. (2008) "Differentiating sulfur compounds: sulfa drugs, glucosamine sulfate, sulfur, and sulfating agents," Institute for Traditional Medicine, (http://www.itmonline.org/arts/sulfa.html), Jan 2008.

Diehl, L. (1976) "New uses for sulfur and pyrites," *Madrid Symposium*: The Sulphur Institute.

Dohner, T.D. (1996) National Chempruf Representative. York, PA, *Durability Problems of Sulfur Concrete*.

Donohue, J. (1965) "The structures of the allotropes of solid sulfur, in Meyer, B., ed., *Elemental sulfur, chemistry and physics*, Interscience Publishers, New York, p. 13-43.

Dorrence, S.M., Barbour, F.A., and Petersen, J.C. (1974) "Direct evidence of ketones in oxidized asphalts," *Anal. Chem.*, 1974, Vol. 46, No. 14, pp 2242–2244.

Dudowicz, J., Freed, K.F., and Douglas, J.F. (2000) "Lattice model of living polymerization. III: evidence for particle clustering from phase separation Properties and rounding of the dynamical clustering transition," *J. Chem. Phys.*, Vol. 113, p. 434.

Dudowicz, J., Freed, K.F., and Douglas, J.F. (1999) "Lattice model of living polymerization. I: basic thermodynamic properties," *J. Chem. Phys.*, Vol. 111, p. 7116.

Duecker, W.W. (1934) "Admixtures improve properties of sulfur cements," *Chemical and Metallurgical Engineering*, Vol. 41, No. 11, p. 583-586.

Dunnette, D.A., Chynoweth, D.P., and Mancy, K.H. (1985) "The source of hydrogen sulfide in anoxic sediment," *Water Resources*, Vol. 19, p. 875-884.

Eary, L.E., Dhanpat R., Mattiod, S.V., Ainsworth, C.C. (1990) "Geochemical factors controlling the mobilization of inorganic constituents from fossil fuel combustion residues," II. Review of the minor elements, *Journal of Environmental Quality*, 19:202.

Easterbrook et al. (1971) "A discussion of some polymerization parameters in the synthesis of EPDM elastomers," XXII IUPAC, *Macro Molecular Preprint*, Vol. 11, p. 712.

Ecker, A. and Steidl, H. (1986) Bundesministerium fur wirtschaftiche Angelenheiten-Strassenforschung (678). Schwefelbton im Strassenbau (in German), "Sulfur Concrete in Road Constructions."

Ekblad, J. (1992) "Forspanda spaltgolvselement as swavelbetonggjutteknik, amercing och egenskaper" Examenarbete 77 Swedish University of Agricultural Science. Department of farm buildings. Division of Research and Education, Uppsala (in Swedish).

El Sarafy, S. (1991) *The Environment as Capital*, Ecological Economics: the Science and Management of Sustainability, Columbia University Press, New York: p. 168-175.

Fairbrother, F., et al., (1955) J. Polym. Sci., Vol. 14, p. 4599.

Feher F. and Hellwig, E. (1958) Z. Anorg. *Allg. Chem.*, Vol. 294, p. 63.

Ferenbaugh, R.W., Gladney, E.S., Soholt, L.F., Lyall, K.A., Wallwork-Barber, M.K., and Hersman, L.E. (1992) "Environnemental interactions of Sulflex pavement," *Environ. Pollut*, Vol. 76, p. 141-145.

FHWA, U.S. Department of Transportation (1990) *Performance Evaluation of Sulfur-extended Asphalt Pavements, Laboratory evaluation*, Publication No. FHWA-RD-90-l 10, Washington, D.C.

Field, C.W. (1972) "Sulfur: element and geochemistry," In: R.W. Fairbridge, ed., *The Encyclopedia of Geochemistry and Environmental Sciences*. Vol. IVA. Van Nostrand Reinhold Co., New York, p. 1142-1147.

Fischer, H. (1984) "Sulphur," in Sander, U.H.F., Fischer, H., Rothe, U., and Kola, R., *Sulphur, sulphur dioxide and sulphuric acid*, London. The British Sulphur Corp. Ltd., p. 415.

Fisher, W.B. and Crescentini, L. (1992) "Caprolactam," In Kroschwitz, J.I., ed., Kirk-Othmer *Encyclopedia of Chemical Technology*, 4th ed., New York, John Wiley & Sons, Vol. 4, p. 827-839.

Funke, R.H. Jr. and McBee, W.C. (1982) "An industrial application of sulfur concretes," *ASC Symposium Series No. 183*, American Chemical Society, Washington, D.C., p. 195-208.

Gabel, D. (1993) *Introductory science skills, Waveland*, Gassy Reactions, Bayer College of Medicine, Houston, TX, 1993, p. 3.10-3.13.

Gagnon, C., Mucci, A., and Pelletier, É. (1996) "Vertical distribution of dissolved sulfur species in coastal marine sediments," *Mar. Chem*, Vol. 52, p. 195-209.

Gallaway, B.M. and Saylak, D. (1974) Paper presented at the *167th National Meeting of the American Chemical Society, Sulfur Use Symposium*, Los Angeles.

Gamble, B. R., Gillott, J.E., Jordaan, I.J., Loov, R.E., and Word, M. A. (1975) "Civil engineering applications of sulfur based materials," *Adv. Chem. Ser. No. 140*, p. 154-166.

Gannon, C.R., Wombles, R.H., Hettinger, W.P., and Watkins, W.D. (1983) "New concepts and discoveries related to the strength of plasticized sulfur," STP 807, ASTM, Philadelphia, PA, p. 84-101.

Garrigues, G. and Vincent, P. (1975) "Sulfur/asphalt binders for road construction," *New Uses of Sulfur. Adv. Chem. Ser., No. 140*, p. 130.

Gaw, W.J. (1976) (with R.A. Burgess, F.D. Young, and H.J. Fromm) "A laboratory and field evaluation of air-blown, low-viscosity waxy asphalts from Western Canada," *Crudes*, Vol. 21, p. 193-224.

Gawel, I. (1989) "Modifikation der bitumen durch Schwefelabfallprodukte der Kokerei und Gasindustrie Strassen," *Tiefbau*, Vol. 43, p. 20.

Gawel, I. (1994) "The selection of additives reducing the evolution of gaseous effluents during the preparation of sulphur-bitumen mixes," *Int. Conf. Catalysis and Adsorption in Environmental Protection*, Szklarska Poreba, PL.

Gawel, I. (2000) "Sulphur modified asphalts," *Asphaltenes and Asphalts*, Vol. 2, *Development in Petroleum Science*, 40 B, Chapter 19, p. 515-535.

Gemelli, E., Cruz, A.A.F., and Camargo, N.H.A. (2004) "A study of the application of residue from burned biomass in mortars," *Materials Research*, Vol. 7, No. 4, p. 545-556.

Georgia Gulf Sulfur Corp. (2000) Products, web site at http://www.georgiagulfsulfur.com/products.htm.

German Patent 1,295,463 (1970).

Gillott, J.E., Jordaan, I.J., Loov, R.E. and Shrive, N.G. (1980) "Sulfur concretes, mortars and the like," U.S. Patent No. 4,188,230.

Gillott, J.E., Jordaan, I.J., Loov, R.E., and Shrive, N.G. (1983) *Durability of building materials*, 1, p. 255-273.

Goode, J.F. and Owings, E.P. (1961) "A Laboratory-field study of hot asphaltic concrete wearing course mixtures," *Public Roads*, Vol. 31, No. 11.

Goodland and Daly (1995) "http://maven.gtri.gatech.edu/sfi/resources/pdf/TR/TR018.PDF.

Goodland, R. (1992) "The case that the World has reached limits," In *Population, Technology, and Lifestyle, The Transition to Sustainability*, ed. R. Goodland, et al., Washington, D.C., Island Press, p. 3-22.

Goodland, R. (1994) "Environmental sustainability: imperative for peace," in *Environmental Poverty. Confict*, eds. N. Graeger and D. Smith, p. 15-46.

Goodland, R. and Daly H. (1996) "Environmental sustainability: universal and non-negotiable," *Ecological Applications*, Vol. 6, p. 1002-1017.

Greer, S. C. (1986) "The dielectric constant of liquid sulfur," *J. Chem. Phys.*, Vol. 84, p. 6984.

Greer, S.C. (1998) "The physical chemistry of equilibrium polymerization," *J. Phys. Chem.* B 102, p. 5413.

Gregor R. and Hackl A. (1978) "A new approach to sulfur concrete," in D.J. Bourne, *New Uses of Sulfur-II: Advances in Chemistry Series 165*, American Chemical Society, Washington, D.C., p. 54–78.

Grun, R. (1937) *Beton ABC*, p. 120.

Hagedorn, C. (2003) *Environmental microbiology*, Virginia Polytechnic Institute and State University, Blacksburg, VA. Available at: http://soils1cses.vt.edu/ch/biol_4684/Cycles/Socidat.html.

Hamilton, W.A. (1998) "Bioenergetics of sulfate-reducing bacteria in relation to their environmental impact," *Biodegradation*, Vol. 9, p. 201-212.

Hammond, R., Deme, I., and McManus, D. (1971) "The use of sand sulfur asphalt mixes for road base and surface applications," *Proc. Can. Tech. Asphalt Assoc.*, Vol. 16, p. 27-52.

Hammons, M.I., Smith, D.M., Wilson, D.E., and Reece, C.S. (1993) "Investigation of modified sulfur concrete as a structural material," *Technical Report CPAR-SL-93-1*, final report to the U.S. Army Corps of Engineers, Washington, D.C. p. 37.

Harrington, S., Zhang, R., Poole, P. H., Sciotino, F., and Stanley, H. E. (1997) *Phys. Rev. Lett.*, Vol. 78, p. 2409.

Hellers, B.G. and Makenya, A.R. (1996) "Sulfur concrete," Hand out paper, Royal Institute of Technology, Stockholm, *Proceeding of the 7DBMC (7th International Conference on Durability of Building Materials and Components)*.

Higmell, E. T., et al. (1972) *Assoc. Asphalt Paving Technol.*, Vol. 41.

Hill, R.C, Bergman, J.G., and Bowen, P.A. (1994) "A framework for the attainment of sustainable construction," In *Proceedings of the First International Conference on Sustainable Construction*, Kibert, C.J., ed. Tampa, FL, November 6-9, CIB TG 16.

Hossein, M., Mohamed, A.M.O., and Hassani, F.P. (1999) "Ettringite Formation in Lime-Remediated Mine Tailings: II. Experimental Results," *CIM Bulletin*, Vol. 92, No. 1029, p. 75-80.

Howard, P.H., ed. (1991) *Handbook of environmental fate and exposure data for organic chemicals, Vol. III: Pesticides*, Lewis Publishers, Chelsea, MI.

Hummel, A. (1960) *Beton ABC*, Vol. 60, p. 12.

International Agency for Research on Cancer, (1987) *Supplement 7, Overall evaluations of carcinogenicity: an updating of IARC monographs*, Volumes 1-42. IARC, Lyon, FR.

Isacsson, U. and Lu, X. (1995) "Testing and appraisal of polymer modified road bitumen-state of the art," *Material and Structures*, Vol. 28.

Ito, K., Moynihan C.T., and Angell, C.A., (1999) "Thermodynamic determination of fragility in liquids and a fragile-to-strong liquid transition in water," *Nature*, Vol. 398, p. 492.

Jander, G. and Spandau, H. (1940) "Short textbook on inorganic chemistry" (in German).

Jones, D.R. (1992) "An asphalt primer: Understanding how the origin and composition of paving grade asphalt cements affect their performance," *SHRP Asphalt Research Program, Technical Memorandum No. 4*, The University of Texas, Austin.

Jong, B. W. & U.S. Bureau of Mines. (1985) *Fiber reinforcement of sulfur concrete to enhance flexural properties [microform]*, B.W. Jong et al., U.S. Dept. of the Interior, Bureau of Mines, Pittsburgh, PA.

Jordaan, I.J. Gillott, J.E., Loov, R.E. and Shrive, N.G. (1978) "Improved ductility of sulphur concretes and its relation to strength," *Proceedings of the International Conference on Sulphur in Construction*, Ottawa, (CANMET, Energy, Mines and Resources Canada), p. 475-488.

Kalb, P.D., Adams, J.W., Milian, L.W., Penny, G., Brewer, J., and Lockwood A. (1999) "Mercury bakeoff technology comparison for the treatment of mixed waste mercury contaminated soils at BNL," in *Proceedings of Waste Management 1999*, Tucson, AZ.

Kalb, P.D. and Colombo, P. (1985) *Modified sulfur cement solidification of low-level wastes*, BNL-51923, Brookhaven National Laboratory, Upton, NY, October 1985.

Kalb, P.D. Heiser, J.H. III, and Colombo, P. (1991a) "Modified sulfur cement encapsulation of mixed waste contaminated incinerator fly ash," *Waste Management*, Vol. 11, No. 3, p. 147-153.

Kalb, P.D., Heiser, J.H. III, and Colombo, P. (1991b) "Modified sulfur cement encapsulation of mixed waste contaminated incinerator fly ash," *Waste Management*, Vol. 11, p. 1-7.

Kalb, P.D., Heiser, J.H. III, Pietrzak, R., and Colombo, P. (1991c) "Durability of incinerator ash waste encapsulated in modified sulfur cement," *Tenth Annual International Incineration Conference*, May 1991, p. 10.

Kalb, P.D., Melamed, D., Patel, B., and Fuhrmann, M. (2002) *Treatment of mercury containing waste*, U.S. Patent No. 6,399,849, June 4, 2002.

Kalb, P.D., Milian, L.W., Yim, S.P., Dyer, R.S., and Michaud, W.R. (1997) "Treatability study on the use of by-product sulfur in Kazakhstan for the stabilization of hazardous and radioactive wastes," *Proceedings of the American Chemical Society Symposium on Emerging Technologies in Hazardous Waste Management IX*, Pittsburgh, PA, Sept.15-17, 1997.

Kalb, P.D., Vagin, S., Beall, P.W., and Levintov, B.L. (2004) "Sustainable development, in Kazakhstan: *Using oil & gas production by-product sulfur for cost-effective secondary end-use products*. REWAS 2004 Global Symposium, September 2004.

Kaplan, I.R. (1972) "Sulfur cycle," in R.W. Fairbridge, ed., *The Encyclopedia of Geochemistry and Environmental Sciences*. Vol. IVA, Van Nostrand Reinhold Co., New York., p. 1148-1151.

Kennepohl, G.J. and Miller, L.J. (1978) "Sulfur-asphalt binder technology for pavements," *New Uses of Sulfur II. Adv. Chem. Ser.*, Vol. 165, p. 113.

Kennepohl, G.J., Logan, A., and Bean, D.C. (1974) *Canadian Sulphur Symposium*, Calgary, Alberta, Canada, p. Q1.

Kennepohl, G.J., Logan, A., and Bean, D.C. (1975) *Proceedings of the Association of Asphalt Paving Technologists*, 44,485.

Kennepohl, G.J., Logan, A., and Bean, D.C. (1978) Canadian Patent No. 1,025,155.

Kennepohl, G.J., Logan, A., and Bean, D.C. (1979) "Apparatus for continuous preparation of sulfur asphalt binders and paving compositions," U.S. Patent 4,155,654.

Kibert, C. (1994) "Establishing principles and a model for sustainable construction," In *Proceedings of the First International Conference on Sustainable Construction*, C. J. Kibert, ed. Tampa, FL, November 6-9, CIB TG 16.

King, J.K., Kostka, J.E., Frischer, M.E., and Saunders, F.M. (2000) "Sulfate-reducing bacteria methylate mercury at variable rates in pure culture and in marine sediments," *Appl. Environ. Mictobiol.* June 2000, p. 2430-2437.

King, R.W., Puzinauskas, V.P., and Holdsworth, C.E. (1980) *Asphalt composition and health effects: A critical review*, API Technical Publication, American Petroleum Institute, Washington, D.C., p. 11.

Kinlaw, D. (1992) *Competitive and green: Sustainable performance in the environmental age*, Pfeiffer & Co., San Diego, CA.

Koh J.C. and Klement, W., Jr. (1970) "Polymer Content of Sulfur Quenched Rapidly from the Melt" *J. Phys. Chem.*, Vol. 74, p. 4280.

Koningsberger, D.C. (1971) "The polymerization in liquid sulfur and selenium, An ESR study," Thesis, Technishe Hogeschool te Eindhoven, Eindhoven.

Korn, J., Prinzler, H.W., and Pape, D. (1966) Zur Kenntnis der Reaktion zwischen Paraffinkohlenwasserstoffen und Schwefel, Erdöl Kohle, Vol. 19, p. 651.

Kozhevnikov, V. F., Viner, J. M., and Taylor, P. C. (2001) *Phys. Rev.* B 64, 214109.

Kramer, D.A. (2002) "Nitrogen, in Metals and minerals," *U.S. Geological Survey Minerals Yearbook 2000*, Vol. I, p. 55.1-55.22.

Kriech, A.J., Kurek, J.T., Osborn, L.V., Wissel, H.L., and Sweeney, B.J. (2002) "Determination of polycyclic aromatic compounds in asphalt and in corresponding leachate water," *Polycyclic Aromatic Hydrocarbons*, Vol. 22, p. 517-535.

Lageraaen, P.R. and Kalb, P.D. (1997) "Use of recycled polymers for encapsulation of radioactive, hazardous and mixed wastes," *BNL-66575 Informal Report*.

Lartaut, M. (1985) *Proceedings of the 3rd Euro bitumen Symposium*, L'Aja, p. 595-600.

Lee, D.Y. (1975) "Modification of asphalt and asphalt paving mixtures by sulfur," *Ind. Eng. Chem., Prod. Res. Dev.*, Vol. 14, No. 3.

Lee, D.Y. (1998) "Manufacturing methods for sulfur concrete sewer pipe," Final report to U.S. Environmental Protection Agency, National Center for Environmental Research, Washington, D.C.

Leffler, W.L. (2000) *Petroleum refining in nontechnical language* 3rd ed. Tulsa, OK., Pennwell Corp., p. 310.

Lepp, H. (1972) "Iron: element in geochemistry," in R.W. Fairbridge, ed., *The Encyclopedia of Geochemistry and Environmental Sciences*, Vol. IVA. Van Nostrand Reinhold Co., New York, p. 599-603.

Leutner, B. and Diehl, L. (1977) "Manufacture of sulfur concrete," U.S. Patent No. 4,025,352.

Levine, R.M. and Wallace, G.J. (2001) "The mineral industries of the Commonwealth of Independent States," USGS, 2001.

Lewis, G. N. (1916) "The atom and the molecule," *J. Am. Chem. Soc.*, Vol. 38, p. 762-785.

Liddle, B.T. (1994) "Construction for sustainability and the sustainability of the construction industry," *Proceedings of the First International Conference on Sustainable Construction*, CIB TG 16, C.J. Kibert, ed., Tampa, FL.

Lin, H., Li, Z., Tohji, K., Tsuchiya, N., and Yamasaki, N. (2004) "Reaction of Sulfur with Water under Hydrothermal Conditions," *14th International Conference on the Properties of Water and Steam*, Kyoto, p. 365.

Lin, S. L., Chian, E.S.K., and Lai, J.S. (1994) "Sustainable technology using sulfur for solidification/stabilization of metal contaminated wastes," *OCESSA J.*, Vol. 11, No. 2, p. 17.

Lin, S.L., Lai, J.S., and Chian, E.S.K. (1995) "Modification of sulfur polymer cement (SPC) stabilization and solidification (S/S) process," *Waste Management*, Vol. 15, Nos. (5/6), p. 441-447.

Lindgren, Å. (1996) "Asphalt wear and pollution transport," *Sci. Tot. Environ.*, Vols. 189/190, p. 281-286.

Liolios, G.C. (1989) "Alkylation—Past, present and future," *National Petroleum Refiners Association Annual Meeting*, San Francisco, CA., March 19-21, 1989, Presentation, p. 11.

Love, G.D. (1979) "Sulfur: potential pavement binder of the future," *Transp. Eng. J. ASCE*, Vol. 105, p. 525.

Lusty, J.R. (1983) "Chemical investigations on leaching of sulfur from sulfur infiltrated concretes," *Cement and Concrete Research*, Vol. 13, No. 2, p. 233-128.

Luther, G.W. and Church, T.M. (1992) "An overview of the environmental chemistry of sulphur in wetland systems," in R.W. Howarth, J.W.B. Stewart, and M.V. Ivanov, eds., *Sulfur Cycling on the Continents*. John Wiley &Sons, Ltd., London, p. 125-142.

Luther, G.W. III, Ferdelman, T.G., Kostka, J.E., Sxamakis, E.J., and Church, T.M. (1991) "Temporal and spatial variability of reduced sulfur species (FeS_2, S_2O_{3-2}) and pore water parameters in salt marsh sediments," *Biogeochem*, Vol. 14, p. 57-88.

Mack, C.J. (1932) "Colloid chemistry of asphalts," *J. Phys. Chem.*, Vol. 36, p. 2901-2914.

Makenya, A.R. (1997) *Composition for durability of a Chempruf-modified sulfur concrete*, Licentiate Thesis, Dept. of Architecture, Royal Institute of Technology, Stockholm, SE.

Makenya, A.R. (2001) *Industrial application of sulfur concrete an environmental-friendly construction material*, Dissertation, Dept. of Architecture, Royal Institute of Technology, Stockholm, SE, 232 p.

Makenya, A.R. and Hellers, B.G. (2000) "European market opportunities: Sulfur concrete in Sweden," *The Sulphur Institute's 7th Biennial Symposium*, Washington, D.C., March 26-28.

Malaiyandi, M., Benedek, A., Holko, A.P., and Bancsi, J.J. (1982) "Measurement of potentially hazardous poly-nuclear aromatic hydrocarbons from occupational exposure during roofing and paving operations," in M. Cooke, A.J. Dennis, and G.L. Fisher eds., *Poly-nuclear Aromatic Hydrocarbons: Physical and Biological Chemistry. Sixth International Symposium*, Battelle Press, Columbus, OH, p. 471-489.

Malhotra, V.M. (1973) "Mechanical properties and freeze-thaw resistance of sulfur concrete," Division Report No. IR 73-18, Energy, Mines and Resources, Ottawa, CA.

Malhotra, V.M. (1974) "Effect of specimen size on compressive strength of sulfur concrete," Division Report No. IR 74-25, Energy, Mines and Resources, Ottawa, CA.

Malhotra, V.M. (1983) "Sulfur concrete and sulfur-infiltrated concrete," *Concrete Technology and Design*, edited by R.N. Swamy, Surrey University Pess, London, Vol. 1, p. 1-40.

Malhotra, V.M. and Winer, A. (1976) "The use of asbestos fiber in Portland cement and sulfur concretes," *Can. Min. Metall. Bull.*, Vol. 69, No. 767, p. 131-138.

Malone, T.F. (1994) "Sustainable human development: A paradigm for the 21st Century," *White Paper for the National Association of State Universities and Land-Grant Colleges*, The Sigma Xi Center, Research Triangle Park, NC, November 6.

Marwil, S.J., Miller, B.R., and Willis, C.G. (1994) *Plasticized sulfur compositions and method of manufacturing same.* U.S. Patent 5,304,238.

Masterton, W.L. and Hurley, C.N. (1997) *Chemistry: Principles and reactions,* 3rd ed., Harcourt Brace College Publishers.

Matos G.R. and Ober. J.A. (2005) "Sulfur end use statistics," *Materials flow of sulfur: U.S. Geological Survey Open file report,* http://minerals.usgs.gov/ds/2005/140/sulfur-use.pdf.

Mattus, C.H. and Mattus, A.J. (1994) *Evaluation of sulfur polymer cement as a waste form for the immobilization of low level radioactive or mixed waste,* OAK Ridge National Laboratory.

Mayberry, J.L., et al. (1993) *Technical area status report for low-level mired waste final waste forms,* Vol. I, DOEMWIP-3, Mixed Waste Integrated Program, Office of Technology Development, U.S. Department of Energy, Washington, D.C.

McBee, W.C. and Sullivan, T.A. (1978a) "Direct substitution of sulphur for asphalt in paving materials," *Report of Investigations* 8303, U.S. Department of the Interior, Bureau of Mines.

McBee, W.C. and Sullivan, T.A. (1978b) "Sulfur utilization in asphalt paving materials," *New Uses of Sulfur II. Adv. Chem. Ser.,* Vol. 165, p. 135.

McBee, W.C. and Sullivan, T.A. (1979) "Development of specialized sulfur concretes," *Bureau of Mines Report No. RI 8346,* U.S. Bureau of Mines, Washington, D.C., 21.

McBee, W.C. and Sullivan, T.A. (1982a) *Concrete formulations comprising polymeric reaction products of sulfur/Cyclopentadiene, Oligmer/Dicyclopenetadiene,* U.S. Patent No. 4,348,313.

McBee, W.C. and Sullivan, T.A. (1982b) *Modified Sulfur Cement,* U.S. Patent No. 4,311,826.

McBee, W.C. and Sullivan, T.A. (1983) *Modified sulfur cement.* U.S. Patent No. 4,391,969.

McBee, W.C., Saylak, D., Sullivan, T.A., and Barnett, R.W. (1976) "Sulfur as a partial replacement for asphalt in bituminous pavements," in *New Horizons in Construction Materials,* Vol. 1, Envo Publishing Co., Lehigh Valley, PA., p. 345-361.

McBee, W.C., Sullivan, T.A., and Fike, H.J. (1985) "Sulfur construction materials," *Bulletin 678.* U.S. Bureau of Mines.

McBee, W.C., Sullivan, T.A., and Fike, H.F. (1986) "Corrosion-resistant sulfur concretes," *Corrosion and Chemical Resistant Masonry Materials Handbook,* Noyes Publications, p. 392-417.

McBee, W.C., Sullivan, T.A., and Jong, B.W. (1981a) "Modified sulfur concrete for use in concretes, flexible paving, coatings, and grouts," *Bureau of Mines Report No. RI 8545*, U.S. Bureau of Mines, Washington, D.C., p. 24.

McBee, W.C., Sullivan, T.A., and Jong, B.W. (1981b) "Modified sulfur concrete technology," *Proceedings, SULPHUR-81 International Conference on Sulfur*, Calgary, p. 367-388.

McBee, W.C., Sullivan, T.A., and Jong, B.W. (1983a) "Corrosion-resistant sulfur concretes," *Bureau of Mines Report No. 8758*, U.S. Bureau of Mines, Washington, D.C., p. 28.

McBee, W.C., Sullivan, T.A., and Jong, B.W. (1983b) "Industrial evaluation of sulfur concrete in corrosive environments," *Bureau of Mines Report No. RI 8786*, U.S. Bureau of Mines, Washington, D.C., p. 15.

McBee, W.C., Sullivan, T.A. and Jong, B.W. (1985) "Industrial evaluation of sulfur concrete in corrosive environment," *Mining Engineering*, Vol. 37, No. 1, Jan. 1985, p. 37-44.

McBee, W.C., Sullivan, T.A., Saylak, D., and Barnett, R.W. (1976) "New horizons in construction materials; sulfur as a partial replacement for asphalt in bituminous pavements."

McBee, W.C., Sullivan, T.A., and Patrick (1977) *Structural materials*, U.S. Patent No. 4,022,626, May 10.

McBee, W.C., Sullivan, T.A., and Tzatt, J.O. (1980) *Publication No. FHWA-IP-80-14*. U.S. Department of Transportation.

McBee, W.C., Weber, H.H., and Dohner, W.T. (1989) "Sulfur polymer concrete for special-purpose applications," *7th Structures Congress*, San Francisco, CA, Sponsor: ASCE, New York, NY, p. 31-43.

McBee, W.C. and Weber, H.H. (1990) "Sulfur polymer cement concrete, in *Proceedings of the 12th Annual Department of Energy Low-Level Waste Management Conference, CONF-9008119*, National Low-Level Waste Management Program, Idaho Natl. Engineering Lab., Idaho Falls, ID.

McKinney, P.V. (1940) "Provisional methods for testing sulfur cements," *ASTM Bulletin* 96-107, p. 27-30.

Mehta, P.K. (1969) "Morphology of calcium sulphoaluminate hydrate," *J. Am. Ceram. Soc.*, Vol. 52 p. 521–522.Mehta, P.K. (1973) "Mechanism of expansion associated with ettringite formation," *Cem. Concr. Res.*, Vol. 3, p. 1–6.

Mehta, P.K. and Wang, S. (1982) "Expansion of ettringite by water adsorption," *Cem. Concr. Res.*, Vol. 12, p. 121–122.

Meyer, B. (1965) "Preparation and properties of sulfur allotropes, in Meyer, B., ed., *Elemental sulfur, chemistry and physics*, Interscience Publishers, New York, p. 45-69.

Meyer, B. (1968) "Elemental sulphur," in Nickless,G., ed., *Inorganic sulphur chemistry*, Elsevier Publishing Company, Amsterdam, p. 241-258.

Meyer, B. (1976) "Elemental sulfur," *Chemical Reviews*, Vol. 76, p. 367-388.

Meyer, B. (1977) *Sulfur, energy, and environment* Elsevier Scientific Publishing Company, Amsterdam, p. 488.

Meyer, B., Oommen, T.V., and Jensen, D. (1971) "The color of liquid sulfur," *J. Phys. Chem.*, p. 75, 912- 917.

Miele, G. M. (1986) "Sulfur Installation construction Methods and Techniques," *International Symposium and Workshop*, Washington, D.C.

Miller, L.J. and Crawford, N.W. (1981) *Preprint to the Annual Meeting of the Canadian Technical Asphalt Association in Montreal.*

Mohamed, A.M.O. (2000) "The Role of Clay Minerals in Marly Soils on Its Stability," *Engineering Geology*, Vol. 57, p. 193-203.

Mohamed, A.M.O. (2002) "Keynote Paper: Geoenvironmental Aspects of Chemically Based Ground Improvement Techniques," *4th International Conference on Ground Improvement Techniques*, March 26-28, 2002, Kuala Lumpur, Malaysia.

Mohamed, A.M.O. (2003a) "Geoenvironmental Aspects of Chemically Based Ground Improvement Techniques for Pyritic Mine Tailings," *Ground Improvement*, Vol. 7, No. 2, p. 73-85.

Mohamed, A.M.O. (2003b) "Ground Improvement of Sulphide Bearing Soils by Inorganic Chemical Processes," *9th Arab Structural Engineering Conference (9ASEC): Emerging Technologies in Structural Engineering*, November 29– December 1, 2003, Abu Dhabi, UAE, p. 837-846.

Mohamed, A.M.O. and Antia, H.E. (1998) "Geoenvironmental Engineering," *Developments in Geotechnical Engineering*, 82, Elsevier, p. 707.

Mohamed, A.M.O., Boily, J.F., Hossein, M., and Hassani, F.P. (1995) "Ettringite Formation in Lime-Remediated Mine Tailings: I. Thermodynamic Modelling," *Canadian Institute of Mining, Metallurgy and Petroleum (CIM) Bulletin*, Vol. 88, No. 995, p. 69-75.

Mohamed, A.M.O. and El Gamal, M. (2006a) "Compositional control on sulfur polymer concrete production for public works," in *Developments in Arid Regions Research*, Vol. 3, p. 27-38.

Mohamed, A.M.O. and El Gamal, M. (2006b) "Compositional control on sulfur polymer concrete production for public works," *Proceedings of the Seventh UAE University Annual Conference*, Al Ain, April 23-25, 2006, p. ENG.

Mohamed, A.M.O. and El Gamal, M. (2007a) "Durability of sulfur concrete manufactured from recycled waste materials," *Proceedings of The 8th UAE University Annual Conference*, Al Ain, April 23-25, 2007, p. ENG.

Mohamed, A.M.O. and El Gamal, M. (2007b) "Durability and leachability characteristics of modified sulfur cement and concrete barriers for containment of hazardous waste in arid lands," *1st Joint QP-JCCP Environment Symposium in Qatar, "Sustainable Development and Climate Change,"* February 5-7, 2007, Doha, Qatar.

Mohamed, A.M.O. and El Gamal, M. (2007c) "Development of modified sulfur cement and concrete barriers for containment of hazardous waste in arid lands," *"Sustainable Development and Climate Change,"* February 5-7, 2007, Doha, Qatar.

Mohamed, A.M.O. and El Gamal, M. (2007d) "Sulfur based hazardous Waste Solidification," *Environmental Geology*, Vol. 53, No. 1, p. 159-175.

Mohamed, A.M.O. and El Gamal, M. (2008a) "Sulfur cement and concrete Production," *Proceedings of the 9th UAE University Annual Conference*, Al Ain, April 23-25, 2008, p. ENG-1-10.

Mohamed, A.M.O. and El Gamal, M. (2008b) *New use of surfactant*, U.K. Patent Application No. 0807612.7, filed by J.A. KEMP & Co., UK, dated April 25, 2008.

Mohamed, A.M.O. and El Gamal, M. (2009a) "Hydro-mechanical behavior of a newly developed sulfur polymer concrete," *Cement and Concrete Composites*, Vol. 31, p. 186-194.

Mohamed, A.M.O. and El Gamal, M. (2009b) *New use of surfactant*, International Patent Application No. PCT/IB2009/005338, filed by J.A. KEMP & Co., UK, dated April 21, 2009.

Mohamed, A.M.O. and El Gamal, M. (2009c) *New use of surfactant*, GCC Patent Application No. GCC/P/2009/13350, filed by J.A. KEMP & Co., UK, dated April 25, 2009.

Mohamed, A.M.O. and El Gamal, M. (2009d) "Evaluation of the Potential Use of Cement Kiln Dust as an Aggregate Material for Manufacturing of Sulfur Polymer Concrete," *Proceedings of the 10th UAE University Annual Conference*, Al Ain, April 13-16, 2009, p. 257-264.

Mohamed, A.M.O., El Gamal, M., and El Saiy, A.K. (2006a) "Thermo-mechanical performance of the newly developed sulfur polymer concrete," in *Developments in Arid Regions Research*, Vol. 3, p. 15-26.

Mohamed, A.M.O., El Gamal, M., and El Saiy, A.K. (2006b) "Thermo-mechanical performance of the newly developed sulfur polymer concrete," *Proceedings of the Seventh UAE University Annual Conference*, Al Ain, April 23-25, 2006, p. ENG.

Mohamed, A.M.O. and Hossein, M. (2004) Solidification/Stabilization of Sulphide Bearing Soils Using ALFA Process," in *Geo Jordan 2004: Advances in Geotechnical Engineering with Emphasis on Dams, Highway Materials, and Soil Improvement: Proceedings of the Conference*, July 12-15, Irbid, Jordan, eds. Khalid Alshibli, Abdallah I. Husein MalKawi, and Mostafa Alsaleh, Published by ASCE, Geotechnical Practice Publication, No. 1, p. 131-144.

Mohamed, A.M.O., Hossein, M., and Hassani, F.P. (2002) "Hydro-mechanical Evaluation of Stabilized Mine Tailings," *Environmental Geology Journal*, Vol. 41, p. 749-759.

Mohamed, A.M.O., Hossein, M., and Hassani, F.P. (2003) "Role of Fly Ash addition on Ettringite Formation in Lime-Remediated Mine Tailings," *Journal of Cement, Concrete and Aggregates*, ASTM, December 2003, Vol. 25, No. 2, p. 49-58.

Mohamed, A.M.O., Hossein, M., and Hassani, F.P. (2004) Keynote Paper: "Evaluation of the Newly Developed ALFA Technology for the Treatment of High Sulphide Content Mine Tailings," *5th Conference on Ground Improvement Techniques*, March 26-28, 2004, Kuala Lumpur, Malaysia, p. 41-54.

Mohamed, A.M.O., Hossein, M., and Hassani, F.P. (2007a) "Evaluation of the Newly Developed Aluminum, Lime, and Fly Ash Technology for Solidification/Stabilization of Mine Tailings," *Journal of Materials in Civil Engineering*, Vol. 19, No. 1, January 1, 2007.

Mohamed, A.M.O., Hossein, M., and Hassani, F.P. (2007b) "Treatment of Mine Tailings Using ALFA® Technology," *Ground Improvement*, Vol. 11, No. 2, p. 77-86.

Mohammadi S. (2006) "Sulfur sewage pipe production methods," Proceeding of ICOMAST. *International Conference on Manufacturing Science and Technology*, August 28-30, Melaka, Malaysia, p. 395.

Moore, A.E. and Taylor, H.F.W. (1968) "Crystal structure of ettringite," *Nature*, Vol. 218, p. 1048-1049.

Moore, W.E. (1972) "Introduction," in *Technology and Social Change*, W.E. Moore, ed., Chicago: Quadrangle Books, p. 3-25.

Morihiro, T., Nakatsuka, Y., Kaminade, T. (2008) *Binder containing modified sulfur and process for producing material containing modified sulfur*, European Patent No. PE1961713A1 published in accordance with Art. 158(3) EPC.

Morishita, T. (2001) *Phys. Rev. Lett.*, Vol. 87, 105701.

Morse, D.E. (1985) "Sulfur," in *Mineral Facts and Problems*, U.S. Bureau of Mines Bulletin 667, p. 783-797.

Morton, M. (1987) "Technology and Engineering," 264p.

Moss, M.R. (1978) "Sources of sulfur in the environment—the global sulfur cycle," in Nriagu, J.O., ed., *The Atmospheric Cycle*, Part 1 of Sulfur in the Environment, New York, John Wiley and Sons, p. 23–50.

Munasinghe, M. and McNeely, J. (1995) "Key concepts and terminology of sustainable development," in *Defining and Measuring Sustainability: The Biogeophysical Foundations*, M. Munasinghe and W. Shearer, eds., Washington, D.C., World Bank.

National Safety Council (1979) "Handling and storage of solid sulfur," Data sheet No. 1-612-79, Chicago, p. 4.

National Science and Technology Council (1994) *Technology for a sustainable future: Framework for action*, Washington, D.C., Office of Science and Technology Policy.

Nellenstyn, J.F. (1924) "The composition of asphalt," *J. Inst. Petr. Technol.*, Vol. 10, p. 311-325.

Nellenstyn, J.F. (1928) "Relation of micelle to the medium in asphalt," *J. Inst. Petr. Technol.*, Vol. 14, p. 134-138.

Nevin, P.J. (1996) *Comparison of performance of sulfur concrete applied by different methods*, Construction Equipment Services, Johannesburg, South Africa. Paper presented at the 1996 fall Convention. American Concrete Institute, New Orleans, LA, November 3-8.

Nimer, E.L. and Campbell R.W. (1983) *Sulfur cement-aggregate–organosilane composition and methods for preparing*, U.S. Patent No. 4,376,830.

Nimer, E.L. and Campbell, R.W. (1985) *Sulfur cement-aggregate compositions and methods for preparing*, U.S. Patent 4,496,659.

Nnabuife, E.C. (1987) "Study of some variables affecting the properties of sulfur-reinforced sugarcane residue based boards," *Indian Journal of Technology*, Vol. 25, p. 363-367.

Norton, B.G. (1992) "A new paradigm for environmental management," in *Ecosystem Health: New Goals for Environmental Management*, R. Costanza, B. G. Norton, and B. D. Haskell, eds. Washington, D.C. Island Press, p. 23-41.

Nuclear Threat Initiative (NTI). (2003) "Reported nuclear, radioisotope, and dual-use materials traffic king incidents involving the newly independent states," http://www.nti.org/db/nistraff/tables/2003%20by%20material.htm>.

Oana S. and Ishikawa H. (1966) "Sulfur isotopic fractionation between sulfur and sulfuric acid in the hydrothermal solution of sulfur dioxide," *Geochemical Journal*, Vol. 1, No. 1, p. 45.

Ober, J.A. (2003) "Materials flow of sulfur," U.S. Geological Survey Open files report 02-298, http://www.usgs.gov/.

Occupational Safety and Health Administration. (1999) Section IV, Chapter 2, Petroleum refining processes, OSHA Technical Manual, website at http://www.osha_slc.gov/dts/osta/otm/otm_iv/otm_iv_2.html. Accessed November, 27, 2001.

Oil and Gas Journal. (1999) "A description of Canadian oil sands characteristics," *Oil & Gas Journal*, Vol. 97, No. 26, June 28, p. 48.

Oil sands CHOA Meeting, Calgary, AB, April 12th, 2006.

Okumura, H.A. (1998) "Early sulfur concrete installations," *Concrete International*, Vol. 20, No. 1, Jan. 1998, p. 72-75.

ONORM. (1969) B 3304 edition January.

Orlowski, J., Leszczewski, M., and Margal, I. (2004) "Stability of polymer concrete with steel reinforcement."

Oyekunle, L.O. and Onyehanere, L.N. (1991) "Studies on Nigerian crudes," *Analysis of sulfurized asphalts, 3, Fuel Sci. Technol. Int.*, Vol. 9, p. 681.

Papirer, E., Fritschy, G., and Eckhardt, A. (1980) "Structural modes of sulphur in sulphur-bitumen composites as studied by electron microscopy," *Fuel*, Vol. 59, p. 617.

Perov, E.I., Moshchenskaya, N.V., Irkhina, E.P., and Smorodinov, V.S. (2001) "Reactivity of various molecular sulfur species in reactions with liquid n-alkanes," *Neftekhimiya*, Vol. 41, No. 5, p. 384-388.

Petersen, J.C. (1981) "Oxidation of sulfur compounds in petroleum residues: reactivity-structural relationships," *Div. Petrol. Chem. Soc.*, Vol. 26, No. 4, p. 898-906.

Petersen, J. C. (2000) "Chemical composition of asphalt as related to asphalt durability," *Asphaltenes and Asphalts, 2, Development in Petroleum Science*, 40 B Chapter 14, p. 363-399.

Petersen, J.C. and Plancher, H. (1981) "Quantitative determination of carboxylic acids and their salts and anhydrides in asphalts by selective chemical reactions and differential infrared spectrometry," *Anal. Chem.*, Vol. 53, p. 786-789.

Petersen, J.C., Barbour, F.A., and Dorrence, S.M. (1974) "Catalysis of asphalt oxidation by mineral aggregate surfaces and asphalt components," *Proc. Assoc. Asphalt Paving Technol.*, Vol. 43, p. 162-177.

Petersen, J.C., Barbour, F.A., and Dorrence, S.M. (1975) "Identification of dicarboxylic anhydrides in oxidized asphalts," Anal Chem., Vol. 47, No. 1, p. 107–111.

Petrossi, U., Bocca, P.L., and Pacor, P. (1972) " Industrial and engineering chemistry reasearch and development," *Ind. Eng. Chem. Prod. Res. Dev.*, Vol. 11, p. 214.

Pfeiffer, J.P. and Saal, R.N.J., (1940) "Asphaltic bitumen as colloid system," *The Journal of Physical Chemistry*, Vol. 44, No. 2, p. 139–149.

Pickard, S.S. (1981) "Sulfur concrete: Understanding/application," *Concrete International*, Vol. 3, Issue 10, p. 57-67.

Pickard, S.S. (1984) "Sulfur Concrete at AMAX Nickel—A Project Case History," *Concrete International*: October 1, Vol. 6, Issue 10, p. 35-41.

Pickard, S.S. (1985) "Sulfur Concrete for Acid Resistance," *Chemical Engineering*, Vol. 92, No. 15, p. 77-78.

Pirages, D. (1994) "Sustainability as an evolving process," *Futures*, Vol. 26, No. 2, p. 197-205.

Plancher, H., Green, E.L., and Peterson, J.C. (1976), "Reduction of oxidative hardening of asphalts by treatment with hydrated lime—a mechanistic study," *Proceedings of the Association of Asphalt Paving Technologists*, Volume 45, p. 1-245. The Association of Asphalt Paving Technologists, Minnesota.

Plato, J. S. (1980), "Sulfur Research and Development," *Sulphur*, Vol. 3, p. 18-20.

Pronk, F. E., Soderberg, A. F., Frizzell, R. T. (1975) "Sulfur modified asphaltic concrete," *Proc. Can. Tech. Asphalt Assoc.*, Vol. 20, p. 135-194.

Pryor, W.A. (1962) *Mechanisms of sulfur reactions*, McGraw-Hill, New York.

Qarles Van Ufford, J.J. and Vlugter, J.C. (1965) *Schwefel und bitumen*, II. Brennst. Chem., Vol. 46, p. 7.

Rahimian, I. (1976) "Reaktion zwischen Bitumen und Schwefel sowie Charakterisierung der entstehenden Verbindungen" 4. Gemeinschaftstagung ÖGEW/DGMK, Salzburg.

Ramsey, J.W., McDonald, F.R., and Petersen, J.C. (1967) "Structural study of asphalts by nuclear magnetic resonance," *Ind. Eng. Chem. Prod. Res. Dev.*, Vol. 6, No. 4, p. 231–236.

Reardon E.J., Czank C.A., Warren C.J., Dayal R., and Johnston, H.M. (1995) "Determining controls on element concentrations in fly ash leachate," *Waste Management & Research*, 13:435.

Rees, W.E. (1990) "The ecology of sustainable development," *The Ecologist*, Vol. 20, No. 1, p.18-23.

Rennie W. J. (1979) "Sulphur asphalts," *New Uses for Sulphur—SUDIC Technology Series No. 2*, 2nd ed.

Roberts, D.V. (1994) "Sustainable development—A challenge for the engineering profession. in the role of engineering in sustainable development," M. Ellis, ed., Washington, D.C., *American Association of Engineering Societies*, p. 44-61.

Rock Binders, Inc. (2002) "Sulfur extended asphalt modifier," Rock Binders, Inc., Stockton, CA. Available at http://www.rockbinders.com/seam.html.

Roger D. Spence and Caijun Shi, eds. (2004) *Stabilization and solidification of hazardous, radioactive and mixed wastes*, CRC Press, Boca Raton, FL, p. 390.

Royal Institute of Technology. (2001) *Industrial application of sulfur concrete.*

Rubbio, G. (1977) *Schwefelzusatz auf die Eigenschaften von Bitumen*. Bitumen, 39:142.

Ruiz-Garcia, Anderson, E.M., and Greer, S.C. (1989) "The shear viscosity of liquid sulfur near the polymerization temperature," *J. Phys. Chem.*, Vol. 93, p. 6980-6983.

Rylova, M.V., Samuilov, A. Y., Sharafutdinova, D. R., Khrapkovskii, G. M., and Samuilov Y. D. (2002) "Interaction of bicyclopentadiene with elemental sulfur initial stages of the reaction," *Chemistry and Computational Simulation, Butlerov Communications*, Vol. 3, No. 9, p. 29-32.

Saal, R.N.J., Bass, P.W., and Heukelom, W. (1946) "The colloidal structure of asphaltic bitumens," *J. Chim. Phys.*, Vol. 43, p. 235-261.

Sander, U.H.F., Rothe, U., and Kola, R. (1984) "Sulphuric acid," in *Sulphur, sulphur dioxide and sulfuric acid*, London, The British Sulphur Corp, p. 257-399.

Sasan, M. (2006) *International Conference on Manufacturing Science and Technology*, August 28-30, Melaka, Malaysia.

Saylak, D, Little, D.N., and Bigley, S.W. (1982) *Beneficial use of sulfur in highway pavements—characterization and analysis of plasticized sulfur Concrete*, A

Minerals Research Contract Report, U.S. Dept. of the Interior, Bureau of Mines, Washington, D.C., p. 57.

Saylak, D. and Conger, W.E. (1982) "A review of the state of the art of sulfur asphalt paving technology," *Sulfur: New Sources and Uses, American Chemical Society Symposium Series*, No. 183, p. 155-193.

Saylak, D. and Gallawy, B.M. (1975) *The use of sulphur in sulphur-asphalt aggregate mixes*, Interam. Conf. Mater. Technol. Caracas, p. 636.

Schmerling, L. (1981) *Organic and petroleum chemistry for non chemists*, Tulsa, Okla., Penn Well Publishing Co., p. 109.

Scientific American Journal. (1970) Vol. 222, No. 5, p. 62-78

Serageldin, M. (1993) "The Use of Land and Infrastructure in the Self Improvement Strategies of Urban Lower Income Families," *14th Commission on Human Settlements and at the Shelter Forum*, Nairobi.

Sereda, P.J. and Beaudoin, J.J. (1980) *Quality control evaluation of repair material*, Building Research Note ISSN 0701-5232.

Shakhashiri, B.Z. (1983) *Chemical demonstrations: Volume One*, University of Wisconsin Press, Madison, p. 53.

Shell International B.V. (2006) "Modified sulfur and product comprising modified sulfur as binder," PCT/EP2006/063220.

Sheppard, W.L. (1975) "Sulfur Mortars," *Corrosion and Chemical Resistance Masonry Materials Handbook*, p. 222.

Shrive N.G. and Ward, M.A. (1977) *Some low temperature ageing and durability aspects of sulphur/asphalt concretes in low temperature properties of bituminous materials*, C. Marek, ed., ASTM STP No. 628.

Shrive, N. G., Loov, R.E., Gillott, J. E., and Jordaan I. J. (1977) *Materials Science and Engineering*, Vol. 30, p. 71.

Siegmann, M.C. (1950) "Manufacture of asphaltic bitumen," in, *The Properties of Asphaltic Bitumen*, J. Pfeiffer (ed.), Elsevier, New York, p. 143.

Simic, M. (1980) *Plasticized sulfur with improved thixotropy*, U.S. Patent No. 4,210,458

Simic, M. (1982) U.S. Patent No. 4,348,233

Sliva, P., Peng, Y.B., Peeler, D.K., Bunnell, L.R., Turner, P.J., Martin, P.F., and Feng, X. (1996) *Sulfur polymer cement as a low-level waste glass matrix encapsulant*, 1996.

Soderberg, A.F. (1983) *A New Construction Material*, Sudicrete, SUDIC, Calgary.

STARcrete Technologies Inc. (2000) "STARcrete—the first commercial sulfur concrete," http://www.starcrete.com/. Accessed September 10, 2001

State of the Art Guideline Manual for Design. (1980) "Quality Control, and Construction of SEA Pavements," FHWA-IP-80, U.S. Department of Transportation (FHWA), Washington, D.C.

Stepanyan, I.V. (1988) *The use of sulfur as a component of asphalt concrete*, Cand. Sci. Dissertation, Moscow.

Steudel, R. (2003) *Topics in Current Chemistry*, Vol. 230, p. 81.

Stevens, J. (1998) "Oil sands projects gather pace," *Sulphur*, No. 254, January-February, p. 27-30.

Sullivan, T.A. (1986) *Corrosion-resistant sulfur concretes—design manual*, The Sulphur Institute, Washington D.C., p. 44.

Sullivan, T.A. and McBee, W.C. (1976) *Development and testing of superior sulfur Concretes*, RI-8160, Bureau of Mines, U.S. Dept. of the Interior, Washington, D.C.

Sullivan, T.A., McBee, W.C., and Blue, D.D. (1974) *Adv. Chem. Ser.*, Vol. 140, p. 55.

Sullivan, T.A., Mc Bee, W.C., and Blue, D.D. (1975) "Sulfur in coatings and structural materials," *Advances in Chemistry Series No. 140*, American Chemical Society, Washington, D.C., p. 55-74.

Sullivan, T.A., McBee, W.C., and Rasmussen, K.L. (1975) *Studies of sand sulfur asphalt paving materials*, U.S. Bur. Mines Rep. Invest. RI-8087.

Sulphur, (2000) "Sulphur removal JV to handle cheaper crude," *Sulphur*, No. 270, September-October, p. 52.

Svenson, A., Edsholt, E., Ricking, M., Remberger M., and Röttorp, J. (1996) "Sediment contaminants and Microtox toxicity tested in a direct contact exposure test," *Environ. Toxicol. Water Qual.*, Vol. 11, p. 293-300.

Svenson, A., Viktor, T., and Remberger, M. (1998) "Toxicity of elemental sulfur in sediments," Environ. Toxicol. Water Qual., Vol. 13, p. 217-224.

Swamy R.N. and Jurjees T.A.R. (1986) "Stability of sulfur concrete beams with steel reinforcement," *Materiaux et Constructions*, Vol. 19, No. 113, p. 351-359.

Syroezhko, A.M., Yu, O., Fedorov, V.V., Gusarova, E.N., (2003) "Modification of paving asphalts with sulfur," *Russian Journal of Applied Chemistry*, Vol. 76, No. 3, p. 491-496.

Tamura, K., Seyer, H.P., and Hensel, F. (1986) Ber. Bunsenges. *Phys. Chem.*, Vol. 90, p. 581.

Texas Gulf Sulfur Co. (1961) *Properties of sulfur*.

Thackray, M. (1970) "Melting point intervals of sulfur allotropes, *J. Chem. Eng. Data*, 15, p. 495-497.

Thamdrup, B., Finster, K., Fossing, H., Hansen, J., and Jørgensen, B.B. (1994) "Thiosulfate and sulfite distributions in pore water of marine sediments related to manganese, iron, and sulfur geochemistry," *Geochim. Cosmochim. Acta*, Vol. 58, p. 67-73.

Thieler, E. (1936) Schwefel (Sulfur), Verlag von Theodor Steinhoff, Dresden und Leipzig (in German).

Tiger Industries Ltd., (1999) *Run with the leader—New Tiger 90CR, quicker breakdown, season-long sulphur nutrition*, Calgary, CA, Tiger Industries Ltd. brochure, p. 8.

Timrot, D. L., Serednitskaya, M. A., and Chkhikvadze, T. D. (1984) *Sov. Phys. Dokl.*, Vol. 29, p. 961-963.

Tobolsky, A.V. and MacKnight W. J. (1965) *Polymeric sulfur and related polymers*, New York, Interscience Publishers.

Tohji, K. (2002) *J. Energy and Resources*, Japan, Vol. 23, p. 15.

Tucker, J.R. and Schweyer, H.E. (1965) *Ind. Eng. Chem. Prod. Res. Dev.* Vol. 4, p. 51.

Tuller, W.N., ed. (1954) *The sulphur data book*, McGraw-Hill Book Co., Inc., New York, p. 143

U.S. Environmental Protection Agency, (1980) *Multi-media assessment of the inorganic chemicals industry*, Vol. III, Chapter 14.

U.S. Bureau of Mines. (1978) "Direct substitution of sulfur for asphalt in paving materials."

U.S. CIA. (2002) "The World Fact Book 2002," Asphalt Institute, *The Asphalt Handbook*.

U.S. EPA. (1991) *R.E.D. facts, Sulfur*, U.S. Environmental Protection Agency, Pesticides and Toxic Substances Branch, Washington, D.C. p. 4.

U.S. National Library of Medicine. (1995) *Hazardous substances databank*, National Library of Medicine, Bethesda, MD.

U.S. Patent No. 3,738,853 (1973).

Vallegra, B.A., White, R.M., and Rostler, K.S. (1970) *Change in fundamental properties of asphalts during service in pavements*, Office of Research and Development, U.S. Bureau of Public Roads.

Van Dalen, A. and Rijpkema, J.E. (1989) *Modified sulphur cement: A low porosity encapsulation material for low, medium and alpha waste*, EUR 12303 EN, Commission of the European Communities, Luxembourg.

Van Oort, W. P. (1956) "Durability of asphalt," *Industrial and Engineering Chemistry*, p. 1196-1201.

Vanegas, J., Du Bose, J., and Pearce, A. (1995) "Sustainable technologies for the building construction industry," *Proceedings of the 1995 Symposium on Design for the Global Environment*, Atlanta, GA, November 2-3.

Voronkov, M.G., Vyazankin, N.S., and Deryagina, E.N. (1979) *Reactions of sulfur with organic compounds*, Nauka, Moscow, p. 367.

Vroom, A. H. (1977) *Sulfur cements, process for making same and sulfur concretes made therefrom*, U.S. Patent No. 4,058,500.

Vroom, A.H. (1981) *Sulfur cements process for making same and sulfur concretes made therefrom*, U.S. Patent No. 4,293,463.

Vroom, A.H. (1998) "Sulfur concrete goes global," *Concrete International*, Vol. 20, No.1, p. 68-71.

Vroom, A.H. (1992) "Sulfur polymer concrete and its application," In *Proceeding of Seventh International Congress on polymers in concrete*, Moscow, p. 606-621.

Vroom, H., and Whitmore, D. W., (1991) "Sulfur concrete for high corrosion resistance," *Proceedings, National Association of Corrosion Engineers*, Saskatoon, Canada, February.

Wallcave, L., Garcia, H., Feldman, R., Lijinsky, W., and Shubik, P.L. (1971) "Skin tumorigenesis in mice by petroleum asphalts and coal-tar pitches of known poly-nuclear aromatic hydrocarbon content," *Toxicol. Appl. Pharmacol.* Vol. 18, p. 41-52.

Warren, C.J. and Dudas, M.J. (1992a) "Acidification adjacent to an elemental sulfur stockpile: I. Mineral weathering," *Canadian Journal of Soil Science*, Vol. 72, p. 113-126.

Warren, C.J. and Dudas, M.J. (1992b) "Acidification adjacent to an elemental sulfur stockpile: II. Trace element redistribution," *Canadian Journal of Soil Science*, Vol. 72, No. 2, p. 27-134.

Watts, R.R., Wallingford, K.M., Williams, R.W., Hours, D.E., and Lewtas, J. (1998) "Airborne exposures to PAH and PM2.5 particles for road paving workers applying conventional asphalt and crumb rubber modified asphalt, *J. Exposure Anal. Environ. Epidemiol.* Vol. 8, p. 213-229.

Weber, H.H. and McBee, W.C., (2000) "New market opportunities for sulphur asphalt," *Sulphur markets today and tomorrow, Biennial International Symposium*, 7th, Washington, D.C., March 26-28, p. 24.

Weber, H.H., McBee, W.C., and Krabbe, E.A. (1990) "Sulfur concrete composite materials for construction and maintenance," *Materials Performance*, Vol. 29, No. 12, 1990, p. 73-77.

West, E. D. (1959) "The heat capacity of sulfur from 25 to 450°C, the heats and temperatures of transition and fusion," *J. Am. Chem. Soc.*, Vol. 81, p. 29.

West, J.R. (1966) "Safety in handling liquid and dry sulfur," *Commercial Fertilizer*, Vol. 112, No. 1, January, p. 15-20.

Weston, R. (1995) "Sustainable development: Integration of free market and natural economic systems," Presented at the *AAES/Engineering Foundation conference "Sustainable Development: Creating Agents of Change,"* Snowbird, Utah. August 4-8.

Whelpdale, D.M. (1992), "An overview of the atmospheric sulphur cycle," in *Sulphur Cycling on the Continents: Wetlands, Terrestrial Ecosystems and Associated Water Bodies*, R.W. Howarth, J.W.B. Stewart and M.V. Ivanov, (eds.), SCOPE report No. 48, John Wiley and Sons, Chichester, UK, p. 5-26.

Whitcomb, J.H., Delaune, R.D., and Parrick, W.H., Jr. (1989) "Chemical oxidation of sulfate to elemental sulfur: its possible role in marsh energy flow," *Mar. Chem.*, Vol. 26, p. 205-214.

Whiteoak, C.D. (1990) *The Shell Bitumen Handbook*," Shell Bitumen, Surrey, UK.

Wilck, K. (1991) "Cement Technology" Danderyd: Cementa AB (in Swedish).

Winter, R., Szornel, C., Pilgrim, W.C., Howells, W.S., and Egelstaff, P.A. (1990) *J. Phys. Condens. Matter*, Vol. 2, p. 8427.

Woo, G. L. (1983) *Phosphoric acid treated sulfur cement-aggregate compositions*, U.S. Patent No. 4,376,831.

Worch, R.J. (1979) "Use of sulfur in bituminous pavements," *Engineering Brief No. 21*. Federal Aviation Administration, Engineering Specifications Division, Washington, D.C.

Wrzesinski, W.R. and McBee, W.C. (1988) "Permeability and corrosion resistance of reinforced sulfur concrete," *Report to the U.S. Bureau of Mines*, Pittsburgh, PA. U.S. Government Printing Office, 605-017/80.001, p. 13.

Yakub, I.A. and Alekseev S.N. (1971) *Korroziya armatury v lyohkih betonah*, Moscov, Stroyizdat.

Yeang, K. (1995) *Designing with Nature*, McGraw Hill, New York, NY.

Yong, R.N., Mohamed, A.M.O., and Warkentin, B.P. (1992) *Principles of contaminant transport in soils*, Elsevier, Amsterdam, The Netherlands, p. 327.

Zheng K.M. and Greer, S.C. (1992) "The density of liquid sulfur near the polymerization temperature," *J. Chem. Phys.*, Vol. 96, p. 2175.

Zwicker, D.A. (1990) "Bright future for brimstone," *The Lamp*, Exxon, Fall, p. 10-15.

INDEX

A

Abrasion resistance, 313
Absorption, changes in, 134–135
    Absorption, sulfur concrete, 204
Accelerated corrosion cell, 320, 321
ACI Committee 548, 141
ACI guide for material selection, 130–135
    absorption, 134–135
    aggregate gradation, 130–131, 132
    corrosion resistance, 131
    moisture absorption, 131
    optimum mix design, 132–133
    voids, 134
ACI mixing scheme, 270–271
AD A/S mixing scheme, 270
Admixtures, durability and, 135–137, 138
Advantages, sulfur concrete and, 205–206
Aggregate dryer/superheated, 282
Aggregate gradation, sulfur concrete, 205
Aggregate type, durability and, 135–137, 138
Aggregates, 122–130, 141
    cement and, 111
    clay containing, 141–142
    coarse, smelting temperature, 291
    density/sulfur content, 125–128
    gradation of, ACI guide and, 130–131, 132
    grain shape, 123–124
    grain size distribution, 124–125, 126, 266
    natural/manufactured, 215
    optimum grain size distribution/sulfur content, 129–130
    sieve analysis of, 124, 126, 128, 129
    strength and, 128–129
    type of, sulfur loading/admixtures and, 135–137
    types of, 292–300
Agricultural chemicals, 51–52, 108. *See also* Fertilizer production
    degradable sulfur, 106
    dusting sulfur, 106
    flowable sulfur, 106
    nitrogen fertilizers, 51
    other agricultural uses, 52
    phosphoric acid/phosphate fertilizers, 51
    plant nutrient sulfur, 51–52
    wettable sulfur, 104, 106
Airport pavements, 369
Aliphatic polysulfide, 150, 372, 373
Allotropes. *See also* Elemental sulfur, properties of
    color of solid, 80–81
    intermolecular sulfur, 108
    most abundant type, 76
    solid sulfur, 76–77
    types of, 75, 107–108
    various forms of, 71–72
Allotropic sulfur forms, 288
Alpha crystallinity, 185
Aluminous cement, 262
American Chemical Society, 110
American Concrete Institute, 144
American Smelting and Refining Company (ASARCO), 260

Anthropogenic sulfur cycle, 38–39
Anti-stripping agents, asphalt and, 340
Aromatic polysulfides, 150, 372–373
ASARCO. *See* American Smelting and Refining Company
Asphalt. *See also* Road paving asphalt; Sulfur-modified asphalt
  aging of, 332, 338–339
  aromatic molecules of, 333
  composition of, 332–334
  fingerprint for, 333
  fractionation of, 334–335
  lime-modified, 339–340
  oxidation of, 332
Asphalt bitumen, 377, 379
Asphalt component interaction, 335–338
  bonding, 335–336
  chemical composition, 336–337
  compatibility, 338
  nonpolar molecules, 338
  polar molecules, 337
  polar vs. non-polar molecules, 337
Asphalt improvement, sulfur and, 359–360
  fatigue life, 360
  low temperature stiffness, 359
  resilient modulus of elasticity, 359–360
  stripping, resistance to, 360
  tensile strength, 359
Asphalt pavements, 55–56
Asphaltenes, 334, 346
Assessment protocol, 327–328
ASTM C1159-98, 145
ASTM C127, 131
ASTM C128, 131
ASTM C1312-97, 145
ASTM C33 specification
  aggregates and, 130, 132, 141
ASTM C457, 134
ASTM C642, 134
ASTM cards, 119
ASTM Committee for Chemical Resistance Materials, 145
ASTM D3515, 132
ASTM specifications, chemical-resistant sulfur mortar, 110
Atmospheric pollutants, Portland cement concrete and, 115
Atoms of sulfur
  liquid sulfur allotropes, 76
  rings and chains, 75–76, 108
  solid sulfur allotropes, 76–77

B

Bacteria, sulfur compounds and, 96–99
Beta crytallinity, 185
Binder requirements, 201–202. *See also* Paving binders
Biocide treatments, 118
Biodiversity, 22, 25
Biodiversity management plan (BMP), 25
Biological cycles, Portland cement concrete and, 118
Biological reactions of sulfur compounds, 96–99
Bitumen, sulfur modified with, 169, 170–183
  chemical reactions, sulfur compounds, 170–171
  composition analysis via FT-IR, 171, 172
  composition analysis via XRD, 171, 172–173
  durability of sulfur bitumen cement, 182–183
  microstructure, 173, 175–179
  physico-chemical composition, 170
  SEM/x-ray mapping, 174–178
  sulfur polymerization, degree of, 173, 180
  thermal analysis, 180–182
  X-ray diffraction patterns, 172
Bitumen-modified sulfur concrete (BMSC), 220–251
  durability of, 237–241
  hydraulic conductivity, 241–243
  leachability, 244–251
  long-term hydromechanical behavior, 234
  microstructure characterization, 225, 226
  production of, 220
  reaction products, 234–237
  strength development for, 227–234
  sulfur ratio/loading, effect of, 223–224
  thermal stability, 221–223
BMP. *See* Biodiversity management plan
BMSC. *See* Bitumen-modified sulfur concrete
BND 60/90, 348
Boiling point, 102
Building construction industry, sustainable development and, 13–15

Index   413

Built environment, ecological systems and, 20
Bureau of Mines, 162

C

Canada
  precast products, 144
  sulfur trade and, 44–45
Canada Center for Mineral and Energy Technology (CANMET), 258
CANMET. *See* Canada Center for Mineral and Energy Technology
Caprolactam, 53
Carbon disulfide, 154
Carcinogenic PAHs, 377–378
Catalytic crackers, petroleum refining, 56–57
Cement, terminology and, 145–146
Cement dust, 254
Cement kiln dust (CKD), 293–300
Cement-based concrete, 111
Cenospheres, 186
Chement 2000 (Chemically Resistant Cement), 261
Chemical loads, 327
Chemical properties of sulfur, 88–99
  biological reactions of sulfur compounds, 96–99
  chemical reactions, elemental sulfur, 93–94
  chemical reactions, sulfur compounds, 95
  chemical reactions, sulfur with olefins, 95
  electronic structure, 89–90
  oxidation states, 90–93
Chemical pulping, 58–59
Chemical reactions, elemental sulfur, 93–94
  acids, reaction with, 94
  air, reaction with, 93
  bases, reaction with, 94
  halogens, reaction with, 94
  iron/zinc, reaction with, 94
  silver, reaction with, 94
  water, reaction with, 93
Chemical reactions, sulfur with olefins, 95
Chemical resistance, 315–317, 318
Chemical stabilizers, 168–169, 191–192, 214
Chemical/industrial sulfur uses, 52–54
  caprolactam, 53
  hydrofluoric acid (HF), 53
  other uses, 54
  titanium dioxide, 53
  vulcanization of rubber, 54
Chemically-bonded sulfur, 343, 344–346
  colloidal structure, 346
  electron paramagnetic resonance, 344
  naphthene-aromatic fraction, 345
  polysulfide linkages, 345
Chemical-resistant sulfur mortar, 110
Chempruf, chemicals and, 314, 315–317
Chempruf mixing scheme, 268, 269, 270
Chempruf modifier, 261
CKD. *See* Cement kiln dust
Claus process, 42–43
Clay containing aggregates, 141–142
Clinkerization, 116
Coarse aggregates handling process, 277–278
Coefficient of thermal expansion, 204
Cold weather placements, 284–285
Colonization of a matrix, 118
Composition, sulfur concrete, 197–201
Compressive strength
  overview of, 111, 112
  reduction, recovery cycles and, 290
  strength development, elemental sulfur concrete, 118–121
  strength development, Portland cement concrete, 112–114
  strength reduction, Portland cement concrete, 114–118
  of sulfur mortar, 152
Compressive strength, sulfur concrete, 204
Concrete mixer, heated, 282, 283
Concrete piping, 264–265
Construction industry
  corrosion resistance materials, 56
  potential applications, 56, 57
  sulfur extended asphalt, 54–56
  sulfur uses in, 54–57
Construction time, 273
Cooling process, energy requirement, 291–292
Corbett asphalt separation method, 335
Corrosion potential, 314, 317, 318–323
  Corrosion resistance
    aggregates and, 131, 141
    materials for, 56
CPDO. *See* Cyclopentadiene oligomer
Crazing resistance, 310

Creep, 310, 311
Critical temperature, 102
Crystal forms of sulfur, 71, 108
Crystalline sulfur, asphalt and, 347–348
Crystallization process, 130, 192
 alpha/beta crystallinity, 185
 for modified sulfur, 147–148
Cyclic penta-sulfide, 154, 156
Cyclic polysulfides, 154
Cyclo-octal sulfur, 76, 77
Cyclooctasulfane, 171
Cyclopentadiene oligomer (CPDO), 212
Cyclpentadiene, sulfur modified with DCPD and oligomer of, 159–164

D

Damage repair, 286
DCPD. See Dicyclopentadiene
DCPD and oligomer of cyclpentadiene, sulfur modified with, 159–164
 chemical reaction, 159–164
 materials/properties of, 162–164
 modifier loadings/physical properties, 164
 test conditions/physical observation, 163
DCPD and styrene, sulfur modified with, 167, 168
Degradable sulfur, 106
Demand for sulfur, 38
Demolition
 reuse and, 33
 waste recycling, 20
Denaturing molten sulfur, 276
Density/sulfur content, aggregates, 125–128
Desulphurization of oil/gas, 121
Development
 reuse of existing, 20
 sulfur concrete, 194–197
Dicyclopentadiene (DCPD), 208. See also Cyclpentadiene, sulfur modified with DCPD and oligomer of
Dicyclopentadiene (DCPD), sulfur modified with, 153–159
 chemical structure of formed product, 154–155, 156
 limitations, 158, 159
 storage effect, 157–158, 159
 sulfur cement composition, 153–159
 surface tension, 155, 156–157
 viscosity, 155, 156

Dicyclopentadiene-cyclopentadiene oligomer-modified sulfur concrete, 212–214
 freeze-thaw, strength and, 212–213
 mix composition, strength and, 212
 percent of DCPD to oligomer in mix, 212–213
Dicyclopentadiene-modified sulfur concrete, 207–212
 DCPD loadings/aggregate types, 208–209
 storage time, 210–211
 thermal stability, 211–212
Differential scanning calorimeter (DSC), 100, 101, 180–182, 221–222
Di-olefins, 95
Dispersive losses of sulfur, 65–66
Dispersive x-ray detector. See EDX
Dissolution/leaching, Portland cement concrete, 115
Dissolved sulfur, asphalt and, 346–347
Di-sulfides, 92
DSC. See Differential scanning calorimeter
DTA pen 60/90, 347
Ductility, 270
Durability, BMSC, 237–241
 compressive strength, acidic environment, 238–239
 compressive strength, saline environment, 240–241
 moisture absorption, 237–238
Durability, elemental sulfur concrete, 135–140
 compressive strength and, 139–140
 cracks/micro-cracks, 138–140
 sulfur loading, aggregate type, different admixtures, 135–137, 138
 water/temperature, effect of, 137–140
Durability, sulfur concrete, 257–258, 292
 abrasion resistance, 313
 chemical resistance, 314, 315–317, 318
 corrosion potential, 314, 317, 318–323
 crazing resistance, 309, 310, 311
 creep, 309, 310
 fatigue strength, 310
 fillers/aggregates, types of, 292–300
 fire load, 307–308, 309, 310
 frost resistance, 305
 reinforcement, 311–313
 service temperature, 306–307
 water absorption, 300–305, 306

Dusting sulfur, 106

E

Ecological effects
  potential, 251–254
  sulfur-modified asphalt, 376–379
Ecological systems
  built environment and, 20
  elemental sulfur effects, 103–104
  sulfur isotopes in, 102–103
Economic feasibility, 329
Economic sustainability, 6–7
  natural system and, 7
EDS. *See* Energy dispersive spectroscopy
EDX (dispersive x-ray detector), 173, 178–179
EIA. *See* Environmental impact assessment
Electrical properties, 102
Electron paramagnetic resonance, 344
Electron spins resonance (ESR) spectra, 87
Electronic structure of sulfur, 89–90
Elemental sulfur
  forms of, 75–77
  uses for, 48
Elemental sulfur concrete, 109–142
  compared to DCPD sulfur concrete/
    Portland cement, 209
  compressive strength, 111, 112
  history, sulfur concrete development, 110
  material composition, 121–135
  overview of, 109–110
  SEM micrograph/EDX spectrum, 120
  strength development for, 118–121
  strength development, Portland cement concrete, 118–121
  strength reduction, Portland cement concrete, 114–118
  summary/conclusion, 140–142
  terminology, 111, 112
  voids in, 121
  X-ray diffraction analysis of, 119
Elemental sulfur, properties of, 77–88, 225
  allotropic transformation, 82–86
  color, 79, 80–81
  density, 79, 80
  melting/freezing points, 77–78
  polymerization, 86–88
  pore sizes, 226

  strength characteristics, 81, 82
  thermal characteristics, 82, 343, 344
  viscosity, liquid sulfur, 79
Emissions, sulfur asphalt and, 358–359
ENB, sulfur modified with, 184–185
  Japan, certification in, 264–265
Energy dispersive spectroscopy (EDS), 121
Energy requirement
  cooling process, 291–292
  heating process, 287–289
  recovery process, 289–291
Environmental impact assessment (EIA), 22
Environmental impacts, avoiding negative, 19–20
Environmental issues, 62–67, 273. *See also*
    Natural resources
  business justification and, 22
  economics, 7
  environmental effects of sulfur, 64
  health effects of sulfur, 63–64
  procurement process and, 25–26
  production/processing, 62–63
  reduce, reuse, recycle, 19–20
  safety, 358
  sustainability, environmental, 5–6, 13
  waste management and, 64–67
Environmental performance standards, 26
Environmental Protection Agency (EPA), 103, 250
EPDP. *See* Ethylene-propylene-diene monomer elastomers
Equilibrium polymerization, 86, 87–88
Equipment, commercial scale application, 273–274
  development of, 271, 272–273
ESR. *See* Electron spins resonance spectra
Ethylene-propylene-diene monomer (EPDM) elastomers, 184
Ettringite crystallization, 116–117
Existing development, reuse of, 20
Expansion joints, 286
Expansive reactions, elemental sulfur/aggregate, 135–140
Expansive reactions, Portland cement concrete, 116–117
  ettringite formation, 116–117
  hydration of CaO and MgO, 116
  thaumasite formation, 117, 118

Explanatory view, modified sulfur solidified concrete production system, 275

F

Fatigue strength, 310
Feasibility, 273–274
Federal Highway Administration (FHA), 366
Fertilizer production. *See also* Agricultural chemicals
   phosphate fertilizer industry, 69
   sulfuric acid and, 49
FHA. *See* Federal Highway Administration
FHWA evaluated sulfur extended asphalt projects, 365
Filler
   cold filler, 270
   composition/properties and, 293
   function of, 142, 146
   grain size distribution, 266
   types of, 292–300
Final modified-sulfur concrete product, 279–281
Fine aggregates handling process, 276–277
Finishing/placing, 284
Fire load, 307–308, 309, 310
Fire retardant additives, 274, 307–308
   5-ethylidene-2-norbornene (ENB), sulfur modified with, 184–185
   5-ethylidene-2-norbornene-modified sulfur concrete, 216–220
   compressive strength, alkaline environment, 218–219
   ignition/biological oxidation, 219–220
   weight loss, alkaline environment, 218
Flame retardancy. *See* Fire load
Flexural strength, sulfur concrete, 204
Flowable sulfur, 106
Flowering process, 74–75
Fly ash, 186
Formed sulfur, 43, 44
Forming/reinforcement, 287
Forms of sulfur, 73
Fractionation of asphalt, 334–335
Framework, sustainable industry, 10–13
   current system, 10–12
   modified system, 12–13
   subsystems in, 11, 13
Frasch mining, 40–41

demise of industry, 62
percentage of mining, 39
Freeze-thaw cycles
   durability, sulfur concrete, 205
   Portland cement concrete and, 117, 118
   sulfur-asphalt bound material, 368–369
Freezing point, elemental sulfur, 77–78
Freshwater sediments, sulfur in, 97
Frost resistance, 305
Fuller distribution, 266
Fuller parabola/Fuller curve, 125, 266–267

G

Gap-grading, 266
Glass fiber reinforcement, 287, 327
Global ecosystems, environmental sustainability and, 5–6
Global sulfur cycle, 36–39
   anthropogenic sulfur cycle, 38–39
   natural sulfur cycle, 37–38
GNP. *See* Gross National Product
Grain shape, aggregates, 123–124
Grain size distribution, aggregates, 124–125, 126
   optimum grain size distribution/sulfur content, 129–130
Grey water systems, 20
Gross National Product (GNP), 6
Guide to joint sealants for concrete structures, 286

H

Health effects
   of sulfur, 63–64
   of sulfur concrete, 251–254, 255
Heat absorption of sulfur, 288
Heated concrete mixer, 282, 283
Heating process energy requirement, 287–289
Heat-jacked Portland cement concrete transit mixers, 282
HF. *See* Hydrofluoric acid
Highways. *See* Asphalt; Paving binders; Road paving asphalt
History, sulfur concrete development, 110
Hot mix. *See* Mixing process
Human body, sulfur in, 60

Human needs/aspirations, sustainability
    and, 17–19
  economic viability and, 18
  healthy built environment, 18–19
  people meeting own needs, 19
  user needs/facility design, matching of,
    18
Hydrated lime treatment for asphalt,
    339–340
Hydration of CaO and MgO, 116
Hydrofluoric acid (HF), 53
Hydrogen bonding, asphalt and, 335
Hydrogen sulfide, 95, 358
  emission control for, 363

I

In situ construction mixing/placing
    techniques, 281–283
Industrial modified sulfur cement, 153–185
  bitumen, sulfur modified with, 169,
    170–183
  DCPD and oligomer of cyclpentadiene,
    sulfur modified with, 159–164
  DCPD and styrene, sulfur modified
    with, 167, 168
  dicyclopentadiene (DCPD), sulfur
    modified with, 153–159
  5-ethylidene-2-norbornene (ENB),
    sulfur modified with, 184–185
  oelfinic hydrocarbon polymers, sulfur
    modified with, 168–169, 170
  styrene, sulfur modified with, 164–167
Industrial sulfur, 106–109, 108
Inorganic plasticizers, 372
ISO 14000, 1, 32
ISO 15686-1, 324–326
Isotopes, 102–103

J

James River Pier, 262
Japan
  ENB-modified sulfur in, 264–265
Japan Petroleum Energy, 274, 275
Joints/joint sealing, 286

K

Key performance indicators (KPIs), 26

L

Land sustainability, 7–8
LCAC. *See* Life-cycle-assessment costs
Leachability, 327
Leachability, BMSC, 244–251
  acidity, influence of, 251
  pH/soil composition, influence of,
    244–247
  salinity, influence of, 248, 249–250
  temperature, influence of, 247–248, 249
Leaching/dissolution, Portland cement
    concrete, 115
Lewis electron diagrams, 90, 92
Life cycle. *See* Project procurement life cycle
Life of product. *See* Service life
Life-cycle-assessment costs (LCAC), 24
Lime-modified asphalt, 339–340
Linear polarization resistance technique
    (LPR), 319, 320
  results, 321, 322–323
Liquid sulfur
  allotropes, 76
  behavior of, 342–343
  polymerization, 72
LPR. *See* Linear polarization resistance
    technique

M

Macro-sulfur crystals, 147
Manufacturing, 274–281
  equipment/methods, 202–204
  joints/joint sealing, 286
  pre-cast mixing/production, 274–281
Marine sediments, sulfur in, 97
Market dynamics, 68
Marshall stabilities, sulfur asphalt binder-
    based mixes, 355–356, 361–362
Material composition, elemental sulfur
    concrete, 121–135, 122
  ACI guide for material selection,
    130–135
  aggregates, 122–130
  elemental sulfur, 121–122
Material recycling, 20
Mechanical loads, 327
Melting point
  of elemental sulfur, 77–78
  typical temperature for, 101–102

Melting sulfur, heating process, 287–289
Microcracking, 270
Microtox assay, 103
Middle Eastern countries, 121
Mineral filler
   function of, 142
   plasticity of mix and, 266
Mix design, 265–267
   for aggregates, 132–133
   critical sulfur content value, 265, 266
Mixing process, 267–271
   fifth scheme, 271
   first scheme, 268–269
   fourth scheme, 270–271
   mix proportions, typical, 268
   second scheme, 269
   third scheme, 269, 270
Mixing/transporting requirements, 283
Modifications, sulfur based cement, 151–152
Modified sulfur, 146, 146–152
   forming/reinforcement, 287
   mechanism for, 146–148
   melting point of, 257
   modification conditions, 151–152
   types of, 148–151
Modified sulfur cement. *See* Sulfur polymer cement
Modified sulfur concrete (MSC), 193–194
Modifying agent handling process, 276
Modulus of elasticity, sulfur concrete, 205
Mohamed and El Gamal mixing scheme, 271, 272
Moisture absorption, aggregates and, 131
Moisture/wind, 285–286
Molten sulfur handling process, 276
Monoclinic sulfur, 76, 108
Morocco, sulfur trade and, 45
MSC. *See* Modified sulfur concrete

N

National Chempruf Concrete, 261
National Research Council (NRC) of Canada, 258
Natural gas, 41
Natural resources. *See also* Environmental issues
   economic system and, 7
   exploitation/use of primary, 11
   input rules for, 5
   management of, 13
   output rule, 6
Natural sulfur cycle, 37–38
Nippon Oil Corporation, 274, 275, 280
Nitrogen fertilizers, 51
NMR. *See* Nuclear magnetic resonance analysis
Non-aqueous-acid-base titration, 337
NRC. *See* National Research Council of Canada
Nuclear magnetic resonance (NMR) analysis, 337
Nucleo-philicity of sulfur, 95

O

Off-spec sulfur, 122
Oil sands, 42
Olefinic hydrocarbon polymers, sulfur modified with, 168–169, 170
Olefins, 58
   di- and tri-olefins, 95
   olefin hydrocarbon polymers, 151
   olefinic hydrocarbon polymer-modified sulfur concrete, 214–216
   olefinic hydrocarbon polymers, 169
On-site construction, *See In situ* construction mixing/placing techniques
Optimum mix design, aggregates, 132–133
Optimum range of sulfur, 266
Ore processing, 57
Organic plasticizers, 372
Organosilane, 302–305
   loadings/effect of, 304, 305, 306
   molten sulfur reactive groups, 303
Orthorhombic sulfur, 108, 147
   bitumen added to, 173
   as most common form, 77
   prohibiting formation of, 153
Oxidation
   biological reactions and, 96–99
   oxidation number, 91
   oxidation states for sulfur, 90–93

P

PAHs, 377–378
Paradigm shift, traditional/sustainable development, 14

Paving binders, 373–374, 379
PCC. *See* Portland cement concrete
PCI. *See Present condition index* (PCI) for pavement
Penetration, sulfur-asphalt blends, 354, 355
  40/50 pen asphalt, 360
  pen 60/90 paving asphalt, 347, 353
Penta-sulfide, 154
Petroleum, 41–42
  alkylation of, 57
  refining/catalytic crackers, 56–57
Pharmaceutical industry, 60–61
Phosphate fertilizer industry, 69
Phosphoric acid/phosphate fertilizers, 51
Photovoltaic (PV) panels, 16
Piers, 262
Pi-pi bonds, asphalt and, 335–336, 337
Piping, concrete, 264–265
Piping/transfer systems, 279
Placing/finishing, 284
Plant nutrient sulfur, 51–52
Plasticity of hot mix, 266
Plasticization of sulfur with mixtures, 95
Plasticized sulfur, 144, 370–376
  case studies, 375–376
  chemicals used in, 372–373
  mixing conditions, 374–375
  plasticization concept, 370–372
  plasticization perceptions, 374
  plasticizing agent requirement, 373–374
Plasticized sulfur/plasticizer, 148–151
Polarization. *See* Linear polarization resistance technique
Pollution, 26
Polymeric polysulfide, 147
Polymerization in sulfur, 86–88, 108
  bond-switching mechanism, 88
Polysulfides, 150
Porous aggregates, 131
Portable machine, 283–284
Portland cement
  aggregates and, 111
  cast-in-place, 57
  compared to elemental/DCPD sulfur concrete, 209
  PC (paste of), 145
Portland cement concrete (PCC), 145. *See also* Strength reduction, Portland cement concrete
  construction time for, 273
  corrosion potential, 321–322, 323
  heat-jacked transit mixers for, 282
  steel reinforcement for, 287
  strength development for, 112–114
  sulfur concrete and, 142
Practical applications, 104–107, 108
  agricultural sulfur, 104, 106
  industrial sulfur, 106–107
  rubber-maker's sulfur, 107
Pre-cast mixing/production, 274–281
  coarse aggregates handling process, 277–278
  final modified-sulfur concrete product, 279–281
  fine aggregates handling process, 276–277
  modified sulfur production process, 278–279
  modifying agent handling process, 276
  molten sulfur handling process, 276
  piping/transfer systems, 279
  sulfur cement production process, 277
Precast sulfur concrete, 144, 259
Present condition index (PCI) for pavement, 366
Processes, sulfur, 74–75
  flouring, 75
  flowering, 74–75
  roasting, 74
  rolling, 75
  sweetening, 74–75
Procurement. *See* Project procurement life cycle; Sustainable procurement process
Production, 258–265
  in the 1970s, 258–260
  in the 1980s, 260–262
  in the 1990s, 263–264
  in the 2000s, 264–265
Products, 104–107
  practical applications, 104–107
  product groups, 104
Project procurement life cycle, sustainability and, 21–33
  business justification, 21–24
  construction process, 29–30
  design and, 27–29
  disposal/re-use of site, 32–33

management/facility operation, 31–32
procurement process, 24–27
Project size, 273
Properties. See Sulfur properties
Pulp and paper, 58–59
PV. See Photovoltaic (PV) panels
Pyrite, 72, 97

Q
Quenched sulfur, 81

R
Recovered elemental sulfur (Claus process), 42–43
Recovery process, energy requirement, 289–291
Recycling, 67
  sustainability and, 19–20
Reduction reactions, sulfur compounds, 96–99
Reduction, sustainability and, 19–20
Reference service life (RSL), 325–326
Refrigerant, 92
Rehabilitation, sustainability and, 16, 20
Reinforcement, 311–313
Reinforcement/forming, 287
Repairing damages, 286
Requirements, sulfur concrete, 201–202, 203, 207
Resource processing/manufacturing, 11
Resource recovery, as sustainability challenge, 13
Retrofitting, sustainability and, 16
Reuse, sustainability and, 16
  existing development, 20
Rhombic sulfur. See Orthorhombic sulfur
Road paving asphalt, 341, 362, 377, 379. See also Asphalt; SEA pavements
Roasting process, 74
RSL. See Reference service life
Rubber-maker's sulfur, 107

S
SAAs. See Sulfur-containing amino acids
SAM. See Sulfur-asphalt mixing module
Sample specimens, 287
Sand-asphalt-sulfur (SAS), 363–364
SC. See Sulfur concrete

Scanning electron microscopy (SEM), 120, 225, 226
Schedule, 274
Scope, 274
SEA. See Sulfur-extended asphalt
SEA binders, 365
SEA pavements, 366, 367, 369
Self-contained machine, 283–284
SEM. See Scanning electron microscopy
Service life, 323–326
  estimation for sulfur concrete, 325–326
  ISO approach, 324–325
Service temperature, 306–307
Shell Thiopave technologies, 342
Silica dust, 254
Smelting process, 62
  for nonferrous sulfides, 41
  sulfuric acid recovery, 68
Social sustainability, 4–5
Solid sulfur allotropes, 76–77
Solidification of elemental sulfur, 140–141
SPC. See Sulfur polymer cement
Splitting tensile strength, sulfur concrete, 204
SRX polymer, 258–259
Stabilizers for sulfur cement, 168–169, 186–187, 191–192
Standard testing, sulfur cement, 189
Starcrete Technologies, 258, 260
Steel reinforcement, 287, 327
Storage of substances, 327
Strength reduction, Portland cement concrete, 114–118
  atmospheric pollutants and, 115
  biological cycles, 118
  dissolution/leaching of, 115
  expansive reactions, 116–117
  freeze-thaw cycles, 117, 118
Stripping, resistance to, 360
Structuring agent, sulfur as, 361
STXTM polymer, 260, 263
Styrene, sulfur modified with, 164–167
  concentration/reaction time, 164–165
  reaction products, 167
  viscosity, 165, 166–167
SUDIC. See Sulfur Development Institute of Canada
Sulfa drugs, 60–61
Sulfate reduction, oxidation and, 96–99

Index **421**

Sulfex, 253
Sulfide ores, 74
Sulf-oxides, 92
Sulfur asphalt binder, rheology of, 351–356
　compressive strength, 353–354
　density, 351
　Marshall stability, 355–356
　penetration, 354, 355
　viscosity, 352–353
Sulfur asphalt binders, preparation apparatus for, 356
Sulfur asphalt development, 363–368
　sand-asphalt-sulfur (SAS), 364
　sulfur-extended asphalt (SEA), 364–368
Sulfur asphalt history, 341–342
Sulfur asphalt interaction, 343–348
　chemically-bonded sulfur, 343, 344–346
　crystalline sulfur, 347–348
　dissolved sulfur, 346–347
Sulfur asphalt mix concepts, 348–351
　mixing temperature/times, 349–350
　sulfur to asphalt ratio, 349
　sulfur-asphalt emulsion, 348–349
　Sulfur asphalt processing technology, 356–357
　construction procedure, 357
　emissions, 358–359
　manufacturing evaluation, 356
　mix production, 357
　premixing, 356–357
　preparation apparatus, sulfur asphalt binders, 356
　safety and the environment, 358
　sulfur asphalt module, 357
Sulfur binder
　aggregates and, 135
　demand for, 122
　modification of, 141
Sulfur bitumen cement, durability of, 182–183
Sulfur cement, 143–192
　advantages of, 190
　development background, 143–145
　disadvantages of, 190–191
　factors controlling formation of, 187–189
　industrial modified sulfur cement, 153–185
　modified sulfur, 146–152
　stabilizers for, 186–187
　standard testing of, 189
　sulfur cement, 186–187
　summary/conclusion, 191–192
　terminology and, 145–146
Sulfur cement formation, factors controlling, 187–189
　cooling rate of, 187–189
　polymer modification of sulfur and, 188
Sulfur cement production process, 277
Sulfur compounds, biological reactions, 96–99
Sulfur concrete (SC), 145, 146
　construction time for, 273
　durability of, 144
Sulfur content/density, aggregates, 125–128
Sulfur crystallization process, 130
Sulfur demand, 38
Sulfur Development Institute of Canada (SUDIC), 259
Sulfur dioxide, 92, 358
Sulfur dust hazards, 103
Sulfur extended asphalt, 54–56
Sulfur loading, durability and, 135–137, 138
Sulfur mortar, compressive strength, 152
Sulfur mortars, 292
Sulfur plasticizer/plasticizer, 148–151
Sulfur polymer cement (SPC), 59, 145, 193–194
Sulfur production/uses, 35–39
　demand for sulfur, 48
　environmental issues, 62–67
　global sulfur cycle, 36–39
　overview of, 35–36
　summary/conclusion, 67–69
　supply of sulfur, 39–44
　trade and, 44–47
　uses for sulfur, 48–62
Sulfur properties, 71–108
　chemical properties, 88–99
　ecological effects, elemental sulfur, 103–104
　for elemental sulfur, 77–88
　elemental sulfur forms, 75–77
　occurrence of sulfur, 72–74
　overview of, 71–72
　processes, 74–75
　product groups/applications, 104–107
　summary/conclusion, 107–108

thermal properties, 99–102
Sulfur supply, 39–44
  production/processes, 39–44
  sulfur trade, 44–47
  Sulfur trade, 44–47
  global market for, 46–47
Sulfur uses, 48–62
  agricultural chemicals, 51–52
  chemical/industrial, 52–54
  construction industry, 54–57
  ore processing, 57
  petroleum alkylation, 57–58
  pharmaceutical industry, 60–61
  pulp and paper, 58–59
  sulfuric acid, 49–50
  waste management and, 59
Sulfur-asphalt mixing module (SAM), 357
Sulfur-containing amino acids (SAAs), 60
Sulfur-extended asphalt (SE), 55–56
  test pavement, public roads, 366
Sulfur-extended asphalt (SEA), 364–368
  adhesion to mineral filler, 367–368
  FHWA evaluated sulfur extended asphalt projects, 365
  freeze-thaw cycling, 368–369
  traditional asphalt materials and, 368–370
  transverse cracking, 369
Sulfuric acid, 49–50
Sulfur-modified asphalt, 331–380
  aging of asphalt, 338–339
  asphalt component interaction, 335–338
  asphalt fractionation, 334–335
  beneficial use of, 341
  characteristics of, 362–363
  composition of asphalt, 332–334
  ecological effects, potential, 376–379
  history of, 341–342
  improved performance, 359–360
  liquid state sulfur behavior, 342–343
  plasticized sulfur, 370–376
  processing technology for, 356–357
  rheology of sulfur asphalt binder, 351–356
  SEA and traditional asphalt materials, 368–370
  sulfur asphalt development, 363–368
  sulfur asphalt interaction, 343–348
  sulfur asphalt mix concepts, 348–351

  sulfur-to-asphalt ratios/properties, 360–363
  summary/conclusion, 379–380
Sulfur-olefin reactions, 155
Sulfur-to-asphalt ratios/properties, 360–363
  hydrogen sulfide emission control, 363
  sulfur-modified asphalt characteristics, 362–363
  sulfur-to-asphalt ratios, 360–362
Sulkret mixing scheme, 268, 269
Sulphur Institute, The (TSI), 258, 259, 261
Sustainable, definition of, 3
Sustainable business justification, 21–24
  economic aspects, 21–22, 23
  environmental aspects, 22, 23
  social aspects, 22, 23, 24
Sustainable construction process, 29–30
  during construction, 29
  before construction begins, 29
  economic aspects, 30
  environmental aspects, 30
  post construction, handover, 29–30
  social aspects, 30
Sustainable design, 27–29
  economic aspects, 27
  environmental aspects, 27–29
  social aspects, 29
Sustainable design/construction, 15–20
  avoiding negative environmental impacts, 19–20
  consumption, minimization of, 15–17
  historical technologies, modification of, 17
  human desires, reshaping of, 17
  human needs/aspirations and, 17–19
  new technologies, creation of, 16–17
  reusing, rehabilitation, retrofitting, 16
  technological efficiency and, 16
Sustainable development, 1–34
  basic elements of, 4
  building construction industry and, 13–15
  definitions of, 3
  economic sustainability, 6–7
  environmental sustainability, 5–6
  framework for, 10–13
  implementation strategies, 15–20
  land sustainability, 7–8
  overview of, 1–2

project procurement life cycle and, 21–33
social sustainability, 4–5
summary/conclusion, 33–34
technology's role in, 9–10
Sustainable disposal/re-use of site, 32–33
  disposal, 32
  economic aspects, 33
  environmental aspects, 33
  reuse, 32–33
  social aspects, 33
Sustainable management/facility operation, 31–32
  economic aspects, 31
  environmental aspects, 31–32
  social aspects, 32
Sustainable procurement process, 24–27
  economic aspects, 24–25
  environmental aspects, 25–26
  social aspects, 26–27
Sweetening process, 74
Swelling clays, sulfur concrete, 205

T

Tank-house flooring, 262
Technology, sustainability and,
  characteristics of technologies, 9–10
  environmental impacts and, 10
  historical technologies, modification of, 17
  human satisfaction and, 10
  minimization of consumption, 10
  new technologies, creation of, 16–17
  technological efficiency and, 16
Temperature, service and, 306–307
Terminology, 111, 112
Testing, recommended, 204–205
Testing, sulfur cement, 189
TGA. *See* Thermal gravimetric analysis
Thaumasite formation, 117, 118
Thermal conductivity values, 306–307
Thermal gravimetric analysis (TGA), 100, 101, 180–181
  of ESC/BMSC, 222–223
Thermal properties, 99–102
  conductivity and, 108
  forms, with temperature, 99–100
  temperatures of interest, 101–102
  thermal analysis techniques, 100

Thiobacillus bacteria, oxidation and, 96, 97–98
Thiols, 92
TIA. *See* Transport impact assessment
Titanium dioxide, 53
Titration test method, EPA, 250
Toxicity characteristics and leaching procedures (TCLP), 376–377
Trade. *See* Sulfur trade
Trans-argonic structure, 91
Transfer/piping systems, 279
Transport impact assessment (TIA), 22
Transport infrastructure, 22
Transporting/mixing requirements, 283
Tri-olefins, 95
TSI. *See* Sulphur Institute, The

U

United States Bureau of Mines, 143
U.S. Bureau of Mines, 261
U.S. Department of Agriculture, 313
Useful characteristics, 255
Uses for sulfur. *See* Sulfur uses

V

Value for Money, 21
Van der Waals forces, 335, 336
Vinyl-norbornene (VNB), 184
Viscosity of liquid sulfur, 79
Viscosity-increasing surface-active finely divided particulate stabilizer, 214–215
VNB. *See* Vinyl-norbornene
Voids
  in sulfur concrete, 134, 204
  in sulfur-asphalt aggregate mixes, 355
Volcanoes/hot springs, 72
Vulcanizing properties, 107

W

Waste minimization/management, 25, 59, 64–67
  dispersive losses, 65–66
  disposal of waste, 13, 66
  recycling and, 19–20, 67
  reduction/reuse, 19–20
Water absorption, 300–305, 306
WAXS. *See* X-ray spectroscopy

Weigh-hopper/scale/conveyor, 283
Wettable sulfur, 104, 106
Whole-Life Costing (WLC), 21, 22
Wind/moisture, 285–286
WLC. See Whole-Life Costing
World War I, 143

X

X-ray spectroscopy (WAXS), 185